FRENCH POSTWAR

TRANSPORT AIRCRAFT

**By
J. CHILLON, J-P. DUBOIS and J. WEGG**

Published by Air-Britain (Historians) Ltd

Registered Office :   1 East Street, Tonbridge, Kent, England

Membership enquiries :   Dept M, 208 Stock Road, Billericay, Essex CM12 OSH

SALES DEPARTMENT :   9 ROOK CLOSE, ELM PARK, HORNCHURCH, ESSEX, RM12 5QH

Front cover :   Caravelle 10B1R EC-BIE of Aviaco on approach to Barajas,
Madrid in September 1975 (A.Camarasa)

Rear cover :   Top: Latécoère 631 F-BDRA flying over Southampton Water
in July 1947 (Flight)
Bottom: Airbus Industrie A300B2 ZS-SDB of South African
Airways (Airbus Industrie)

ISBN 0 85130 078 2

CONTENTS

ENGLISH PREFACE

The idea for this monograph goes back to late 1973, when the two French compilers considered writing a work in French to be published either as a special issue of the publication of the French branch of Air-Britain, "Le Trait d'Union", or in English as an Air-Britain monograph. Initially dealing with the Caravelle only, it was soon felt that other French airliners, be they successful or unsuccessful, also deserved coverage. An Air-Britain monograph devoted to both the Caravelle and the Comet (on the grounds that they had similar nose sections!) was later considered for some time, but this project was not proceeded with.

Then, by 1976, it was decided to conduct preliminary work on a monograph devoted to all French transport aircraft, this being submitted for approval only after preliminary research had shown that sufficient information was available to embark on such a large project. With the final decision for a go-ahead in April 1977, work began. It was felt to be a good opportunity to pay a tribute to French commercial transports, which were not very well known on the international aviation scene, and with the advent of international co-operation it now appears likely that purely French airliners will never be designed again.

Thanks to the extensive research carried out by Jacques Chillon at the Bureau Véritas and DGAC offices, it is also possible to have in print much previously unpublished information.

The Caravelle was of course to be the backbone of the monograph, and during the early stage of preparation John Wegg, who had already done extensive research on the subject, offered to join the team. It was immediately decided that, with his wide experience, he could take over responsibility of the whole Caravelle section, and all preliminary notes and data already prepared in France were forwarded to him for checking against his own information.

The two French compilers were left with all the remaining aircraft, and from the start it was clear that dealing in traditional Air-Britain monograph style would result in an enormous book. So a choice was made: the Caravelle, being the major French airliner, would be dealt with in the usual way, publishing as much data as possible. On the other hand, for the lesser-known types more basic information would be given, ie a short history and a production list. A constant difficulty for the French compilers was what to omit, and this was often a problem. Jacques Chillon has compiled most of the production lists with Jean-Pierre Dubois providing the background text. But on the whole this publication may be said to be a result of close co-operation between three people, one living in the Paris area, one in California and the other on the shores of the Mediterranean. A difficult job indeed!

To define the scope of this monograph, it has been decided to exclude all single-engined aircraft but to include those interesting prototypes which failed to make commercial reality. Furthermore many aircraft were used by the French Air Force or French Naval Aviation. Tracing the complete military service histories of French aircraft is always difficult except in a few instances owing to administrative regulations which put severe limitations on research into public records when the period concerned is too recent. This was the case for the military aircraft at least, and therefore no attempt has been made to record everything known on the military types; and of course it must be remembered that military codes and identities are changed every time the aircraft is moved to a new unit. Thus in the absence of full data being available, only the basic history will be mentioned in the production list of types widely used by the military. On the other hand, those types that were basically commercial airliners and have also found a way into military service will include the full military history (as far as it is known).

We are of course indebted to many people and organisations.

First it must be remembered that both French contributors prepared their typescript directly in English. This being the first monograph published by Air-Britain based on a manuscript provided by authors whose native language was not English, the painstaking work of rewriting must be credited to Chris Chatfield and Peter Marson, who have endeavoured to amend the style and grammar to a more anglicised format where necessary.

Many individuals have helped in the compilation, and we should like to thank specifically our team of friends and correspondents of the French Branch of Air-Britain and contributors to its magazine, "Le Trait d'Union", especially Bernard Chenal (for his invaluable knowledge of military histories), Jean Delmas (editor of "Le Trait d'Union"), Patrick-Xavier Henry, Claude Petit and Marc Suavet. Thanks are also due to Giorgio Apostolo for his help with the Vema 51 details and Matthew E Rodina for a wealth of detail on the SCAN.30s.

The following have also provided generous assistance in compiling the Caravelle story, and their help is grateful acknowledged: Antonio Camarasa, Ron Clark, Jaime Escobar Corradine, Bruce Drum, Hu K Gagos (previously with Douglas Co), T Lakmaker, Rolf Larrson, Markku Nokkala, Eino Ritaranta, J Sherlock, Jack T Womack (AiResearch Co), Eric Falk (General Electric) and Stanley Booker (International Air).

The help afforded by various agencies and airlines has also been outstanding, and special thanks should go to Col J Bouley (SIRPA), M Bourlanges (Bureau Veritas), Messrs Esperou, Ferry, Lecoq and Pachet (DGAC), M Eldin (Air France) and M Richon (UTA). The following airlines and companies have also helped considerably, particularly with the Caravelle section: Air Algérie, Air France, Alitalia, Atlanta Skylarks, Iberia, Indian Airlines, LTU, MEA, Lufthansa, Rolls-Royce, Sabena, SAS, Sterling, Swissair, SNECMA, SOGERMA and Transavia.

Considerable use of previously published information has been made, including various Air-Britain publications, Aviation Letter and French and British magazines such as Aviation Magazine,

Aérospatiale, Air et Cosmos, Air Pictorial, Flying Review and Air International, and also John Stroud's basic reference work "European Transport Aircraft since 1910" (Putnam).

In closing, the authors wish to acknowledge the dedicated work of various enthusiasts in the United Kingdom and elsewhere, particularly Jim Birch, Charles Cain, Mike Gradidge, Don Hannah, Bernard Martin, Reimar Wendt and Peter Watson. The patience of their wives during the preparation of this monograph should also not go unrecorded.

Considerable thanks are also due to the indefatigable Peter Gerhardt who not only confirmed but also added so much to the German data included in this Monograph.

The authors will of course be glad to receive any comments, additions and amendments to the various texts comprising this publication.

Jacques Chillon
3 Rue des Morillons
95130 Franconville
France

John Wegg
PO Box 3371
Ontario
California
U S A 91761

Jean-Pierre Dubois
28 bis rue Trachel
06000 Nice
France

December 1979

FRENCH PREFACE

Cette monographie vient compléter une liste déjà importante d'ouvrages de ce type publiés par Air-Britain, Association Internationale des Historiens de l'Aviation.

Jusqu'à présent, seuls des avions de transport anglo-saxons avaient été traités, et ce pour des raisons évidentes, qui tiennent autant à la production massive de ces appareils qu'au nombre élevé d'historiens aéronautiques de langue anglaise.

L'aéronautique française, elle n'a pas toujours pu s'implanter solidement sur le marche mondial de l'aviation de transport. Et ce n'est pas faute d'avoir essayé. De nombreux avions de ligne - ou plus généralement de transport - ont été dessinés et construits en France depuis 1945.

Il semblait donc utile de leur rendre justice et de les faire figurer en bonne place dans une collection de monographies rédigées en langue anglaise et destinées a être lués par les "aérophiles' du monde entier.

Que le lecteur ne s'y trompe pas, donc: le but poursuivi par les auteurs a bien été de dresser un bilan global des avions de transport français, et autant que faire se pouvait, un bilan qui puisse être suffisamment clair pour l'amateur non francophone, qu'il s'agisse d'un Néerlandais, d'un Brésilien, Néo-Zélandais, Americain, Italien ou Japonais. Ce qui explique certaines simplifications ou, en revanche, dans bien des cas, des explications qui pourraient paraître inutiles au lecteur français.

Chaque appareil a fait l'objet d'un historique succinct - Caravelle exceptée - suivi d'une liste de production, document de travail essentiel pour tout historien de l'aviation. Ces listes ont été soigneusement préparées, puis minutieusement vérifiées, et sont aussi complètes que possible. Une exception toutefois: les appareils militaires, non prévus initialement, et en fin de compte ajoutés tardivement au moment où cette monographie allait être 'bouclée', ont été généralement présentés sous une forme simplifiée - et ce délibérément. Les informations dans ce domaine sont souvent encore incomplètes, non disponibles, et les recherches historiques toujours en cours.

Cette monographie est le résultat d'un travail de longue haleine, plus de deux ans. Mais il faut dire qu'il reflète un travail beaucoup plus long, et mené non pas par les seuls auteurs mais aussi par bien des amis et correspondants passionnés eux aussi des choses de l'air quelles que soient leurs activités professionnelles. Qu'ils soient tous ici à nouveau remerciés.

Toute addition ou correction à cet ouvrage sera accueillié avec reconnaissance par les auteurs. Leur adresse figure au bas de la préface rédigée en langue anglaise.

Qu'il soit permis, pour terminer, de dire quelques mots sur l'Association qui a publie cet ouvrage. Air-Britain est une association a but non lucratif qui regroupe plus de 3500 membres dans le monde entier, tous amateurs de l'histoire de l'aviation.

La cotisation annuelle permet de recevoir tout ou partie des publications et bulletins publiés régulièrement en langue anglaise, et ouvre l'accès à de nombreux autres services (réseau de spécialistes, service diapositives, tarifs préférentiels pour les monographies etc).

Des branches locales ou nationales se sont également constituées. Pour la France, la Branche Française d'Air-Britain, regie par la Loi de 1901 sur les Associations, publie tous les deux mois un bulletin en langue française et qui couvre à peu près tous les aspects de l'histoire de l'aviation française. Association à but non lucratif, entièrement animée par des bénévoles et des amateurs, elle n'exclut pas pour autant un travail de recherche de haute qualité.

Pour plus de renseignements contacter:

Air-Britain (Historians) Ltd
Department M
208 Stock Road
Billericay
Essex CM12 OSH
Grande Bretagne

Préciser simplement, avec votre adresse, "I would like to receive further information on Air-Britain."

Ou, en ce qui concerne la Branche Française et son bulletin "Le Trait d'Union":

"Le Trait d'Union"
M Jean Delmas
Résidence Colbert
107 Avenue D Casanova
93320 Pavillons-sous-Bois
(France)

(English-speaking readers may obtain information on the publication of the French Branch of Air-Britain, "Le Trait d'Union", at the above address. "Le Trait d'Union" is published six times a year, in French, and includes articles and information on all aspects of French aviation.)

The abbreviations are presented in two parts: a list of those common to the production lists, followed by a general list of those found in French aviation and referred to in the texts.

Common Abbreviations in Production Lists

A/L	Airlines
A/W	Airways
canc	cancelled
Cie	Compagnie
CofA	Certificate of Airworthiness issue
c/s	call sign
cub m	cubic metres
cv	converted, conversion
dbf	destroyed by fire
dbr	damaged beyond repair
del	delivered
Dept	Département (French administrative geographical division)
Ets	Etablissements
f/f	first flown, first flight
i/s	in service
kg	kilogram
kgp	kilogram static thrust (rate for jet engines)
km	kilometre
lsd	leased
m	metre
nm	nautical mile
NTU	not taken up
regd	registered
retd	returned
r/o	rolled out
soc	struck off charge (applies to military aircraft of l'Armee de l'Air)
Soc	Société
sq m	square metre
Ste	Société
TT	total time
wfs	withdrawn from service
WFU	withdrawn from use

General Abbreviations

The non-French reader should remember that France is a highly centralised country with many administration services and agencies and a developed civil service. This is also true of French aviation, hence the many abbreviations which appear below (Centre, Direction, Office, Service, but also Centre, Ecole etc). The organisation is too complex to be given here in detail, even in the form of short comments - and only a few comments have been included in brackets where it was felt necessary. Others should be self-evident to a non French-speaking reader.

AF	Air France
AIA	Atelier Industriel de l'Air (military industrial facilities)
AMD	Avions Marcel Dassault
AMD-BA	Avions Marcel Dassault-Bréguet Aviation
AMSR	Advanced medium-short range
ASECNA	Agence pour la Sécurité et le Controle de la Navigation Aérienne
ASI	Air Service International
ATESA	Aero Taxi Ecuatoriana SA
ATT	Atterrissage Tout Temps
BA	Base Aérienne (French AF)
BE	Base Ecole
CAA	Civil Aviation Agency
CADAF	Compagnie Aérienne des Alpes Françaises
CAS	Continentale Air Service
CATA	Compagnie Algérienne de Transports Aeriens
CATI	Compagnie Aérienne de Transports Indo-chinois
CATRE	Centre Aéro-Technique de Réception et d'Entraînement (CEV)
CEP	Centre d'Expérimentations du Pacifique
CEAM	Centre d'Expérimentation des Armes Aériennes (French AF)
CEPM	Centre d'Exploitation Postale Metropolitain (French domestic air postal service)
CEV	Centre d'Essais en Vol (equivalent to British Royal Aircraft Establishment)
CGT	Compagnie Générale Transsaharienne
CGTA	Compagnie Générale de Transports Aériens
CGTM	Compagnie Générale de Turbo-Machines
CIET	Centre d'Instruction des Equipages de Transport (French AF)
CN	Centre National
COGESAT	Compagnie de Gestion et des Services Administratifs et Techniques
c/s	call sign ("indicatif radio" in French)
COTAM	Commandement du Transport Aérien Militaire (French AF Transport Command)
DAC	Direction de l'Aviation Civile
DEFA	Direction des Etudes et Fabrications d'Armement
DGAC	Direction Générale de l'Aviation Civile (ex SGACC, ex SGAC)
DGER	Direction Générale des Etudes et Recherches?
DIT	Déclassé instruction technique (ie permanently grounded for instructional purposes)
DLT	Deutsche Luft Transport
Domaines	Administration des Domaines - responsible for all government-owned property and organiser of public auction sales of surplus property
DTA	Direction des Transports Aeriens
DTCA	Direction Technique des Constructions Aeronautiques
DTI	Direction Technique Industrielle
EARS	Escadrille Aerienne de Recherches et de Sauvetage (SAR unit)
ECMA	(not traced)
EFS	Europe Falcon Service
ENAC	Ecole Nationale de l'Aviation Civile
ENSA	Ecole Nationale Supérieure d'Aéronautique
EPNER	Ecole du Personnel Navigant d'Essais et de Réception (CEV test pilot school)
ET	Escadron de Transport (French AF)
FAA	Federal Aviation Administration/Agency
FAP	Força Aerea Portuguesa
French AF	French Air Force (Armée de l'Air)
Fr Govt	French Government
Fr Navy	French Navy (Marine Nationale, French Naval Aviation or Aeronautique Navale - usually abbreviated to Aeronavale)
GAM	Groupe Aerien Mixte (French AF)
GAMD	Générale Aéronautique Marcel Dassault (later AMD)
GIE	Groupement d'Intérêts Economiques

GLAM	Groupement de Liaisons Aériennes Ministériel (French AF VIP unit)
GT	Groupe de Transport (French AF)
ICAO	International Civil Aviation Organisation
IFG	(not traced)
IGN	Institut Géographique National
JAT	Jugoslovensko Aero Transport
LASO-France	Lignes Aériennes du Sud-Ouest de la France
LCA	Lake Central Airlines
LOT	Lotnicze Polskie Airlinie (Polish Airlines)
OFEMA	Office Français d'Exportation des Matériels Aériens Office Français d'Exportation de Materiel Aéronautique
ONERA	Office National d'Etudes et de Recherches Aéronautiques, later Aérospatiales (French equivalent to American NACA/NASA)
PTT	Postes, Télégraphes et Télécommunications (French postal administration)
RAK	Rikets Allmoanna Kartverk (Swedish Air mapping and survey dept)
RF	République Française (often quoted as owner of government-operated aircraft)
SAA	South African Airways
SABENA	Société Anonyme Belge d'Exploitation de la Navigation Aérienne
SAGETA	Société Anonyme de Gérance et d'Exploitation des Transports Aériens
SALS	Service de l'Aviation Légère et Sportive (light aviation administration)
SANA	Société Auxiliare de Navigation Aérienne
SAR	Search and Rescue (French SAMAR also used)
SATA	Société Anonyme de Transports Aeriens
SATI	Société des Avions Taxis d'Indochine
SATT	Société Africaine des Transports Tropicaux
SCAN	Société de Constructions Aéronavales du Port-Neuf
SCRAS	(not traced)
SEMAF	Société d'Exploitation du Materiel Aéronautique Francais
SERAC	Société pour l'Etude et la Réalisation de l'Aviation Commerciale
SERIMA	(not traced)
SFA	Service de la Formation Aéronautique
SFCA(-Lignel)	Société Française de Constructions Aéronautiques
SFERMA	Société Française d'Entretien et de Réparation de Matériel Aéronautique
SGACC	Secrétariat Général a l'Aviation Civile et Commerciale (later SGAC only, now DGAC)
SIAL	Société Industriel des Avions l'Atécoère

SNCA	Société Nationale des Constructions Aéro-nautiques du .. (French prefix for nationalised aeronautical industries, as follows)
SNCAC	.. du Centre (type number prefix: NC)
SNCAM	.. du Midi (type number prefix: none)
SNCAN	.. du Nord (type number prefix: N)
SNCASE	.. du Sud-Est (type number prefix: SE)
SNCASO	.. du Sud-Ouest (type number prefix: SO)
SNECMA	Société Nationale d'Etude et de Construction de Moteurs d'Aviation
SNIAS	Société Nationale des Industries Aéro-nautiques et Spatiales, also shortened to Aérospatiale
SOCOTRA	Société Coopérative de Transport Aérien
STA	Société Transatlantique Aérienne
STA	Service Technique de l'Aéronautique
STA	(Algeria) Société de Travail Aérien
STAEO	(also STA/EO) Société des Transports Aériens d'Extrême-Orient
STAR	Société Tourisme et Transports Aériens Rapides
TAI	Transports Aériens Intercontinentaux
TAM	Transports Aériens du Midi
TEA	Trans European Airways
UACL	United Aircraft of Canada Ltd
UAT	Union Aéromaritime de Transports
UTA	Union des Transports Aériens (TAI + UAT merger)
VHI	Vought Helicopter Inc
WGAF	West German Air Force

For a very short period after World War II the French nationalised industries (eg SNCAC, SNCAN, SNCASE etc) were also referred to as Aérocentre, Aéronord, Aérosudest, but this practice was soon discontinued.

THE METRIC SYSTEM

As all Western European countries use the metric system for weights, lengths, distances and liquid capacity, we present here for the benefit of the American reader (and probably for some British readers as well) the approximate equivalent of metric measures in the American and British systems.

Length

```
1 metre     =    3.281 ft
25 metres   =   82.02 ft
50 metres   =  164.0 ft
75 metres   =  246.1 ft
100 metres  =  328.1 ft
```

Area

```
1 sq metre    =   10.76 sq ft
25 sq metres  =  269.1 sq ft
50 sq metres  =  538.2 sq ft
75 sq metres  =  807.3 sq ft
100 sq metres = 1076 sq ft
```

Liquid Capacity

```
1 litre     =   0.220 Imp Gall/0.264 US Gall
25 litres   =   5.500 Imp Gall/6.600 US Gall
50 litres   =  11.00 Imp Gall/13.20 US Gall
75 litres   =  16.50 Imp Gall/19.80 US Gall
100 litres  =  22.00 Imp Gall/26.40 US Gall
```

Distance

```
1 kilometre     =   0.621 st m/0.540 nm
25 kilometres   =  15.54 st m/13.49 nm
50 kilometres   =  31.07 st m/26.98 nm
75 kilometres   =  46.60 st m/40.47 nm
100 kilometres  =  62.14 st m/53.96 nm
```

Weight

```
1 kilo     =    2.205 lbs
25 kilos   =   55.11 lbs
50 kilos   =  110.2 lbs
75 kilos   =  165.3 lbs
100 kilos  =  220.5 lbs
```

Introduction

During the late fifties and early sixties, with the growing costs of aeronautical research and development, particularly in designing large aircraft, international collaboration was launched on many programmes. This was especially so in Europe, where closer political involvement was a prime moving force. Private ventures being usually too costly, government financing was obvious. Concorde was a typical example, with company-to-company arrangements being ratified immediately by the French and British governments.

Clearly, in a monograph devoted to French airliners, it would have been unfair to mention Concorde and the A300 Airbus as totally national products, which is in any case untrue. Thus these two aircraft have been included in an International Co-operation section. The Dassault Mercure has been included with the French airliners proper, even though some contribution from other European partners was made to the programme.

In this context, other aircraft must also be mentioned: the Transall is a Franco-German aircraft but has been included in a chapter on military transport aircraft. The American Mohawk 298 conversion of the Nord 262 represents another aspect of international co-operation, as would have been the case if American derivatives of the Breguet 941 had been built.

Curiously, this opening chapter of the monograph may well serve as an epitaph for totally French airliners, as it now seems unlikely that purely French transport aircraft will appear in the future.

A300

EMERGENCE OF A EUROPEAN WIDE-BODY TRANSPORT

The origin of the first European wide-body airliner, the Airbus Industrie A300, or simply Airbus, may be traced back to 1965, when a Franco-British committee drew up plans for a large capacity, short-haul, low-cost transport aircraft. This was confirmed in October of that year when a symposium was held in London by the major European airlines to prepare specifications for such an aircraft, known as the Aerobus or Airbus, the latter name ultimately being preferred.

Many proposals were offered; Bréguet entered the competition with the Bréguet 124 project, announced in July 1965, featuring a double-bubble fuselage (fuselage bilobé) reminiscent of the Bréguet 1150 Atlantic maritime patrol aircraft and using the double-deck passenger accommodation already put into practice on the Bréguet Deux Ponts. The Br 124 would have carried 240 to 265 passengers and would have been powered by four Rolls-Royce Spey engines positioned in pairs under the wings. Other power-plants were considered (such as the JT8D), and twin-engined versions with 13,600 kgp Rolls-Royce RB-178s or 15,000 kgp Pratt & Whitney JT9s were also announced. Empty and loaded weights for the four-engined Br 124 were given as 46,600 and 95,000 kg respectively, range being only 1,500 km with 25,000 kg disposable load. It appears that the double-deck feature did not attract much enthusiasm, and by the end of 1965 Bréguet joined with other challengers to work on the common HBN-100 project.

A Bréguet 946 derivative also deserves mention; this project was a Br 124 fuselage mated to a four turboprop blown wing (aile soufflée) similar to the wing of the Br 940/942 design. Unlike the high-wing Bréguet 944

project, the 946 would have displayed a mid-wing layout.

Curiously enough, a contemporary Nord Aviation proposal also featured a double-bubble fuselage but with transverse layout, the cabin being divided by a lengthwise bulkhead. Also a four-engined project (with 5,800 kgp Spey 50 turbofans), the 200/250 seat Nord 600 would have flown like the Br 124 at Mach 0.8. Gross weight was 80,000 kg.

On 1 June 1965 Bréguet and Nord Aviation signed an agreement for co-operation on the Airbus programme, and when joined later in November by Hawker-Siddeley Aviation (who had designed the HS-132 high-density project based on a Trident design) embarked upon the design of a combined proposal known as the HBN-100. A 226/252 seat aircraft with circular fuselage section, the HBN-100 (Hawker + Bréguet + Nord) was a twin-engined wide-body.

Early in 1966 Sud Aviation also entered the competition with various projects. The generic name of Galion was adopted for the Sud proposals, the Galleon being an obvious choice of name for an aircraft intended as a successor to the Caravelle. With operation being considered by eight major European airlines, the aircraft was to be a 225-seater with a typical range of 2,200 km. Power-plants recommended were to be either the Pratt & Whitney JT9D (18,600 kgp) or Rolls-Royce RB-178 (19,960 kgp).

Dassault had an agreement to share work with Sud if the 241-269 seater was selected, and both French concerns were apparently ready to make the Galion an all-French programme had European co-operation prospects failed.

Loaded weight of the final Galion project was 95,000 kg, and an estimated maximum range was given as 1,850 km. With a fuselage diameter of some six metres, the Galion was also a break-through into the wide-body category and was generally similar to the 101,000 kg auw HBN-100 competitor.

With joint talks initiated in September 1966 the Airbus became a major European issue, and decisions taken in the following month at government level left three concerns to proceed with work on the project. HSA was the British choice, and the German counterpart was the Deutsche Airbus organisation set up specifically for the project. In France Sud Aviation was selected as design leader, thus taking on the responsibility of design work previously under-taken by Nord, Bréguet and Dassault.

By December 1966 many projects were being readied, and wind-tunnel tests began on scale models of the various proposals. The agreement was however still only in principle, and a go-ahead decision was still to be ratified.

In the spring of 1967 Britain, Germany and France agreed on a scaled-up 277-seat design with the same operational performance as previous projects. This new design, which basically grew out of the HBN-100 and Sud Galion proposals, became the first European wide-body airliner, the A300.

The new aircraft was to have two engines only, choice being limited to the RB-207 or JT9D. With all previous designs merged into a single project, contributing shares were also announced: 37.5% each for HSA and Sud and 25% for the Arbeitsgemeinschaft Airbus. Finally, by a Memorandum of Understanding dated 25 July 1967, the three countries involved (Britain, France and Germany) at last gave the go-ahead decision, settling on the uprated 26,000 kgp RB-207 as the final power-plant choice.

The future of the A300 still however remained uncertain, and after some hesitation (and an unexpected French crisis in May 1968 which was viewed by some specialists as a threat to the programme), a redesigned version appeared in December 1968.

The new A300B (the designation retained when production was launched) was chosen as a basic airframe with stretched variants considered from the start, and the Pratt & Whitney JT9D, General Electric CF-6 or Rolls-Royce RB-211 power-plants all now came under consideration.

The 1969 Le Bourget air show offered an opportunity to take another step in the launching of the programme. Since the British had dropped out of any further participation in the development of the Airbus (this having been announced shortly before the show at a Ministerial Meeting held on 10 April 1969), the French and German governments decided to pro-ceed on their own, with the French aerospace industry nominated as design and production leader. The A300B1 can be said to have been born at the historic meeting of 26 May 1969.

With such a major programme ahead, GIE Airbus Industrie was set up on 18 December 1970 with specific liabilities involving risk-sharing partnership under French law. The French SNIAS concern and Deutsche Airbus GmbH were joined by associates Hawker-Siddeley of Great Britain and VFW-Fokker (the latter in December 1970) under the easily translated

Groupement d'Interêt Economique (hence GIE). Ten days later the Dutch Government agreed on the participation of the Netherlands in the programme. The Spanish firm Construcciones Aeronauticas (CASA) joined the Airbus programme on 23 December 1971 for a 4.2% share in pro-duction. It was by now clear that this was a truly important European venture, not only in terms of technical co-operation but backed by government decisions implying political involvement.

Hopes in the Airbus were finalised with the first production order on 9 November 1971, when six aircraft of the A300B2 version were ordered by Air France, with ten more on option.

ATLAS, AEROFORMATION AND SUPER GUPPY OPERATION

It is appropriate to pause in the development history at this stage to mention three aspects all directly connected with the A300 programme. It must be realised that designing, building and operating this large airliner, even at European level, was a huge undertaking, and the three following items neatly illustrate the complexity.

The Groupe Atlas was in fact unrelated to the A300 in its origination but was later to be largely involved with its design. In June 1967 Air France had set up the so-called Montparnasse Committee to study inter-airline co-operation for large aircraft maintenance, thus reducing costs and storage of spares. This decision was of course made in view of the introduction into European service of the large American Boeing 747s.

The organisation was later enlarged into the European Airlines Montparnasse Committee with some twenty airline members. The Groupe Atlas itself was born in July 1968 out of an agreement between Air France, Alitalia, Lufthansa and Sabena (and later Iberia) to share investment costs on maintenance, overhaul and flight-simulator training on the Boeing 747. This became effective during the following year with the definition of a new division of maintenance work between members.

By 1970 it was clear that the DC-10 and A300 wide-bodies were likely to be ordered and used by airline members, and in January 1971 co-operation between the Groupe Atlas and Airbus Industrie was announced. This took the shape of technical meetings and the study of RFCs (Request for Change Specifications), no fewer than nine technical teams having been set up by Atlas, each dealing with a specific aspect of A300 operation. For example, Team 1 was involved with Performance, Economy and Noise (PEN); another team dealt with the flight deck, another with air-conditioning, and so on. Many master changes were incorporated in the design of the A300B at the request of the Atlas teams. The Atlas group also organised their own maintenance plans for the aircraft when it was delivered to member operators.

The Atlas/Airbus Industrie talks reflect an interesting example of co-operation between two international organisations: on the one side several airlines, and on the other aircraft manufacturers of no fewer than five nations. With so many members involved and, inevitably, conflicting views, this could have led to insuperable negotiation problems. In fact, it proved a general success.

Aeroformation was a similar venture, this time dealing specifically with A300 crew training.   Like Airbus Industrie, Aeroformation was set up as a grouping of economic interests (GIE), also under French law.   It was created on 21 September 1971 by Airbus Industrie and the US-based Flight Safety Inc, which already possessed wide experience in the problems of crew training.   With a fifty-fifty share basis, Aeroformation established its training centre in France, near Toulouse-Blagnac, this being conveniently located close to the A300 assembly line.   Aeroformation was staffed by personnel from both Airbus and Flight Safety, with extra members from SNIAS and Deutsche Airbus.   Full instruction programmes developed included the use of both a systems trainer and a cockpit procedure trainer, apart from a flight simulator.   The simulator used is a six-axis version built by LMT (Le Matériel Téléphonique) in conjunction with Canada's CAE Electronics and the LINK division of the US Singer Company, with French CII computer support.

The third aspect of the Airbus support programme was the introduction to European skies of the <u>Super Guppy</u>, two of which were purchased by Airbus Industrie for operation by Aeromaritime, thus enabling airborne delivery to the assembly line in Toulouse of large components and fuselage sections (mostly from MBB-HFB at Finkenwerder).   Super Guppy operations were started on 29 September 1971. Both aircraft are French registered, F-BTGV c/n 0001 ex N211AS and F-BPPA c/n 0002 ex N212AS. The aircraft have also been used for other programmes such as the Dassault Mercure.

INTO PRODUCTION

In October 1969 Airbus Industrie finally settled for an American power-plant, the 22,690 kgp General Electric CF6-50, also developed for the McDonnell-Douglas DC-10-30, using experience gained on the earlier CF6-6.   The CF6-50 was test-flown in 1971 on a B-52E-55-BW serial 57-119 (0-70119), and certification of the CF6-50A now retained for production aircraft was granted by the FAA in March 1972, being rated at 22,225 kgp (49,000 lbs).

As already stated, the first A300B order came from Air France in November 1971, and was followed in January 1972 by one from Iberia. The Spanish company intended to purchase four aircraft of the A300B4 variant (with eight more on option), and c/n 12, 14, 16 and 17 were allocated.   In fact, the order was later cancelled, and the aircraft were re-allocated to other customers.   At this time the cancellation was considered a severe setback to the whole programme, especially with the Spanish CASA being a member of the European Airbus consortium.   Ironically, Iberia subsequently ordered the A300.   Sterling was the third customer with three aircraft (also A300B4s), but again this early order was cancelled.

No real prototype was built, production starting with c/n 1, an A300B1 registered F-WUAB, which was rolled out at Toulouse on 1 August 1972, ground tests beginning the next day.   During the same period fatigue tests were being organised at Ottobrun (near Munich) by the German IABG (Industrie Anlagen Betriebgesellschaft GmbH) and also at Bremen and Hamburg by MBB-HFB.   The EF (Essais de Fatigue) airframe was split into four sections for separate tests (with of course appropriate lengthening of each section to simulate full-length fuselage tests).   Target for the IABG

was 96,000 simulated flights.   This was reached on 22 August 1975, and tests continued to the 120,000 final mark.   The other airframe, ES (Essais de Structure), was tested in Toulouse by CEAT in the static test programme.

Ground vibration tests were applied to F-WUAB by ONERA shortly after its roll out in August 1972.   Maiden flight of the first aircraft took place on 28 October 1972.   The second A300 (F-WUAC), which was also initially an A300B1, followed in February 1973.   The next aircraft, c/n 3 and 4, were A300B2 variants similar to those ordered by Air France, with 137,000 kg auw and a typical range of 1,200 nm and introducing a fuselage stretched slightly by 2.61m.   The B2 was intended for short-range operation and was powered by uprated CF6-50C engines.   Both were used by Airbus Industrie (including use on the certification programme) until the delivery of c/n 4 to Air Inter in January 1977, with the unexpected change of registration from F-WUAA to F-BUAE.

The A300B3 was a long-range version but was not developed, and the A300B4 was next to appear. A heavier design, the A300B4 was a medium-range version (2,000 nm) with extra fuel capacity. This version was originally intended for Iberia, but deliveries to the Spanish company were not finalised following cancellation of their order.

The B4 prototype was c/n 9, registered F-ODCY (originally F-WLGA), following the well-established practice of using the French Overseas register for sales tours abroad.   Similarly the original A300B1 F-WUAB had also appeared on the same register as F-OCAZ.   Extensive demonstration tours had in fact begun with the first aircraft as early as September 1973, first to the Americas and then to the Middle East, India and Pakistan.   In November and December 1973 the A300 was also flown to Africa, many countries throughout the continent being visited. In the middle of 1974 further tours were organised to the Far East and Australia.   A special 134-seat Press layout was chosen with appropriate bar equipment for these tours.

In the meantime, after an efficient test programme conducted with the first four aircraft, French and German certifications (including autoland CAT.II operation) were granted on 15 March 1974, FAA certification being issued on 30 May of the same year.   CAT. IIIA autoland certification was obtained on 30 September 1974, and the B4 type certification (with CAT.III provisions) was granted on 26 March 1975.

The initial production go-ahead programme was gradually increased to some fifty aircraft as the A300 began proving itself a highly reliable aircraft with a satisfactory low-noise level and low-cost operation.   It must be said however that sales were relatively slow in the initial years, first because the A300 appeared at a time of worldwide economic slump and second because of the inherent difficulty of convincing airline operators to buy a type launched by a new manufacturer (as Dassault-Bréguet was sadly to experience with the Mercure).

A300 IN SERVICE

By 1974 the unnamed A300 became more and more regularly referred to as the Airbus, which was originally only the name of the programme specification.   Air France was the first operator to start services with the A300, flying the busy London-Paris route for the first time on 23 May 1974 with the fifth aircraft, F-BVGA. Only one aircraft had been delivered at that time (delivery taking place only 12 days before) but an impressive success was achieved immediately. Air France passenger accommodation was for 26 first-class and 222 'touriste' (economy class). Two more aircraft followed in June and August, the remainder of the B2 order for six being delivered throughout 1975, the full Air France designation being A300B2-1A. The aircraft entered service on such routes as Paris to Nice, Milan and Dusseldorf and later on the Algiers, Athens and Tunis sectors.

The next Air France aircraft were A300B4-2Cs. The first of these, F-BVGG, was fitted with a special all-economy 270-seat layout and was flown to the French West Indies for operation on the New York route, thus becoming the first revenue-earning A300 to land at New York, achieving this on 20 November 1975. The aircraft was soon to make the headlines worldwide when, after the end of the peak period on the Antilles service, it was flown back to Paris and used on Middle East routes. On 28 June 1976, on flight AF139 from Tel Aviv to Paris, it was hijacked with a crew of twelve and 258 passengers shortly after take-off from Athens. After a stop-over in Benghazi, F-BVGG landed at Entebbe, and there followed the daring Israeli raid to liberate the hostages, of which much has been written (and filmed) elsewhere.

Other early A300 operators included Air Siam and TEA, the former using a leased aircraft for twelve months on the Bangkok-Hong Kong-Singapore route. This was a B2, c/n 8, pending delivery of the A300B4 ordered. Owing to local difficulties however the order was cancelled. The leased aircraft, HS-VGD, was returned to Airbus Industrie and then leased to Transavia who, as the major inclusive tour Dutch operator, had one A300 on order. But late in 1976 this order was cancelled, first evidence of the strong competition between the Franco-German concern and Boeing. In January 1977 Boeing Vice-Chairman, Clarence F Wilde, complained of unfair Airbus competition, the European airliner being supported at government level.

Shortly after the Air Siam lease, another A300, the second aircraft built, was leased to Air Algerie via the Belgian operator TEA to operate Muslim pilgrim flights to Saudi Arabia for two months with its maximum passenger all-economy layout. The aircraft was then returned to TEA and was subsequently used with a second aircraft, B4 c/n 17, both aircraft having 320-seat accommodation.

Germanair followed with the twelfth A300, beginning services on 1 June 1975, a second aircraft following in May 1977 with both aircraft flying mostly to Mediterranean resorts.

The next scheduled operators to use the A300 were Korean Air Lines and Lufthansa. The Far East introduction of the type by Korean took place in the summer of 1975 (six B4s being ordered), an interesting breakthrough for a European airliner. Lufthansa's B2 and B4 aircraft were put into service on 1 April 1976, initially on services to London and Paris out of Frankfurt.

Late in 1976 more orders were fulfilled, and the A300 entered service with Indian Airlines, South African Airways and the French domestic airline Air Inter. The first Indian B2 was delivered on 10 November 1976, and the three aircraft fly domestic routes, acting as Caravelle replacements. The SAA order was for a special variant, fitted with inboard leading edge Kruger flaps to improve take-off performance and designated A300 B2K. The first South African A300, ZS-SDA, arrived on 23 November 1976.

During the same period the French domestic airline, Air Inter, began A300 operations, first on the busy Paris-Lyon route, one of the aircraft being A300 c/n 8 which had previously been leased to Transavia and Air Siam.

Sales had slowed by early 1977, until an order from Thai International was announced in April for four B4 variants, with the first delivery to take place in October.

As already stated, the A300 had to win credibility on the worldwide market and, in this intensely competitive sphere, had to face strong rivalry from the US manufacturers. By 1977 it was felt that the future of the A300 might well lie in US sales, especially since US certification for the B4 had been granted on 30 June 1976.

A300 AND THE UNITED STATES

By August 1977 total sales of the airliner had reached 45, with 24 more on option. In retrospect, this represented an undeniably slow rate of sales, although this was never admitted publicly at the time. Hopes were however soon revived with the disclosure of a lease of four aircraft by Eastern Airlines, news which was immediately hailed by many as a major opportunity to boost the A300 sales programme.

The four Eastern B4s were c/n 41, 42, 43 and 44 registered N201EA to N204EA, and the first Airbus Industrie-owned aircraft (c/n 44) landed in New York on 24 August 1977, official handover taking place five days later. With a 26 first-class and 211 economy seat layout, these aircraft were intended for use on the heavy traffic winter routes between New York and Florida, and proving operations were planned for six months, starting early in December 1977. From the start, A300 operation by Eastern was a success, with excellent reliability results. Apart from a few experimental services flown from 18 November out of Newark, regular services were inaugurated on 13 December from both Newark and La Guardia to four Florida terminals.

By this time more orders had been obtained, reaching 82 including options. But then came the La Guardia affair; the runways at La Guardia being on land-fill, there were restrictions on the weights at which aircraft could be operated, and the Port Authority of New York told Eastern to limit its A300B4 weight to 240,000 lb. This had the effect of making A300 operation uneconomical. With DC-10s and TriStars landing at La Guardia at weights up to 370,000 lb, a new confrontation began to develop to rival that which had recently culminated in the granting of Concorde landing rights at New York Kennedy. The main technical argument was that the design of the undercarriage was not suitable for operation into and out of La Guardia - the A300 mainwheel track being shorter than that of the DC-10 and TriStar by approximately 1.5m (or five feet).

Needless to say, European reaction - particularly in the French Press - was considerable. The more reactionary newspaper commentators and politicians began mentioning retaliation should the European aircraft be denied right of operation in the United States through what was described as an unconvincing technical argument.   The furore was unnecessary as a compromise was found by March 1978, involving some reinforcement of runway piles at La Guardia and short-term modifications to the A300 undercarriage.

On 6 April 1978 Eastern Airlines finally announced an order for 19 new aircraft and for the purchase of the four already in use.   The variant ordered was the B4 with 230 seats, with three aircraft to be delivered before the end of 1978 and four to be delivered each year from 1979 to 1982.   Options were simultaneously disclosed for nine 245-seat B4s and 25 200-seat B10s.

This order was naturally met with acclaim, particularly in France, representing the most significant order from an American customer since the 1960's United Airlines order for Caravelles or, more broadly in European terms, since the Viscount breakthrough on the American market.

With total orders standing at 74 (with 34 on option), the Eastern order was felt as a decisive boost to future A300 production, particularly with a 95% American near-monopoly of the world airliner market.

Several American operators had already evinced interest in the new airliner, including Allegheny, American, Continental, Pacific Southwest, TWA and United.   A line of argument soom promoted by Airbus Industrie involved the fact that some 314 American firms were involved in some way or other in the A300 programme and that American participation in every A300 built reached 33% of the aircraft cost (mostly in power-plants and electronics), the German and French shares both being 25%, with 17% for the remaining European partners.

As expected, American reaction was prompt, with fears that a huge market could be lost, particularly for the new Boeing designs.   But hopes of a more extensive breakthrough on the American market received a first blow in July when United Airlines ordered the Boeing 767, the competitor to the proposed shortened version of the basic A300.   By mid-November 1978 American Airlines and Delta Airlines had also ordered large numbers of the new Boeing aircraft, leaving only Eastern still holding options on the shortened variant.

1978, THE DECISIVE YEAR

As already mentioned, sales of the A300 by August 1977 were still disappointing.   But late in December a significant new customer was added to the list with the order of two A300B2s (in 242-seat layout with basic 3,000 km range) by SAS, ten more aircraft being held on option. By late February 1978 SAS announced that the aircraft ordered were to be fitted with Pratt & Whitney JT9D-59A engines, providing commonality with its convertible 747s and ease of maintenance at Stockholm-Bromma.   At the same time, flight simulators were ordered from the French firm LMT.

This order probably signalled the end of the decline in sales, as events were soon to prove. By February and March several orders were rumoured, not the least being Japanese interest in developing an agency for the import of the European airliner.   Iran Air was the first to place an order, and sales had reached 53 (plus 41 options) in late 1977, twenty aircraft having been sold that year.   A first step had been taken in December 1977 with the lease of two aircraft by Iran Air, and late in February 1978 the airline finally ordered six A300 B2Ks for delivery between late 1979 and mid-1981, the leased aircraft being used pending delivery of the brand new machines.

In March Olympic Airways followed with a letter of intent for the purchase of two A300s, converted to a firm order in June.   Early in April Indian Airlines and Thai International announced orders for more aircraft while, as already mentioned, Eastern Airlines were in the process of making a final decision on a larger order.

Following the Eastern order, Thai's order was increased to eight early in July, bringing the total number of A300s on order and in operation in south-east Asia to 21.   This number was set to be increased when Taiwan-based China Airlines announced in the same month their intention of buying four B4s with an option on four more. This order has remained unratified by the Taiwan Government, who have been under strong pressure to buy the American DC-10.

This letter of intent was obtained only twelve days after Pakistan International Airlines had ordered the B4-200 variant on 16 July with longer range and 165 tonnes gross weight, fitted with 52,200 lb st CF6-50C2 engines.

The summer of 1978 was definitely a good season for A300 sales.   On 31 July the first order for an A300C4 was placed by Hapag-Lloyd, and on 4 August another major European operator, Alitalia, ordered the A300 with a letter of intent for four aircraft, later changed into a firm order for eight plus three on option.

On 30 August Malaysian Airline System ordered three B4s with an option on one more, by which time orders had increased to 113 with 49 more on option.   Late in October the disappointment of not securing an order from El Al was balanced by the final decision of Alitalia to purchase the European airliner.   A noteworthy development occurred on 27 October, with British Aerospace becoming a member of Airbus Industrie, the final signing taking place on 1 January 1979.

The year 1978 ended with an order from Iberia for four A300B4-100s with options on four more, delivery to begin in February 1981.   The Spanish airline became the twenty-first confirmed A300 customer (including the single aircraft option for an undisclosed customer), and of the major Western European flag carriers only British Airways, KLM and Sabena had yet to place an order.

With seventy sales in the year, 1978 will probably be remembered by aeronautical historians as the decisive year, not necessarily because of the several orders secured during that period, but because of the many problems and issues that developed and were overcome.   The major success of the initial US order indeed began to resemble the occurrences of twenty years earlier when United Airlines ordered the Caravelle, and it became clear that the A300 family would develop into a realistic competitor to the US manufacturers.

On 31 December 1978 total ratified orders for the A300 had reached 131, with 52 more on option. The year 1979 witnessed a continuation of sales impetus, so that by late July the order and delivery book was as follows:

Customer	Orders B2	Orders B4	Options B2	Options B4	Delivered B2	Delivered B4	Total Delivered
Aerocondor	-	1	-	-	-	1*	1
Air Afrique	-	1	-	-	-	-	-
Air France	9	15	12		7	7	14
Air Inter	7	-	-	-	5	-	5
Alitalia	-	8	-	3	-	-	-
Bavaria-Germanair	-	4	-	-	-	4	4
Cruzeiro	-	2	-	2	-	-	-
Eastern Airlines	-	23	-	9	-	7	7
Egyptair	-	3	-	4	-	2 lsd	2
Garuda**	-	6	-	6	-	-	-
Hapag Lloyd	-	2+	-	-	1	-	1
Iberia**	-	4	-	4	-	-	-
Indian Airlines	8	-	3	-	5	-	5
Iran Air	6	-	3	-	2 lsd*	-	2
Korean Air Lines	-	8	-	-	-	8	8
Laker	-	10	-	-	-	-	-
Lufthansa	5	6	9		5	5	10
Malaysian Airline System	-	3	-	1	-	-	-
Olympic	-	2	-	3	-	2	2
Pakistan International	-	4	-	6	-	-	-
Philippine Airlines	-	2	-	2	-	-	-
Scandinavian Airline System**	-	4	-	8	-	-	-
Singapore International	-	6	-	6	-	-	-
South African Airways	4	1	-	-	4	-	4
Thai International	-	10	-	-	-	6	6
TOA Domestic	6	-	-	-	-	-	-
Trans European	1(B1) 1		-	-	1(B1) 1		2
	46	126	6	54	30	43	73

* Since returned to Airbus Industrie

** JT9D

+ One B4, one C4 (B4 only delivered)

A300 Designations

In mid 1978 a new designation system was set up to rationalise the growing number of A300 variants:

A300B: passenger only aircraft

A300B1 - retained for two original aircraft, one in service with TEA.

A300B2 - three-digit suffixes were added to the original B2 variant:
A300B2-100 - slotted wing leading edge with sub-variants according to power-plant:

-101 to -119 CF6-50	-101	CF6-50C
	-102	CF6-50C1
	-103	CF6-50C2
-120 to -139 JT9D	-120	JT9D-59A
	-121	JT9D-59B
-140 to -159 RB-211		

Further two-digit suffixes refer to maximum gross weight, eg A300B2-103-03 142 tonnes.

A300B2-200 - leading edge Kruger slats giving improved field performance (previously A300B2K).

A300B4 - three-digit suffixes were added to the original B4 variant, as with the A300B2.

A300B4-100 - former B4 variant.

A300B4-200 - former A300B4 Stage III variant (gross weight 165 tonnes).
Further suffixes apply as with the A300B2, eg A300B4-120-06 refers to an A300B4 powered by JT9D-59A and with 165 tonnes gross weight.

A300C: convertible aircraft (passenger and cargo)

A300F: freighter aircraft

Specifications

A300B1

Span: 44.84m
Length: 50.97m
Height: 16.56m
Wing area: 260 sq m
Empty weight: 83,000 kg
Maximum payload: 27,000 kg
Maximum take-off weight: 132,000 kg
Maximum usable fuel: 34,500 kg
Range with 257 passengers: 2,690 km
Range with maximum fuel: 3,930 km
Take-off field at maximum weight: 1,950 m
(ISA + 15°C at sea level)

A300B2

Span: 44.84m
Length: 53.62m
Height: 16.53m
Wing area: 260 sq m
Empty weight: 84,740 kg
Maximum payload: 31,760 kg
Maximum take-off weight: 137,000 kg
Maximum usable fuel: 34,500 kg
Range with 281 passengers: 2,600 km
Range with maximum fuel: 3,700 km
Take-off field at maximum weight: 1,965m
  (ISA + 15°C at sea level)

A300B4

Span: 44.84m
Length: 53.62m
Height: 16.53m
Wing area: 260 sq m
Empty weight: 86,790 kg
Maximum payload: 35,210 kg
Maximum take-off weight: 150,000 kg
Maximum usable fuel: 45,400 kg
Range with 281 passengers: 3,890 km
Range with maximum fuel: 5,190 km
Take-off field at maximum weight: 2,800m
  (ISA + 15°C at sea level)

The above figures apply to initial production aircraft, and increases in weight were granted to the B2 in
1975 and to the B4 in 1976.

Power-plants were initially two 22,226 kgp General
Electric CF6-50As, with two 23,133 kgp General Electric
CF6-50Cs from 1975.

The performance figures quoted above are given as an
indication only, as they depend on the conditions of aircraft operation.   Maximum cruising speed is Mach 0.84
(917 km/hr) at 9,145m, and economical cruising speed is
847 km/hr.

With the mid-1978 re-designation, the major details
of the four major airframe variants are given below for
completeness.

A300B2-100

Power-plants: 2 x 51,000 lb GE CF6-50C
Span: 44.84m
Length: 53.57m
Height: 16.53m
Wing area: 260 sq m
Maximum seating: 336
Take-off weight: 142,000 kg
Landing weight: 130,000 kg
Empty weight: 89,910 kg
Maximum cruising speed: 935 km/hr
Cost-economical cruising speed: 869 km/hr
Long-range cruising speed: 839 km/hr

A300B2-200

Power-plants: 2 x 51,000 lb GE CF6-50C
Span: 44.84m
Length: 53.57m
Height: 16.53m
Wing area: 260 sq m
Maximum seating: 336
Take-off weight: 142,000 kg
Landing weight: 130,000 kg
Empty weight: 85,900 kg
Maximum cruising speed: 935 km/hr
Cost-economical cruising speed: 869 km/hr
Long-range cruising speed: 839 km/hr

A300B4-100

Power-plants: 2 x 51,000 lb GE CF-50C
Span: 44.84m
Length: 53.57m
Height: 16.53m
Wing area: 260 sq m
Maximum seating: 336
Take-off weight: 157,000 kg
Landing weight: 134,000 kg
Empty weight: 88,100 kg
Maximum cruising speed: 910 km/hr
Cost-economical cruising speed: 869 km/hr
Long-range cruising speed: 841 km/hr

A300B4-200

Power-plants: 2 x 52,500 lb GE CF6-50C1
Span: 44.84m
Length: 53.57m
Height: 16.53m
Wing area: 260 sq m
Maximum seating: 336
Take-off weight: 165,000 kg
Landing weight: 134,000 kg
Empty weight: 88,400 kg
Maximum cruising speed: 910 km/hr
Cost-economical cruising speed: 869 km/hr
Long-range cruising speed: 849 km/hr

Airbus Developments and Projects

While A300 sales were only slowly   increasing in the
early years, Airbus Industrie announced several developments, of which the more important are described below.

A300F and A300C

Both projects were announced in the autumn of 1973,
the A300F being a 150 ton freighter and the A300C a
convertible variant.   These projects had a large side-
loading door.

By 1976 the A300B4/FC (fret convertible) proposal had
emerged, possessing basic compatibility with current
freighters such as the 707-321C, DC-8-63F and 727-25QC.
A total capacity of 294 sq m was envisaged (186 sq m in
upper cabin, 108 sq m for the two lower holds), with a
maximum payload of 40,000 kg.

During 1979 a B4 model, c/n 83, was under modification
to A300C4 standard at VFW-Lemwerder for Hapag-Lloyd
delivery in January 1980, and this aircraft will be able
to change from 281 passenger layout to 41 tonnes freighter.

A300B9

The B9 project was a stretched variant (overall
length 62.1m) with an all-economy layout for 336
passengers (or 309 seat accommodation with first-class and
economy cabins) for operation stages of over 4,000 km.
With a slightly increased wing span (48.44m), the maximum
take-off weight would be around 170 metric tons (374,000
lb).   Power-plants considered were the CF6-50M (rated at
24,950 kpg/55,000 lb) and the JT9D-59A.

A300B10

See next section.

A300B11

This was a proposal for a long-range Airbus (10,000
km) with four engines and passenger accommodation for
205/218 passengers.   Choice of power-plants was
uncertain, the CFM-56 and JT-10D being proposed, with the
CF6-32 and RB211-534 as alternatives.   With a
comparatively short fuselage but extended wing span, the
B11 could sustain long-range cruising speeds of Mach 0.82.

Dimensions given in 1978 were as follows:

Length (overall): 47.49m
Span: 51.74m
Wing area: 315 sq m
Empty weight: 90,700 kg (199,500 lb)

Maximum take-off weight: 195,000 kg (429,000 lb)

A200

By mid 1977 a new short- and medium-haul transport with accommodation for between 120 and 160 passengers was considered by the French SNIAS. With the future of the ASMR Douglas-Dassault project uncertain, the smaller A200 could represent another opportunity for Airbus Industrie. Target date for introduction of the new type into scheduled service was planned for 1982.

With a basic range of 3,000 km (1,600 nm) and CFM-56 power-plants, the A200 is intended to be a Boeing 737, 727 and Douglas DC-9 replacement, with two variants designated A200A and A200B. Current information on the two designs is as follows:

A200A

Span: 34.15m
Length: 34.82m
Height: 11.87m
Empty weight: 36,495 kg
Maximum take-off weight: 66,000 kg
Power-plants: 2 x 22,000 kgp
Accommodation: 12 + 108
Hold capacity: 30.6 sq m
Estimated range of long-range version: 4,925 km
                                        (2,660 nm)

A200B

Span: 34.15m
Length: 40.91m
Height: 11.87m
Empty weight: 39,662 kg
Maximum take-off weight: 72,000 kg
Power-plants: 2 x 22/24,000 kgp or 2 x 25,000 kgp
Accommodation: 16 + 144
Hold capacity: 45.2 sq m
Estimated range of long-range version: 3,740 km
                                        (2,020 nm)
                              or   4,185 km
        depending on power-plant   (2,260 nm)

FROM THE A300B10 TO THE A310

Among the Airbus derivatives was the A300B10 proposal which, after some hesitation, received growing attention from Airbus Industrie design teams early in 1978. This was a projected shortened fuselage version of the basic air-liner for stage lengths of 3,950-5,600 km in the respective SR (short-range) and MR (medium-range) versions, revealing a slightly increased range over initial B10 projects.

On 8 March 1978 it was announced that a Memorandum of Understanding had been signed between the four European companies - SNIAS, British Aerospace, MBB and VFW-Fokker - on joint programmes for the development of a family of short/medium-haul airliners. The family of aircraft was to be the Airbus line on one side (with particular mention of the A300B10 derivative) and a new project to be called JET, in two basic versions for 130 and 160 passengers. The initials JET stood for Joint European Transport, or also Joint Engineering Team. On the continent this announcement was promptly hailed as an increased commitment by the recently nationalised British Aerospace industry to major European programmes. Until then the British contribution to the A300 programme had been limited to that of a private firm, Hawker-Siddeley.

Soon however British Aerospace made it clear that this was not a firm commitment to proceed with a joint European programme but only an agreement on the shape things would take if the project were proceeded with.

During the spring of 1978 Airbus Industrie announced that they soon hoped to take a final decision to launch the B10 version. This was, one should remember, during a booming period for A300 sales, with the Eastern Airlines order and several others under negotiation. With options on the B10 already taken by Eastern, Lufthansa and Air France, a decision had to be taken as soon as possible. In the meantime, the A300B10 had become the A300-10 under an attempt to rationalise the designation system of the growing number of A300 variants. Finally, on 7 July, Airbus Industrie announced the go-ahead decision for development and production of the shortened fuselage version. The new airliner, it was announced, would be designed to carry about 234 passengers, and its most important feature would be a smaller wing using an advanced aerofoil section.

The programme was backed by letters of intent from several major operators, including Eastern Airlines (25), Lufthansa (10), Swissair (six) and Air France (four), with deliveries to begin in 1983.

A few weeks later, during Franco-Spanish talks in Madrid, Iberia also evinced an interest in buying four aircraft. It was decided that the final sizing of the A300-10 would have to be frozen by September or October to ensure fleet introduction by 1983.

By the time the French and German governments were confirming their approval on 13 July for the go-ahead on the A300-10, clouds were gathering, however. First, on 14 July, a new Boeing airliner, the 767, was given the go-ahead following an order for thirty by United Airlines. As a 197-seater (in the United configuration) and with a typical 3,540 km range, the 767 was immediately viewed by most specialists as a direct competitor to the A300-10 programme.

Matters were further complicated in the following month, when British Airways was given government approval to order 19 Boeing 757s to be powered by Rolls-Royce RB211-535 power-plants.

But Eric Varley, the British Minister of Industry, also announced that British Aerospace was willing to acquire 20% of the shares in Airbus Industrie, thus becoming a major partner in the Franco-German concern. The French Minister of Transport, Joël le Theule, immediately declared in a press conference on 31 August that the French Government was not seeking British co-operation with the European manufacturers if the country was also buying American at the same time, a viewpoint which had gradually become clearer since July. The French daily press went into a furore, but curiously the French aeronautical press probably better understood the British attitude, which indeed appeared complicated. On the one side was co-operation with the United States in the order for Boeing 757s and a consequent boost to future sales of RB211 jet engines, and on the other side the joining of a major European aero-nautical programme, the A310 (as the A300-10 had become in the meantime). This was not to mention government approval for full-scale development of the HS/BAe 146, a decision taken in July and which was felt in some quarters as another blow to European co-operation, being in competition with similar VFW-Fokker designs. At the same time, the British order for 757s was a serious

Airbus Industrie A300B2 F-WUAA in Air France colours
(Airbus Industrie)

Concorde F-BVTA of Air France (Aerospatiale via J-P Dubois)

Top   :   SO.30P Bretagne F-OAIX of Air Algerie
            with two underwing Palas jets (via Charles W Cain)

Bottom :   Close-up of Palas auxiliary jets on
            SO.30P F-OAIX (via Charles W Cain)

Top   : SO.30P Bretagne F-BAYZ of Air Atlas-Air Maroc
          at Bordeaux-Merignac in April 1955 (via JMG Gradidge)

Centre : SO.30P Bretagne No.43 of Aeronavale at Le Bourget
          (Blandin/Regnier)

Bottom : SO.30P Bretagne No.29 of the Armee de l'Air at
          Blackbushe in September 1955 (F Hudson)

blow to the proposed JET programme.

In fact the only problem regarding the A310 programme was that British Aerospace would not be a particularly welcome partner unless there was an accompanying British order. While the French remained very strict on this aspect, the German Government was more conciliatory, being anxious not to see the whole British aerospace industry go American and probably also to counter-balance the influence of their only major partner, the French.

Fortunately a compromise was soon to be found on 27 September, following Anglo-French talks in Paris. The French delegation was authorised to express German views, and the difficulties were settled.

First, British Aerospace was to become a full partner, taking a 20% share in the group. But it was also mentioned that British Airways was not to order a direct competitor - which clearly meant the Boeing 767. It became evident that the 180-seat Boeing 757 was not a danger to the A310 early in September when Eastern Airlines, following an order for 21 Boeing 757s on 31 August, confirmed that the options for 25 A310s were to be maintained.

French Government approval of BAe joining Airbus Industrie was finally confirmed on 24 October, to take effect from 1 January 1979. The British partner, by now a welcome major manufacturer, is expected to design and manufacture the wing of the A310 as its 20% share in the programme, as well as its continuing commitment to A300 wing production.

Early in the summer of 1979 Airbus Industrie signed an agreement with a Belgian consortium, Belairbus, for participation in the A310 programme. Belairbus will be responsible for production of the slats, slat tracks and Kruger flaps, apart from development, finance and tooling investment. Belgium's associate status, like that of the Netherlands, made it the sixth participant in Airbus production.

It was too early, of course, to foresee the future of the new A310 airliner, but what was clear was that the European A310 would have a formidable rival in the Boeing 767, which by mid November had been ordered by American Airlines (30 plus 20 on option) and Delta Airlines (with an order for 42), this reducing chances for American A310 orders, the Eastern options being the sole exception.

The fuselage of the A310 was to possess much in common with the A300 but would be thirteen fuselage frames fewer with a length of 46.65m. Compared with the A300, the tail fins would be smaller, but, with the La Guardia problem still remembered the A300B4 undercarriage would be retained.

Power would be provided by two derated General Electric CF6-45s, which had already been chosen by Lufthansa in view of their near 100% commonality with the CF6-50s fitted to the airline's B2/B4s, DC-10s and 747s. Choice of the CF6-45B2 would not require adoption of new engine nacelles, although Airbus Industrie was also working on variants to be powered by the CF6-80, JT9D-7 and Rolls-Royce RB211-534. Obviously the aim is commonality for customers operating JT9D-powered 747s, Rolls-Royce powered L-1011s and Pratt & Whitney-powered DC-10s.

Short and medium range versions are planned (with -100 and -200 suffixes respectively), and the designation system of the A300 is also used (eg -101 and -201 for aircraft powered by CF6-45s, -102 and -202 for CF6-80s, -120 and -220 for JT9D-7(46), -140 and -240 for RB211s). Typical ranges with a full payload of 20,500 kg are 1,750 nm (3,238 km) for the short-range version and 3,000 nm (5,550 km) for the medium-range. Provisions will be made for improved freight operation, including LD3 and also LD-6F standard containers in the lower hold.

The first flight of the A310 is planned for March 1982, with certification following a year later. Orders and options to mid 1979 are given below:

	Orders	Options
Air Afrique	2	–
Air France	5	10
KLM	10	10
Lufthansa	25	25
Swissair*	10	10
	52	55

*JT9D

Specifications

Proposed specification data for the short and medium-range A310 variants are given below:

A310-101

Maximum take-off weight: 267,860 lb
Maximum landing weight: 254,630 lb
Manufacturer's empty weight: 149,438 lb
Seating: 213
Maximum fuel capacity: 64,820 lb

A310-201

Maximum take-off weight: 288,800 lb
Maximum landing weight: 262,350 lb
Manufacturer's empty weight: 151,431 lb
Seating: 210
Maximum fuel capacity: 99,200 lb

Production

In the following A300 Production List, the first 100 aircraft are detailed, bringing information current to late 1979.

A brief note about delivery dates is in order : those quoted after del in the text refer to the dates given in the official paperwork and may differ from the actual handover or fly-away dates. Where these are known, they are also quoted, together with the airport of arrival in the airline's home country.

C/n D01   B1. F/f 28 Oct 72 as F-WUAB. Re-regd
           F-OCAZ 13 Sep 73 still to Airbus Industrie.
Final flight 31 Aug 74 and WFU.

C/n D02   B1. F/f 05 Feb 73 as F-WUAC. Painted in
           Iberia colours before acquisition by Trans
European Airways as OO-TEF. Lsd immediately and del
direct to Air Algerie 25 Nov 74. Retd to TEA 19 Jan
75.

C/n D03   B1/B2-1C. F/f 28 Jun 73 as F-WUAD. F-ODCX
           allocated but NTU. Re-regd F-BUAD
78, still to Airbus Industrie, for use as training
and demonstration a/c. Cold weather trials in Sweden
Jan 79.

C/n D04   B2-1C. F/f 20 Nov 73 as F-WUAA. Flown in
           Air France colours prior to del 12 Jan 77 to
Air Inter as F-BUAE. Left Toulouse 26 Jan 77.

C/n N05   B2-1A/1C. F/f 15 Apr 74 as F-WVGA. To Air
           France as F-BVGA del 08 May 74, ferried Tou-
louse-Paris 11 May 74.

C/n N06   B2-1A/1C. F/f 23 Jun 74, carrying F-WVGB
           underwing only. To Air France as F-BVGB del
28 Jun 74, ferried Toulouse-Paris same day.

C/n N07   B2-1A/1C. F/f 06 Aug 74, carrying F-WVGC
           underwing only. To Air France as F-BVGC del
11 Aug 74 and ferried Toulouse-Paris same day.

C/n N08   B2-1C. Originally painted as F-WNDB but roll
           out and first flight as HS-VGD. F/f 02 Oct
74. To Air Siam del 17 Oct 74, and left Toulouse 21
Oct 74. Ent serv same day, named "Chao Praya". Retd
to Airbus Industrie 13 Oct 75. Intended to become
F-BDHC but NTU and re-regd from F-WNDB to F-ODHC 16 Nov
75. Lsd to Transavia 08 May 76 - 15 Jan 77, del to
Amsterdam 11 May 76 as PH-TVL, named "Apollo 76". Retd
to Airbus 17 Jan 77 and re-regd F-WNDB. To Air Inter
as F-BUAF (CofA 22 Feb 78), del 11 Mar 78.

C/n N09   B4-2C. F/f 26 Dec 74 as F-WLGA. Intended to
           become HS-VGF but NTU. Re-regd F-ODCY 04 Jun
76 (CofA 31 May 76). Lsd to Air France 16-27 Jul 76.
To Bavaria-Germanair as D-AMAP del 20 Feb 78. Lsd to
Hapag-Lloyd by 24 Mar 79.

C/n N10   B2-1A/1C. F/f 07 Mar 75 as F-BVGD. To Air
           France del 21 Mar 75 and ferried Toulouse-
Paris same day.

C/n N11   B2-1A/1C. F/f 23 Apr 75 as F-BVGE. To Air
           France del 30 Apr 75 and ferried Toulouse-
Paris next day.

C/n N12   B4-2C. F/f 20 May 75 as F-WLGC. To Germanair
           as D-AMAX del 23 May 75 and left Toulouse same
day. Airline renamed Bavaria-Germanair from 01 Jan 77,
and a/c named "Maximilian". Airline bought by Hapag-
Lloyd   Apr 77, in Hapag-Lloyd titles by May 79, but in
BV titles Jun 79.

C/n N13   B2-1A/1C. F/f 31 May 75 as F-BVGF. To Air
           France del 11 Jun 75 and ferried Toulouse-
Paris same day.

C/n N14   B4-2C. F/f 23 Jul 75 as F-WLGB. Originally
           allocated to Iberia order, but subsequently
cancelled. To Korean Air Lines as HL7218 del 31 Jul 75.
Ferried Toulouse-Seoul 09 Aug 75, ent serv 28 Aug 75.

C/n N15   B2-1C. F/f 18 Jun 75 as F-WLGC. To Air Inter
           as F-BUAG del 15 Oct 76, ferried Toulouse-Paris
25 Oct 76. Lsd to Air Charter International Jun 79.

C/n N16   B4-2C. F/f 22 Aug 75 as F-WLGB. Originally
           allocated to Iberia order, but subsequently
cancelled. To Korean Air Lines as HL7219 del 31 Aug 75
and ferried Toulouse-Seoul 09 Sep 75.

C/n N17   B4-2C. F/f 07 Oct 75 as OO-TEG. Originally
           allocated to Iberia order, but subsequently
cancelled. To Trans European Airways del 16 Oct 75,
left Toulouse 22 Oct 75, named "Adrianus Andreas Jr".
Lsd to Zaire Aero Service, still as OO-TEG, Nov 76 -
11 Dec 76. Retd to TEA and then lsd to Egyptair   Apr
77. Sold to Hapag-Lloyd as OO-TEG 02 Jul 79, still lsd
to Egyptair and re-regd SU-BBS

C/n N18   B4-2C. F/f 23 Oct 75 as HL7220. To Korean Air
           Lines del 11 Nov 75 and ferried Toulouse-Seoul
next day.

C/n N19   B4-2C. F/f 11 Nov 75 as F-BVGG. To Air France
           del 17 Nov 75, and ferried Toulouse-Paris same
day. Flown to New York 18 Nov 75 for Air France West Indies service. Hijacked 27 Jun 76 and flown to Entebbe,
Uganda. Retd 22 Jul 76 to Paris.

C/n N20   B4-2C. F/f 19 Dec 75 as D-AMAY. To Germanair
           del 30 Mar 76, named "Ludwig I", left Toulouse
01 Apr 76. Airline renamed Bavaria-Germanair from 01 Jan
77 and bought by Hapag-Lloyd   Apr 77. A/c still in BV
colours and titles Apr 79.

C/n N21   B2-1C. F/f 29 Nov 75 as F-WNDA. To Lufthansa
           as D-AIAA del 02 Feb 76, named "Garmisch-Parten-
kirchen". Ferried Toulouse-Frankfurt 08 Feb 76.

C/n N22   B2-1C. F/f 23 Jan 76 as F-WNDC. To Lufthansa
           as D-AIAB del 17 Mar 76, named "Rüdesheim-am-
Rhein". Ferried Toulouse-Frankfurt 19 Mar 76.

C/n N23   B4-2C. F/f 12 Feb 76 as F-WVGH. To Air France
           as F-BVGH del 13 Apr 76, ferried Toulouse-Paris
14 Apr 76.

C/n N24   B4-2C. F/f 03 Mar 76 as F-WNDD. To Korean Air
           Lines as HL7221 del 21 Apr 76, ferried Toulouse-
Seoul 23 Apr 76.

C/n N25   B4-2C. F/f 23 Mar 76 as F-WNDA. To Bavaria-
           Germanair as D-AMAZ del 29 Apr 77. Left Toulouse
10 May 77. Lsd to Egyptair as D-AMAZ from 10 May 77. Re-
regd SU-AZY 19 Jan 78 on lease-purchase.

C/n N26   B2-1C. F/f 11 Mar 76 as F-WNDB. To Lufthansa
           as D-AIAC del 24 Apr 76, named "Lüneburg", and
ferried Toulouse-Frankfurt next day.

C/n N27   B2-1C. F/f 03 May 76 as F-WLGB. Re-regd F-WLGC,
           still to Airbus Industrie, 06 Nov 76. To Air
Inter as F-BUAH del 28 Sep 76 and ferried Toulouse-Paris
same day. Lsd to Air Charter International Jun 79.

C/n N28   B4-2C. F/f 16 Apr 76 as F-WNDC. To Korean Air
           Lines as HL7223 del 06 Jul 76, and ferried Toul-
ouse-Seoul 08 Jul 76.

C/n N29   B4-102. F/f 08 May 76 as F-WNDD. Lsd to Aero-
           Condor Colombia, and carried F-WNDD and HK-2057X
on proving flights. Del 10 Dec 77 as HK-2057 and flown
Toulouse-Barranquilla same day, named "Ciudad de Barranquilla". Repossessed by Airbus Industrie and retd to Toulouse   79. Re-regd F-ODJU to Airbus Industrie.

C/n N30   B4-2C. F/f 04 Jun 76 as F-WNDB. To Korean Air
           Lines as HL7224 del 24 Feb 77, and ferried Tou-
louse-Seoul 03 Mar 77.

C/n N31   B4-2C. F/f 08 Jul 76 as F-WUAY, although first
           allocated F-WLGC but NTU. Intended for conver-
sion to A300C4 standard for demonstration at 1978 Hannover
Air Show, but conversion not undertaken. Re-regd F-WZEQ
02 Jun 78. To Korean Air Lines as HL7238 del 09 Aug 78
and ferried Toulouse-Seoul 10 Aug 78.

C/n N32   B2K-3C. F/f 30 Jul 76 as F-WLGA. To South
           African Airways as ZS-SDA del 12 Nov 76, named
"Blesbok". Ferried Toulouse-Johannesburg 23 Nov 76.

C/n N33   B4-2C.   F/f 21 Sep 76 as F-WNDC. To Thai Air-
           ways International as HS-TGH del 25 Oct 77,
named "Srimuang". Ferried Toulouse-Bangkok 26 Oct 77.

C/n N34   B2-1C.   F/f 30 Aug 76 as F-WLGB. To Indian
           Airlines as VT-EDV del 28 Oct 76. Ferried
Toulouse-Bombay 10 Nov 76.

C/n N35   B4-2C.   F/f 02 Feb 77 as F-WLGA. To Thai Air-
           ways International as HS-TGK del 14 Dec 77,
named "Suranaree". Ferried Toulouse-Bangkok 15 Dec 77.

C/n N36   B2-1C.   F/f 05 Oct 76 as F-WUAT, although first
           allocated F-WLGB but NTU. To Indian Airlines
as VT-EDW del 27 Nov 76, and ferried Toulouse-Bombay 30
Nov 76.

C/n N37   B2K-3C.   F/f 02 Nov 76 as F-WUAU. To South
           African Airways as ZS-SDB del 22 Dec 76, named
"Gemsbok". Ferried Toulouse-Johannesburg 23 Dec 76.

C/n N38   B2-1C.   F/f 16 Nov 76 as F-WUAV. To Indian
           Airlines as VT-EDX del 28 Dec 76, and ferried
Toulouse-Bombay 30 Dec 76.

C/n N39   B2K-3C.   F/f 01 Dec 76 as F-WLGB. To South
           African Airways as ZS-SDC del 24 Jan 77 and
ferried Toulouse-Johannesburg next day, named "Waterbok".

C/n N40   B2K-3C.   F/f 13 Dec 76 as F-WUAX. To South
           African Airways as ZS-SDD del 14 Feb 77, and
ferried Toulouse-Johannesburg 16 Feb 77, named "Rooi-
bok".

C/n N41   B4-2C.   F/f 11 Jan 77 as F-WUAZ. Lsd to
           Eastern Airlines as N201EA, del 03 Dec 77 and
ferried Toulouse-New York same day. Purchased by Eastern Airlines

C/n N42   B4-2C.   F/f 24 Feb 77 as F-WUAU. Lsd to
           Eastern Airlines as N202EA, del 19 Nov 77 and
ferried Toulouse-Miami same day. Purchased by Eastern
Airlines

C/n N43   B4-2C.   F/f 13 May 77 as F-WUAT. Lsd to
           Eastern Airlines as N203EA, del 29 Oct 77 and
ferried Toulouse-New York same day. Purchased by Eastern Airlines

C/n N44   B4-2C.   Originally F-WNDB, then painted as
           HS-TGH, and repainted as F-WUAX in Eastern
colours for f/f 13 Jul 77. Lsd to Eastern Airlines as
N204EA, del 24 Aug 77 and ferried Toulouse-New York same
day. Purchased by Eastern Airlines

C/n N45   B4-2C.   F/f 18 Feb 77 as F-BVGI. To Air
           France del 23 Mar 77 and ferried Toulouse-
Paris next day.

C/n N46   B4-102.   F/f 28 Mar 77 as F-WLGB. Intended
           for Thai Airways International but NTU and re-
regd F-WZER   78. To Olympic Airways as SX-BEB
del 31 Jan 79, named "Odysseus". Ferried Toulouse-
Athens 22 Feb 79.

C/n N47   B4-2C.   F/f 22 Apr 77 as F-BVGJ, although
           first allocated F-WUAX but NTU. To Air France
del 20 Oct 77 and ferried Toulouse-Paris same day.

C/n N48   B2-1C.   F/f 14 Mar 77 as F-WNDB. To Lufthansa
           as D-AIAD del 16 Apr 77, named "Westerland-
Sylt". Ferried Toulouse-Frankfurt 16 Apr 77. Renamed
"Westerland" by 1979.

C/n N49   B2-202.   F/f 10 May 77 as F-WUAV. Re-regd
           F-ODHY (CofA 22 Feb 78) and lsd to Iran Air del
07 Mar 78. Ferried Toulouse-Istanbul-Tehran same day,
named "Kermanshahan". Retd to Airbus Industrie and
stored at MBB Hamburg 06 Apr 79. Re-regd F-WZES prior
to Feb 78. To Eastern Airlines as N291EA del   Jan 80,
ferried Toulouse-

C/n N50   B2-1C.   F/f 16 Jun 77 as F-WNDB. Re-regd
           F-WZET   , and again F-GBEA   Jun 78.
To Air France del 21 Jun 78 and ferried Toulouse-Paris
same day.

C/n N51   B2-202.   F/f 03 Feb 78 as F-ODHZ, although first
           allocated F-WZEA but NTU. Lsd to Iran Air as
F-ODHZ (CofA 08 Mar 78), del 16 Mar 78 and ferried Toulouse-
Tehran same day, named "Hormozgan". Retd to Airbus Industrie and stored at MBB Hamburg 06 Apr 79. To Eastern Airlines as N292EA del   Jan 80, and ferried Toulouse-

C/n N52   B2-1C.   F/f 17 Nov 77 as F-WZEB. To Lufthansa
           as D-AIAE del 07 Jan 78, named "Neustadt an der
Weinstrasse", and ferried Toulouse-Frankfurt same day.

C/n N53   B4-2C.   F/f 16 Aug 77 as F-WZEE. To Lufthansa
           as D-AIBA del 29 Sep 77, named "Rothenburg ob der
Tauber", and ferried Toulouse-Frankfurt same day.

C/n N54   B4-2C.   F/f 05 Jan 78 as F-WZED. To Thai Air-
           ways International as HS-TGL del 17 Feb 78, named
"Srisoonthorn", and ferried Toulouse-Bangkok same day.

C/n N55   B4-2C.   F/f 01 Mar 78 as F-WZEC. To Thai Air-
           ways International as HS-TGM del 14 Apr 78, named
"Thepsatri", and ferried Toulouse-Bangkok 18 Apr 78.

C/n N56   B4-102.   F/f 19 Dec 78 as F-WZEF. Originally
           intended for Pakistan International but instead
to Olympic Airways as SX-BEC del 13 Feb 79, named "Achilles".
Ferried Toulouse-Athens 22 Feb 79.

C/n N57   B4-2C.   F/f 23 Jan 78 as F-WZEG. To Lufthansa
           as D-AIBB del 23 Mar 78, named "Freudenstadt-
Schwarzwald", and ferried Toulouse-Frankfurt same day.

C/n N58   B4-102.   F/f   as F-WZEH. Originally
           intended for Philippine Airlines but instead
to Olympic Airways as SX-BED del   , named

C/n N59   B2-1C.   F/f 29 Mar 78 as F-WZEI. To Indian
           Airlines as VT-EDY del 11 May 78, and ferried
Toulouse-Bombay 12 May 78.

C/n N60   B2-1C.   F/f 19 Apr 78 as F-WZEJ. To Indian
           Airlines as VT-EDZ del 08 Jun 78, and ferried
Toulouse-Bombay next day.

C/n N61   B2-203.   F/f   as F-WZEK. Originally
           scheduled for del to Iran Air May 79, but del
suspended due to revolution. Del re-scheduled as EP-IBR
for Feb 80

C/n N62   B2-1C.   F/f 18 Aug 78 as F-BUAI. To Air Inter
           del 13 Oct 78, and ferried Toulouse-Paris same
day. Lsd to Air Charter International Jun 79.

C/n N63   B4-103.   F/f 10 Oct 79 as F-WZEL. Originally
           intended for 1se to Pakistan International, then
painted in Olympic colours Jun 78, and finally sold to
Philippine Airlines as RP-C3001 del 28 Nov 79. Ferried
Toulouse-Manila 29-30 Nov 79.

C/n N64   B4-103.   F/F 27 Feb 79 as F-WZEM. Ordered by
           Bavaria-Germanair, but order taken over by Hapag-
Lloyd Apr 77. Del to Hapag-Lloyd 02 Apr 79 as D-AHLA(2),
and ferried Toulouse-Hannover 04 Apr 79.

C/n N65   B4-2C.   F/f 13 Sep 78 as F-WZEN. Re-regd
           F-GBNA   . To Eastern Airlines del 10 Nov
78. Ferried Toulouse-Atlanta as F-GBNA 13 Nov 78, and re-
regd N205EA at Atlanta.

C/n N66   B4-2C.   F/f 06 Oct 78 as F-WZEO. Re-regd
           F-GBNB   . To Eastern Airlines del 30 Nov
78. Ferried Toulouse-New York as F-GBNB same day, and re-
regd N206EA at Atlanta.

C/n N67   B4-2C. F/f 20 Oct 78 as F-WZEP. Re-regd
          F-GBNC   . To Eastern Airlines del 07 Dec
78. Ferried Toulouse-New York 10 Dec 78 as F-GBNC, and
re-regd N207EA at Atlanta.

C/n N68   B4-103. Originally intended for Philippine Air-
          lines with f/f regn F-WZEA but NTU. Instead
f/f 06 Aug 79 as F-GBND. H/o to Eastern Airlines at
Toulouse 05 Oct 79, ferried to Atlanta as F-GBND 09 Oct
79 and officially del 12 Oct 79. Re-regd N208EA at
Atlanta.

C/n N69   B4-103. F/f 23 Oct 79 as F-WZEB. To Philippine
          Airlines as RP-C3002 del

C/n N70   B4-203. F/f 22 Feb 79 as F-BVGK. To Air France
          del 26 Apr 79 and ferried Toulouse-Paris next
day.

C/n N71   B4-2C. F/f 18 Dec 78 as F-WZEC. To Thai Air-
          ways International as HS-TGN del 02 Mar 79,
named "Sudawadi". Ferried Toulouse-Bangkok 06 Mar 79.

C/n N72   B4-2C. F/f 22 Jan 79 as F-WZED. To Thai Air-
          ways International as HS-TGO del 14 Mar 79,
named "Srichulalak". Ferried Toulouse-Bangkok 16 Mar 79.

C/n N73   B4-203. F/f 12 Jul 79 as F-WZEE. To Malaysian
          Airline System as 9M-MHA del 30 Oct 79, and
ferried Toulouse-Kuala Lumpur 03 Nov 79.

C/n N74   B4-203. F/f 27 Mar 79 as F-BVGL. To Air France
          del 14 May 79, and ferried Toulouse-Paris next
day.

C/n N75   B4-2C. F/f 31 Jan 79 as F-WZEG. To Lufthansa
          as D-AIBC del 23 Mar 79, named "Lindau-Bodensee".
Ferried Toulouse-Frankfurt 23 Mar 79.

C/n N76   B4-2C. F/f 19 Feb 79 as F-WZEI. To Lufthansa
          as D-AIBD del 30 Mar 79, named "Erbach-
Odenwald". Ferried Toulouse-Frankfurt 30 Mar 79.

C/n N77   B4-2C. F/f 08 Mar 79 as F-WZEJ. To Lufthansa
          as D-AIBF del 19 Apr 79 and ferried Toulouse-
Frankfurt 22 Apr 79, named "Krohnberg-Taunus". Operated
by Condor from del to 20 Jan 80, named "Philharmonie
Hamburg". Retd to Lufthansa and renamed "Krohnberg-
Taunus".

C/n N78   B4-203. F/f 06 Apr 79 as F-BVGM. To Air France
          del 31 May 79 and ferried Toulouse-Paris same
day.

C/n N79   B2-320. Painted as LN-RCA in SAS colours for
          publicity photos before f/f 28 Apr 79 as F-WZEN.
First A300 with Pratt & Whitney JT9D-59A engines. To
Scandinavian Airline System as LN-RCA del   ,
named "Snorre Viking".

C/n N80   B2-203. F/f   as F-WZEO. Originally
          scheduled for del to Iran Air May 79, but del
suspended due to revolution. Del re-scheduled as EP-IBS

C/n N81   B4-   F/f 09 May 79 as F-WZEP. To Korean
          Air Lines as HL7246 del 27 Jun 79, and ferried
Toulouse-Seoul 29 Jun 79.

C/n N82   B2-203. F/f   as F-WZEQ. To TOA
          Domestic as JA

C/n N83   C4-203. F/f 16 May 79 as F-WZES. Originally
          built as a B4 without a cargo door, but cvtd to
C4 standard by a VFW/Lemwerder cargo door retrofit. To
Hapag-Lloyd as D-AHLB del

C/n N84   B4-2C. F/f 08 Jun 79 as F-WZET. To Thai Air-
          ways International as HS-TGP del 09 Aug 79, named
"Srisubhan". Ferried Toulouse-Bangkok 09 Aug 79.

C/n N85   B4-2C. F/f 03 Jul 79 as F-WZEC. To Thai Air-
          ways International as HS-TGR del 06 Sep 79, named
"Thepamat", and ferried Toulouse-Bangkok next day.

C/n N86   B4-103. F/f 26 Jul 79 as F-GBNE. H/o to
          Eastern Airlines at Toulouse 08 Nov 79, ferried
to Atlanta as F-GBNE 12 Nov 79 and officially del 14 Nov
79. Re-regd N209EA at Atlanta.

C/n N87   B4-103. F/f 22 Aug 79 as F-GBNF. H/o to
          Eastern Airlines at Toulouse 08 Nov 79, ferried
to Atlanta as F-GBNF 12 Nov 79 and officially del 15 Nov
79. Re-regd N210EA at Atlanta.

C/n N88   B2-1C. F/f 03 Sep 79 as F-WZED. To Indian
          Airlines as VT-EFV del 25 Oct 79, ferried
Toulouse-Bombay

C/n N89   B2-203. F/f   as F-WZEF. To TOA Domes-
          tic as JA

C/n N90   B2-203. F/f   as F-WZEG. To TOA Domes-
          tic as JA

C/n N91   B4-103. F/f 26 Sep 79 as F-GBNG. H/o to
          Eastern Airlines at Toulouse 26 Nov 79, ferried
to Atlanta as F-GBNG 28 Nov 79 and officially del
. Re-regd N212EA at Atlanta.

C/n N92   B4-103. F/f 17 Oct 79 as F-GBNH. H/o to
          Eastern Airlines at Toulouse 06 Dec 79, ferried
to Atlanta as F-GBNH 09 Dec 79 and officially del
. Re-regd N213EA at Atlanta.

C/n N93   B4-203. F/f 18 Oct 79 as F-WZEI. To Malaysian
          Airline System as 9M-MHB del

C/n N94   B2-320. F/f 06 Dec 79 as F-WZEJ. To SAS as
          SE-DFK del   . Ferried Toulouse-
                   , named "Sven Viking".

C/n N95   B4-203. F/f 20 Nov 79 as F-WZEM. To Malaysian
          Airline System as 9M-MHC del

C/n N96   B4-203. F/f 30 Nov 79 as F-WZEP. To Pakistan
          International Airlines as AP-BAX del

C/n N97   B2-1C. F/f   as F-BUAJ. To Air Inter
          del

C/n N98   B4-203. F/f   as F-WZER. To Pakistan
          International Airlines as AP-BAY del

C/n N99   B4-103. F/f   as F-WZET. To Pakistan
          International Airlines as AP-BAZ del

C/n N100 B4-203. F/f   as F-BVGN. To Air France
          del

C/n N101   F/f   as   .

C/n N102   F/f   as   .

C/n N103   F/f   as   .

C/n N104   F/f   as   .

CONCORDE

So much has been written on Concorde that it was decided to keep this introduction deliberately short in order to avoid duplication. It is felt that even with the small number of aircraft currently flying, a whole monograph could be devoted to this single subject, such was the impact of the supersonic airliner on world aviation, at least as far as technology and controversy were concerned.

The beginnings may be traced back to the individual design work undertaken in Britain by the Bristol Aeroplane Company and by Sud Aviation in France. The original 1959 joint Sud-Dassault proposal was called the Super Caravelle, and the modified French programme of late 1961 grew closer to the British work produced under the Bristol 198 contract. With previous experience of a friendly exchange of data over the Channel, international co-operation between the two countries soon became inevitable. An agreement was signed for the joint development of the SST on 25 October 1962, with government ratification from both sides of the Channel following in November.

Work was shared equally between the two countries, with two separate final assembly lines being adopted. Production work was agreed on 40% and 60% shares, the latter part going to Sud Aviation, which had been nominated design leader. Regarding the power-plants, Bristol-Siddeley (later a division of Rolls-Royce Limited) had already signed an agreement on 28 November 1961 with the French SNECMA to start collaboration on the Bristol Olympus engines selected for the supersonic transport. Bench testing was also improved by in-flight tests with Vulcan XA903. Engine work development was shared with a British majority (60% for Bristol-Siddeley and 40% for SNECMA).

The design and test programme was huge, including research aircraft such as the BAC 221 (the rebuilt Fairey FD.2 WG774) and the HP-115, built to investigate the handling characteristics of the future airliner. French experience had been gained from such aircraft as the SE-212 Durandal and SO-9000/9050 Trident, which with the Nord 1500 Griffon had pioneered supersonic flights in France, not to mention data obtained from the successful line of Dassault fighters, including a specially modified Mirage IIIB.

The technological jump required to produce a supersonic airliner was enormous and will not be detailed here. When the American proposals (such as the Boeing 2707) were abandoned, only the Russian TU-144 was left as a competitor but of course with no hopes of being marketed outside the Soviet Union.

Thus the Franco-British (or Anglo-French) supersonic airliner remained virtually alone on the western scene, with world attention focused on it. Cost escalation was high, and this and other arguments soon gave rise to controversy in both countries (but possibly less in France).

By 1967 however there were already 74 Concorde option holders. Following initial Air France and BOAC contracts for eight aircraft each, the American airlines had shown interest in the SST. Pan American first had six reservations in June 1963 (later eight),

immediately followed by Continental (three) and American (four and later six) and TWA with a similar number. Many more customers joined the option line: MEA, Qantas, Air India, JAL, Sabena, United, Eastern, Braniff, Lufthansa, Air Canada and, at a later date, Communist China. But the conversion of these options into firm contracts was never finalised.

When the first Concorde was flown in March 1969, with a specially allocated registration F-WTSS (for Transport Supersonique, the British prototype being G-BSST), the event was seen as a major step in scientific progress, although accompanied by growing doubts on the feasibility of profit-making commercial supersonic air transport. With the successful progress of the flight test programme, a reduction in the planned production became inevitable by 1973 with the decisions of several airlines to drop their options. But apparently it was too late to consider cancellation of the whole costly programme, which was therefore proceeded with, causing even more controversy. The Concorde programme became not only a matter of national prestige but also a serious gamble.

By early 1974 six Concordes had been flown, including 001, which had made its final flight in October 1973 to be delivered to the French Musée de l'Air at Le Bourget. The fact that this was done without publicity shows that there were growing doubts about the programme, so great was the fear of irritating public opinion with so much money seemingly wasted on an aircraft relegated to a museum.

As far as French production is concerned, the first production aircraft was flown in December 1973 (the first British production aircraft flying two months later), and certification trials were undertaken during 1974. Then the Supersonic Transport started making more frequent headlines with the controversy over US landing rights, which were won only after a long legal battle. With certification granted in 1975, route proving flights were immediately started, and at long last the debut of supersonic commercial air transport began on 21 January 1976 with the departure of a British Airways SST from London-Heathrow for Bahrain and an Air France Concorde from Paris-Roissy to Dakar and Rio de Janeiro. On 9 April the Air France route to Venezuela was inaugurated, and on 24 May aircraft of both airlines landed almost simultaneously at Washington-Dulles.

Reliability was immediately praised by both airlines, but profitability remained a major problem and is said to start from a 2,750 hours yearly operation and a ten-year amortisation period, provided the regular load factor of 60% is fulfilled. With the third anniversary of scheduled Concorde services already celebrated, landings at New York being supplemented by route extensions to Dallas-Fort Worth and Singapore (in conjunction with Braniff and Singapore Airlines respectively), slow headway was being made as 1979 began. Humourists of the year joked that Concorde was making scarcely more noise than the environmentalists who had fought against it. However with only Chinese and Iranian interest, officially at least, and no more than 16 aircraft authorised for production, this history of Concorde development is likely to remain as an outstanding - if costly - first for the two European countries involved, France and Great Britain.

Specification

Pre-Production Concorde

Span: 25.6m
Length: 58.83m
Height: 11.58m
Wing area: 358.2 sq m
Operating empty weight: 72,120 kg
Maximum take-off weight: 166,465 kg
Payload: 12,700 kg
Power-plants: 4 x Bristol-SNECMA Olympus 593B turbojets
              rated at 14,890 kgp (stage 0) and 15,900
              kgp (stage 1 with 9% afterburning)
Maximum cruising speed: Mach 2.2
Service ceiling: 16,765-18,900m
Take-off distance: 2,805m

Production Concorde

Span: 25.56m
Length: 61.74m
Height: 11.32m
Wing area: 358.25 sq m
Operating empty weight: 78,830 kg
Maximum take-off weight: 181,440 kg
Basic payload: 10,000 kg depending on fuel weight
Certification: 128 passengers plus 5/6 crew
Power-plants: 4 x Rolls-Royce/SNECMA Olympus 593 Mk 610
              with nominal static thrust of 17,260 kgp
              at sea level with afterburning or 4,550
              kgp at Mach 2 cruise, 53,000 ft ISA + 5°C

Production

French

C/n 001 F-WTSS. F/f 02 Mar 69. Last flt 19 Oct 73
        when del to Musée de l'Air, Le Bourget, Paris.

C/n 02 F-WTSA. F/f 10 Jan 73. Last flt Apr 76 when
        del to Orly, Paris, for permanent static
exhibition.

C/n 1   F-WTSB. F/f 06 Dec 73. Currently oper for
        crew training and French presidential journeys.

C/n 3   F-WTSC. F/f 31 Jan 75. To F-BTSC SNIAS (CofA
        22 May 75). Cvtd to Air France standards 1977/
78.

C/n 5   F-BVFA. F/f 25 Oct 75. To Air France, del 19
        Dec 75.

C/n 7   F-BVFB. F/f 06 Mar 76. To Air France, del 09
        Apr 76.

C/n 9   F-BVFC. F/f 09 Jul 76. To Air France, del 03
        Aug 76.

C/n 11 F-BVFD. F/f 10 Feb 77. To Air France, del 26
        Mar 77.

C/n 13 F-WJAM. F/f 26 Jun 78. To F-BTSD SNIAS (CofA
        20 Sep 78). To Air France del   1978. Lsd
Oct 78 to Braniff International Airways for crew
training, still in Air France colours.

C/n 15 F-WJAN. F/f 26 Dec 78. Re-regd F-BVFF Jun 79.

British

C/n 13520/002 G-BSST. F/f 09 Apr 69. To RNAS Yeovilton
                       for exhibition on behalf of
Science Museum.

C/n 13522/01 G-AXDN. F/f 17 Dec 71. To Duxford
                      for exhibition on behalf of Imperial War
Museum.

C/n 13523/02 G-BBDG. F/f 13 Feb 74.

C/n 100-004 G-BOAC. F/f 27 Feb 75. To British
                      Airways Board, del 13 Feb 76, ent serv 16
Feb 76. Re-regd G-N81AC 05 Jan 79 and regd N81AC for
Braniff International Airways' oper.

C/n 100-006 G-BOAA. F/f 05 Nov 75. To British
                      Airways Board, del 14 Jan 76, ent serv 21
Jan 76. Re-regd G-N94AA 12 Jan 79 and regd N94AA for
Braniff International Airways' oper.

C/n 100-008 G-BOAB. F/f 18 May 76. To British
                      Airways Board, del 30 Sep 76. Re-regd
G-N94AB 12 Jan 79 and regd N94AB for Braniff International
Airways' oper.

C/n 100-010 G-BOAD. F/f 25 Aug 76. To British
                      Airways Board, del 06 Dec 76. Carries
Singapore Airlines colours on one side. Re-regd G-N94AD
09 Jan 79 and regd N94AD for Braniff International
Airways' oper.

C/n 100-012 G-BOAE. F/f   . To British
                      Airways Board, del 20 Jul 77. Re-regd
G-N94AE 05 Jan 79 and regd N94AE for Braniff International
Airways' oper.

C/n 100-014 G-BFKW. F/f 21 Apr 78. To British
                      Airways Board.

C/n 100-016 G-BFKX. F/f 20 Apr 79. To British
                      Airways Board.

Whichever way the historian approaches a study of world air transport, he will find the subject filled with the names of Douglas Commercial series, the Boeing jetliners and Lockheed transports.   Justice would also be done to the British aircraft industry by mentioning the Comet and Viscount.   By then, very little space remains for French airliners.

France had been a leading aeronautical country pre-war, but so much time and experience had been wasted during the occupation years of 1940-44, followed by the uncertain post-war political and economic stability, that the country never regained any superiority.   Only from the early sixties, with great economic progress achieved at national level, was the aircraft industry able to compete seriously on world markets.

However, from the start France continued to design and build airliners which had apparently few sales prospects when compared to their American and British equivalents.   This was often undertaken with high financial losses. The French highly-centralised system and early nationalised industries with government subsidies allowed only eight different types of airliners to enter commercial service.   Whether the system is good or bad remains a matter of opinion, but the aircraft for study are listed below in brief.

The four-engined SE-161 transport was a pre-war design but was plagued with teething troubles for many years; yet 100 were built. The SE-2010 - hailed with abuse by the French aeronautical press - was no match for its American four-engined competitors.   The Latécoère 631 was a stubborn venture into the long-gone era of commercial flying-boat operation.

Then there was the Caravelle....

It is true that the aircraft was acclaimed as a major success by the French, but was it really? By the then prevailing standards in France. admittedly yes; but what if compared to world sales of American jetliners?

The Nord 262 was apparently a promising short-haul airliner but did not sell very well, although it has now been used in many countries the world over.   Ironically, as a revival of interest was being shown with the United States Mohawk 298 conversion, the French production line was being closed down.   Insufficient capacity may be blamed, as had been the case with the SO-30 Bretagne years earlier.

The two airliners not designed by the nationalised industry were the Breguet Deux-Ponts and Dassault Mercure.   Both were apparently good aircraft but failed to sell well.

An unrealistic government policy in the Deux-Ponts programme accelerated the decline of Bréguet and its disappearance from the French aeronautical scene, the name now only surviving in the company AMD-BA (Avions Marcel Dassault-Bréguet Aviation, usually referred to as 'Dassault-Bréguet').   The Mercure is a different story - related elsewhere - and Dassault were in no danger with their successful range of military and business jets.

Last but not least, a major problem with French airliners has always been the problem of cost: when aircraft are produced in small numbers the unit cost will be increased auto-matically, and sales become more difficult.

Mention should also be made of the Aérospatiale AS-35 project, announced at the 1979 Le Bourget Salon.   This is intended as a high-wing twin turbo-prop commuter airliner for 36-44 passengers, but it is naturally too early at the present stage to pre-dict its future.

SO-30P BRETAGNE

During the occupation of France the "Groupe Technique de Cannes", an offshoot of SNCASO, had begun work on a modern twin-engined air-liner, the SO-30N, which was to be called Bellatrix under the new programme for aircraft names issued in March 1943 by the Secretariat d'Etat à l'Air, which stated that the names for all SNCASO designs should begin with the letter B.   When the "Zone Sud" (or unoccupied southern France) was invaded in November 1942, the prototype had already been prepared for its first test flight, but permission to initiate this was not granted by the Armistice Commission, so the Bellatrix was promptly dis-mantled and the components hidden in various places near Draguignan, then the capital of the Département du Var, a hilly and scarcely populated area.   (One recalls that a Latécoère 631 was also preserved in similar conditions during the same period, but with German consent.)

Preventing the SO-30N from falling into enemy hands was a success, and after the liberation

of the area starting in August 1944, the components were recovered and the Bellatrix was flown in February 1945.   While on the subject, it must be remembered that another transport, the SO-90, designed and built in Cannes by the same team, made a remarkable escape across the Mediterranean on its unauthorised first flight!

Meanwhile design office work on the SO-30 had been continuing, and an improved version appeared late in 1945 as the SO-30R, the R suffix denoting change of engines.   (The SO-30R had been named Bellatrix A in the 1943 Directions for Aircraft names.)   The first SO-30R proto-type had a single fin-and-rudder tail assembly which was not retained for production variants. French aeronautical design had been out of touch with modern aircraft development during the dark years of occupation, and airliners appearing after the war soon had to be adapted to the new ICAO requirements.   This was the case for the large SE-2010 Armagnac airliner as well as the SO-30P which was consequently modified.

American Pratt & Whitney power-plants were chosen for the production batch, with resulting change of designation from SO-30R (Gnome & Rhone 14R-5) to SO-30P (Pratt & Whitney R-2800 B-43). The maiden flight of the production SO-30P c/n 1 took place in 1947, with the new name of Bretagne (Brittany) adopted by March 1949 - conforming to a tradition of naming airliners after French provinces (as evidenced with the SE-2010 Armagnac, SO-95 Corse, SE-161 Languedoc and Br-763 Provence).

Air France, an obvious customer for the SO-30P, had been a consultant in the design of the production variant, but the national airline lost interest in the type. The early production machines with registrations already reserved and allocated fell by the wayside except for c/n 2, which was later converted for experimental purposes. The next SO-30P to appear from the Saint-Nazaire factory was c/n 8 in 1948. With no sales arranged, the Bretagne transports now in production were taken over by the DTI (Direction Technique et Industrielle de l'Aéronautique), this being a government body organised in 1946 to provide for the technical and industrial needs of French aviation. Two aircraft also went to the SGACC, the French civil aviation authority (a third SO-30P, c/n 11, also being transferred later from the DTI).

With a somewhat large production batch at hand, it was necessary to find operators for the type, and various arrangements were made. In 1950-51 three aircraft were temporarily used on loan from the DTI: one by Aéro-Cargo (c/n 11), another abroad with Iranian Airways as EP-AAM, while c/n 12 was operated by Air France for some time. The leased Iranair machine was operated from November 1950 to May 1951 on the Teheran-Bombay run, and a firm order for several aircraft was considered but not finalised owing to financial difficulties. EP-AAM was flown back from Iran to France after 400 hours of operation.

France still had a large world influence and commitments in North Africa (Algeria, Tunisia and Morocco), and there existed a possibility of using the new aircraft there. Air Algerie became a large operator of the type in 1951-52 with as many as nine aircraft on strength, immediately followed by Air Maroc with twelve. But France was waging a bitter and costly war in Indochina where French influence still predominated, and between 1952 and 1954 seven Bretagne transports were delivered to STA-EO. Then, with the battle reaching a climax early in 1954, six aircraft went to another operator, Aigle Azur, most of these being flown with Air Laos colour schemes. Aigle Azur variants, which incorporated a larger loading door, were also equipped with 150 kg s.t. Turboméca Palas auxiliary jets on a trial basis.

In 1952 two SO-30Ps had been specially modified for use as VIP aircraft by the President of the Republic and the President du Conseil (equivalent to the Prime Minister).

Two more had become jet aircraft as flying test-beds. The SO-30R-02 (second prototype) was used as a Rolls-Royce Nene test-bed (with trials initiated in 1951 and used until 1954), and the second SO-30P production aircraft which flew in 1953 with two ATAR jet engines. Both aircraft had a single-fin tail assembly.

Another production change took place with the introduction of the Pratt & Whitney R-2800 CA-18 power-plants, first fitted to aircraft c/n 10. The power of each engine was raised from 1620 hp on the R-2800 B43 to 1800 hp, and a few more aircraft of the CA-18 sub-type followed, as shown in the following production list.

The story of the SO-30 would not be complete without mention of the unsuccessful SO-30C freighter designed in the late forties to meet an Armée de l'Air specification for a military transport, other competitors being the Bréguet 891R (qv) and the Nord 2500 Noratlas, which won the production contract. The single SO-30C, flown in 1950, had belly loading hatches.

Finally by the mid-fifties most existing SO-30Ps were transferred to the French Air Force or Navy, with half a dozen going to the CEV for communications and various experimental purposes. Aéronavale was the largest operator of the type, with up to 18 on strength by 1960 used for transport and various fleet duties, including target-towing for warship AA guns. The last two Navy SO-30Ps were phased out in 1968.

Two CEV aircraft have been preserved, one - F-ZABI - at La Roche-sur-Yonne, and the other - F-ZABZ - being stored in the open at the test flight centre of Istres. Two ex-Armée de l'Air aircraft, c/n 19 and probably 22, had also been preserved earlier in a curious combination where two fuselages made up a coffee house and bar in the vicinity of Ambérieu-en-Bugey in the Ain département. Unfortunately this combination was later destroyed by fire.

Judging from the apparently safe record of Bretagne operation under various skies and by various operators, one may wonder why it was so difficult to find customers for the aircraft (total production reached 45, including the prototypes). A major explanation can probably be found in the fact that this short- and medium-haul transport was generally comparable to the famous Douglas DC-3 as far as size and performance (apart from speed) were concerned, and the cheaper American aircraft was usually preferred by operators.

Specifications

1 SO-30R Bellatrix (two prototypes)

Span: 25.61m
Length: 18.43m
Height: 5.85m (with single-fin tail)
Empty weight: 9,870 kg
Loaded weight: 16,400 kg
Payload: 3,255 kg
Power plants: 2 x 1,700 hp Gnôme-Rhône 14 R-5
Maximum speed: 545 km/hr at 7,700m or 436 km/hr at sea-level
Cruising speed: 430 km/hr
Service ceiling: 8,450m

Passenger accommodation was divided into three cabins, seating (as a 30 passenger day-plane operating over 2,000 km range) nine, six and fifteen respectively. Night flying over 3,000 km range was also considered, the aircraft becoming a sleeper with berths for 16 passengers and 1,800 kg disposable load for freight or mail.

2 SO-30P Bretagne (production aircraft)

Span: 26.90m
Length: 18.95m
Height: 4.90m
Wing area: 86.17 sq m
Empty weight: 12,700 kg
Loaded weight: 18,500 kg
Power plants: 2 x 1,620 hp Pratt & Whitney R-2800 B-43

Cruising speed: 400 km/hr
Service ceiling: 8,000m

   Crew of three, 30 passengers ('luxe' version), 34 ('confort' version) or 43 ('coach' variant).

3   SO-30P CA-18 variant

C/n 10, 18, 19, 20, 22, 23, 24, 25, 26.

Power plants: 2 x R-2800 CA-18 rated at 1,927 hp/2,600 rpm or 2,434 hp on take-off with water injection
Loaded weight: raised to 19,500 kg or even 20,000 kg under specific conditions

4   SO-30P B43 with auxiliary jet engines (two Turboméca Palas)

Loaded weight: increased to 20,150 kg on low-altitude airfields, authorised up to 20,480 kg for aircraft operated in Indochina, depending on local temperature

Production List

C/n 01   SO-30N F-BALY built Nov 42 and stored until f/f 26 Feb 45.   To F-WALY.   WFU 1951 at Orly and stored dismantled until early 60s.

C/n 01   SO-30R F-BAYA built 1945 at Courbevoie.   F/f 06 Nov 45.   To F-WAYA.

C/n 02   SO-30R F-WAYB built 1946 at Courbevoie.   Cvtd with twin fins and rudders Oct 46.   Cvtd 1950 to SO-30 Nene and f/f 15 Mar 51.   Used as engine test-bed with two Rolls-Royce Nenes.   WFU 1954 at Brétigny.

C/n 01   SO-30C F-WAYN f/f 06 Jan 50. Last flown 12 Dec 50.

C/n 1   SO-30P F-WAYC built 1947 at St Nazaire.   Type B-43.   F/f 11 Dec 47.

C/n 2   SO-30P F-WAYD built 1947/48 at St Nazaire.   Type B-43.   Cvtd 1952 to SO-30 Atar and f/f 27 Jan 53.   Used as engine test-bed with two SNECMA Atars. Still active 1965.

C/n 3-6   SO-30.   Believed not completed (F-WAYE/H NTU).   One of them probably destroyed in static tests Apr 47.   The production order was for 40, supposedly c/n 7 to 45, plus the SO-30C.

C/n 7   SO-30P F-WAYI.   Type B-43.   Built 1948.   Believed loaned to French AF 14 Sep 49.   To F-BAYU DTI/Air Algérie (CofA 16 May 52).   To F-OAQG Aigle Azur (CofA 27 Apr 54).   WFU 1954.   To French Navy.

C/n 8   SO-30P F-WAYJ.   Type B-43.   Built 1949.   F/f Aug 49.   To F-BAYJ SGACC (CofA 25 Oct 51). Based at Le Bourget, then at Casablanca 1954.   Canc Sep 56.   To French Navy.

C/n 9   SO-30P F-WAYK.   Type B-43.   Built 1949.   To F-BAYK DTI (CofA 1 Sep 50).   Loaned to Air Algérie Dec 51 after overhaul.   To F-OAQH Aigle Azur (CofA 27 Apr 54).   WFU 1954.   To French Navy.

C/n 10   SO-30P F-WAYL.   Type CA-18.   Built 1949.   F/f 15 Sep 49.   To French Navy.

C/n 11   SO-30P F-WAYM.   Type B-43.   Built 1948/50.   To F-BAYM DTI/Aérocargo del 29 Mar 50 (500 hrs operating tests).   To SGACC Nov 52 (or earlier).   Canc Sep 56.   Intended for French Navy but not traced.

C/n 12   SO-30P F-WAYO.   Type B-43.   Built 1950.   F/f Jun 50.   To F-BAYO DTI (CofA 06 Sep 50). Loaned to Air France Sep 50.   To F-OAIY Air Algérie (CofA 23 Mar 51).   Written off 30 Oct 51 at Orly.

C/n 13   SO-30P F-WAYP.   Type B-43.   Built 1950.   To F-OAIT DTI/Air Algérie (CofA 11 Jan 51).   Ret to manufacturer 1953.   Canc late 1956.

C/n 14   SO-30P F-WAYQ?   Type B-43.   Built 1950.   To F-BAYQ DTI (CofA 28 Nov 50).   To EP-AAM Iranian A/W del 30 Nov 50.   To F-DAAC DTI/Air Maroc (CofA 04 Jul 51).   To F-OAMA Air Algérie (CofA 13 Oct 52).   To Aigle Azur Apr 54.   Written off 30 Apr 54 at Saigon (operated by Air Laos).

C/n 15   SO-30P F-WAYR.   Type B-43.   Built 1950.   To F-OAIX DTI/Air Algérie (CofA 03 Feb 51).   Tested with additional Turboméca Palas jet engines.   To Aigle Azur Apr 54.   WFU 1954.   To French Navy.

C/n 16   SO-30P F-WAYS.   Type B-43.   Built 1950/51.   To F-OALG STAEO "Bach Hao" del 24 Jul 52.   WFU 1955. To CEV.

C/n 17   SO-30P F-WAYT?   Type B-43.   Built 1951/53.   To F-OALH STAEO (CofA 18 Feb 53).   To F-BAYT STAEO (CofA 04 May 53).   To F-OALH again   STAEO (CofA 12 Apr 54).   To DTI 1955.   To SFERMA by Feb 56, canc and over-hauled for French Navy.

C/n 18   SO-30P F-BAYU.   Type CA-18.   Built 1951.   To DTI/Air Algerie (CofA 27 Nov 51).   To Air Atlas-Air Maroc Apr 54.   WFU 1955.   To CEV briefly as F-ZXRR and then as F-ZABI.   On display in the open at La Roche-sur-Yon (Dept de la Vendée).   Steps have been taken to prevent the aircraft from becoming derelict.   Following a visit of the Director of the Musée de l'Air, it has been decided that the aircraft should be transferred to the National Museum at Le Bourget.   Transfer planned to take place in spring 1979.

C/n 19   SO-30P F-BAYV.   Type CA-18.   Built 1951.   To DTI/Air Algérie (CofA 05 Mar 52).   To Air Atlas-Air Maroc Jul 54.   WFU 1955.   To French AF Jul 55.

C/n 20   SO-30P F-BAYX.   Type CA-18.   Built 1951.   To SGACC (CofA 13 May 53).   Used by High Commissioner at Brazzaville from 1955 until Apr 60.   To French AF Nov 62.

C/n 21   SO-30P F-WAYY.   Type B-43.   Built 1951.   Loaned to Air Algérie as F-WAYY?   To F-OALI STAEO "Van Hong" del 24 Jul 52.   To DTI 1955.   To SFERMA by Feb 56, canc and overhauled.   Intended for French AF/GLAM but to French Navy instead.

C/n 22   SO-30P F-BAYZ.   Type CA-18.   Built 1951/52.   To DTI/Air Maroc (CofA 23 Apr 52).   WFU 1954.   To French AF Apr 56.

C/n 23   SO-30P F-BEHA.   Type CA-18.   Built 1951/52.   To DTI/Air Maroc (CofA 27 May 52).   Ret to SNCASO 05 Dec 52 after accident and WFU.   Intended for French Navy Sep 56 but already dbr.

C/n 24   SO-30P (F-BEHB).   Type CA-18.   To French AF "France" as Presidential aircraft del 29 Jul 52.

C/n 25   SO-30P (F-BEHC).   Type CA-18.   To French AF "Anjou" as Prime Ministerial aircraft del Apr 52.

C/n 26   SO-30P F-BEHD.   Type CA-18.   Built 1951/52.   To DTI/Air Maroc (CofA 16 Jun 52).   WFU 1954.   To French AF.

C/n 27   SO-30P (F-BEHE).   Type B-43.   Built 1953.   To F-OALJ STAEO (CofA 28 Sep 53).   To DTI 1955. To SFERMA by Feb 56, canc and overhauled for French Navy.

C/n 28   SO-30P (F-BEHF).   Type B-43.   Built 1948/52.   To F-DAAX Air Maroc (CofA 31 Jul 52).   WFU 1954. To CEV.

C/n 29   SO-30P F-WEHG.   To CEV Brétigny (by Feb 56) (ex F-DAAY? NTU).   To French AF del Aug 56.

C/n 30 SO-30P F-WEHH. To CEV Brétigny by Feb 56 (ex
        F-DAAZ? NTU). To French AF del Aug 56.

C/n 31 SO-30P (F-BEHI). Type B-43. Built 1952. To
        F-DABA Air Maroc (CofA 27 Aug 52). WFU 1954.
To French Navy.

C/n 32 SO-30P (F-BEHJ). Type B-43. Built 1948/52.
        To F-DABB Air Maroc (CofA 20 Aug 52). WFU 1954.
To French Navy.

C/n 33 SO-30P (F-BEHK). Type B-43. Built 1951/52.
        To F-DABC Air Maroc (CofA 29 Oct 52). WFU 1954.
To French Navy.

C/n 34 SO-30P (F-BEHL). Type B-43. Built 1952. To
        F-DABD Air Maroc (CofA 22 May 53). Written off
Oct 53 at Tangiers.

C/n 35 SO-30P (F-BEHM). Type B-43. Built 1951/52.
        To F-DABE Air Maroc (CofA 05 Mar 53). WFU 1954.
To French Navy.

C/n 36 SO-30P F-WEHN. To French Navy del 01 Feb 55.

C/n 37 SO-30P F-WEHO. F/f 02 Apr 53. To CEV
        Brétigny by Feb 56. Later to military marks,
mostly as F-ZABZ. Last flew 08 Jul 71. Wfu and preserved within precinct of CEV/BA125 at Istres (Dept
des Bouches-du-Rhône), then transferred to St. Nazaire. A preservationist organisation, 'Phénix', based
in SE France has reportedly been involved with the
aircraft recently.

C/n 38 SO-30P F-BEHP. Type B-43. Built 1953. To
        DTI/STAEO (CofA 12 Feb 54). WFU 1954. To
French Navy.

C/n 39 SO-30P F-BEHQ. Type B-43. Built 1953. To
        DTI/STAEO (CofA 06 May 54). Fitted with Palas
jet engines. WFU 1954. Intended for French Navy but to
CEV instead.

C/n 40 SO-30P F-WEHR. Type B-43. Built 1953.
        Reported as "F-OAKS" of Air Laos but NTU. To
F-BEHR DTI/Aigle Azur (CofA 18 Mar 54). To French Navy
del 28 Aug 54.

C/n 41 SO-30P F-WEHS. Type B-43. Built 1954. To
        F-BEHS DTI/Aigle Azur (CofA 17 Jun 54). Written
off 30 Aug 54 at Hanoi.

C/n 42 SO-30P F-WEHT. To French Navy del 24 Feb 55.

C/n 43 SO-30P F-BEHU. Type B-43. Built 1954. To
        STAEO (CofA 27 Aug 54). WFU 1955. To French Navy.

C/n 44 SO-30P F-WEHV. To CEV Brétigny (by Feb 56).
        Later to military marks, Dbr on beach in
Camargue, when oper by CEV/EPNER.

C/n 45 SO-30P F-WEHX. To French Navy del 15 Mar 55.

SE-2010 ARMAGNAC

Early in 1942 a SNCASE project had been
designed under the reference 052. The 100-
seat airliner would have been powered by four
Hispano-Suiza 12Z or Gnôme-Rhône 14R engines,
but with 1,200 hp engines such as these, the
aircraft would have been underpowered, and the
next step, in September 1942, was the SE-2000
project, with twin-decks, a pressurised
fuselage and four 2000 hp Gnôme-Rhône 18R
engines. With a loaded weight of 43 metric
tons and 40 passengers, estimated performance
on the South Atlantic route was 400 km/hr at
7000m. Early in 1943, however, the SE-2000
outline was changed to a conventional
cylindrical fuselage, and a prototype was
ordered on 4 August 1943. With the liberation
of France in the following year, an Air France
specification for a transatlantic airliner was
drawn up, and the project was further modified.
Now referred to as the SE-2010, with operation
by the national airline specifically in mind,
the prototype was ordered on 12 March 1945 and
the name Armagnac (a French province) adopted.

Rolled out late in 1948, the SE-2010 prototype, with modified tail fin to suit ICAO
standards, was at long last flown at Toulouse-
Blagnac on 2 April 1949. At the controls was
Chief Pilot Pierre Nadot, who was to become the
Caravelle test pilot a few years later. The
Armagnac was demonstrated at the 1949 Paris Air
Show which, at that time, was still being held
at Orly. The provisional Wasp Major VSB-11G
power-plants (3,040 hp each) were abandoned by
mid-December 1949 after 145 hours of test
flights and replaced by the planned 3,550 hp
R-4360s. Flight tests were resumed during the
spring of 1950, but the prototype was short-
lived and crashed on 30 June 1950 while landing
at Blagnac following the in-flight opening of an
unlocked leading-edge panel. There were two
fatalities on board and one on the ground, and
the aircraft was a total write-off.

In what can only be described as the
optimistic mood prevailing at that time, fifty
SE-2010 production aircraft had been ordered by
the French Government, work being shared by the
three SNCASE plants at Toulouse, Marignane and
La Courneuve as well as the ill-fated SNCAC
works at Colombes. The unrealistic and daring
order was soon reduced to 25 machines and then
fifteen. The latter production run was to
include eight aircraft for Air France, three for
TAI, one Allison T40 turboprop conversion (the
fifteenth airframe) and three for potential
export sales. Allocations for the prototype
and 15 production aircraft were reserved in the
F-BAVA block. In fact, only nine aircraft were
eventually laid down, and the first production
Armagnac flew on 30 December 1950, then being
used to complete the trials left uncompleted
owing to the early loss of the prototype.
Manufacturing problems were encountered, coupled
with a scarcity of the appropriate alloys, and
this resulted in a slight increase in weight.

The standard Air France colour scheme was
painted on early aircraft, but in fact the airline was never to fly the Armagnac. By that
time proven US airliners had been or were
entering commercial service and were preferred.
Over-enthusiastically described as a "flying
palace" by the French Press, the SE-2010 was
probably a pleasant airliner but not truly
competitive with the DC-6. From the passenger
point of view only the cabin layout left something to be desired; there were two compartments separated amidships by what was described
as a pantry and bar. In fact, the main wing
spar went right across the cabin and one had to
walk over it, with a resulting split deck and
appropriate stairs. Not the most efficient
design!

In spite of a major campaign led in favour of the new French transport, particularly by the aeronautical magazine "Les Ailes", the whole batch of eight production aircraft remained unsold after Air France had lost its interest in the type.

France was thus left with a fleet of 90/100-seat airliners still under construction. In mid-1952 an agreement with TAI was reached, and four Armagnacs (c/n 3, 4, 5, 6) entered service with the airline, flying from Casablanca, Abidjan and Tananarive to Jeddah on pilgrim flights and also on a regular route from Paris to Morocco up to 1953. But after being re-placed in TAI service by DC-6Bs they were returned in July 1953.

A solution had to be found, and late in 1953 a new operator was formed under the name of SAGETA (Société Auxiliaire de Gérance et d'Exploitation de Transports Aériens). This was a joint venture between the three major French airlines (TAI, UAT and Air France), Les Messageries Maritimes, Air Algérie, Aigle Azur and also SNCASE, with deliveries scheduled between January and July 1954. Initially based at Toulouse, where storage had begun in 1953, the newly-formed company was later moved to the Paris area in 1956, with its main base at Orly. A TAI colour scheme was retained but with the lion emblem of the Province of Armagnac, evidence that SAGETA had been formed only to operate the Armagnac. In fact, the agreement was extremely unrealistic, SAGETA being only the subsidiary of the pool with no right to use the aircraft on its own without previous agreement from the airline members.

France was however waging a bitter war in Indochina, and operations on the long run from Toulouse to Saigon were soon launched via Beirut, Karachi and Calcutta. The first SE-2010 flight to Indochina took place on 22-24 December 1953, the aircraft being back in Toulouse on 28 December. Seven aircraft were used in 1954 on this 11,455 km route without major problems, the line being closed down in 1955 after the 163rd Armagnac trip to Indo-china. With the end of the war in French Indochina, operation of the type was slowed down. SAGETA was a non-scheduled operator, an early example of French charter work with large aircraft on long distance flights over-seas, including on some occasions chartered trips on behalf of the Armée de l'Air which then lacked appropriate transport aircraft.

Hours logged by the SE-2010 fleet slowly rose, total times between 1,300 and 2,700 hours being reached by the aircraft by mid-1955, with the 4,000-5,000 hours mark not being passed until mid-1958. The sixth aircraft crashed at Orly in poor weather in January 1957 on a flight from Tunis with 70 on board. There were no fatalities except for an elderly passenger who later died from a heart attack. On 20 December 1957 the seventh Armagnac was damaged in flight by a time-bomb explosion over France but landed safely at Lyon Airport with a large hole in the fuselage. The same aircraft (F-BAVH) was substantially damaged at Djibouti in June 1958 when the nose gear re-tracted during take-off. All 85 passengers escaped unhurt. The airliner was repaired and flown back to France but not used again.

SAGETA operations ended in October 1958, and the fleet reverted to government ownership two months later. Five aircraft were stored at Orly and one at Bordeaux (F-BAVH), until

between April and June 1959 F-BAVD, F-BAVE, F-BAVI and F-BAVC were successively ferried back to Bordeaux on a last flight and the whole fleet was scrapped in 1961, apart from the last Armagnac (F-BAVI) which was put on show for a few years before also being scrapped.

Of interest is the career of the first production aircraft. Transfer to the CEV for experimental purposes took place in 1953, and the aircraft was extensively modified as an engine test-bed, being finally delivered in April 1955. Redesignated SE-2060, it was widely used for podded SNECMA Atar jet engine development programmes and other research work. It was further modified and fitted with multi-purpose pods – flying again on 23 February 1962. Another modification programme brought back the SE-2060 to the CEV for more experimental work (f/f again 27 September 1965) including high-altitude parachute tests. This last survivor of the Armagnac breed was withdrawn from use in 1968 after a successful career on experimental work but unfortunately was not preserved.

Thus ended the early venture of SNCASE into the modern airliner market. The SE-2010 had been an affaire de coeur with most French aero-nautical newsmen, and probably over-much emphasis was placed on the quality of the air-liner. There were many drawbacks too, and it must be accepted that the Armagnac just didn't quite...

The next SNCASE attempt was much more successful and was called ... the Caravelle (qv).

Specification

SE-2000 (early project)

Span: 42.26m
Length: 39.45m
Loaded weight: 44,000 kg
Power plants: 4 x 2,000 hp Gnôme-Rhône 18R

Eightyseven passengers over 1,000 km or 32 passengers over 3,000 km. Project modified after the French Liberation when data on modern airliners became available. Weight increased to 64,000 kg and still later to 70,000 kg. Span: 49m. Four Pratt & Whitney R-4360 Wasp Major (3,500 hp each).

SE-2010 Armagnac

Span: 49m
Length: 40m
Height: 14m
Wing area: 230 sq m
Empty weight: 44,300 to 47,131 kg
Maximum weight: 75,000 to 77,000 kg
Payload: 6,869 to 7,700 kg
Power plants: 4 x Pratt & Whitney R4360-B 13 (2,687 hp at 2,550 rpm, sea level) With injection, 3,550 hp at 2,700 rpm on take-off (Prototype initially flown with 3,040 hp VSB-11G Wasp Majors)
Propellors: Curtiss Electric C644-5-B-306
Maximum speed: 580 km/hr
Cruising speed: 450 km/hr
Range: 5,120 km

Four-engined airliner with cantilever mid-wing. Crew of five to fifteen, 91 to 104 passengers.

Production List

C/n 01   F-WAVA prototype. F/f 02 Apr 49. Written off 30 Jun 50 near Toulouse.

C/n 1   F-WAVB.   F/f 30 Dec 50.   To CEV 1953.   Coded
        "Q" (F-ZWSQ) and modified 1954 as jet engine
test-bed (pods under wings).   Initially used with
SNECMA Atars.   Modified with multi-purpose pods 1962.
WFU Mar 68 at Melun-Villaroche.

C/n 2   F-WAVC.   Built 1952.   To F-BAVC SAGETA (CofA 22
        Jun 54).   WFU at Orly Jun 59 and later stored at
Bordeaux until scrapped 1961.

C/n 3   F-WAVD.   F/f 02 Jan 52.   To F-BAVD TAI (CofA 05
        May 52).   To SAGETA Jan 54.   WFU at Orly Apr
59, later stored at Bordeaux until scrapped 1961.

C/n 4   F-BAVE.   F/f 11 Jul 52?   To TAI (CofA 01 Aug
        52).   To SAGETA Mar 54.   WFU at Orly Apr 59
and later stored at Bordeaux until scrapped 1961.

C/n 5   F-BAVF.   To TAI (CofA 17 Nov 52).   To SAGETA
        Jul 54.   WFU at Orly Feb 59 and scrapped 1961.

C/n 6   F-BAVG.   Built 1952.   To TAI (CofA 06 Feb 53).
        To SAGETA 1954.   Written off 29 Jan 57 at Orly.

C/n 7   F-BAVH.   Built 1953.   To SAGETA (CofA 01 Mar
        54).   Damaged 27 Jun 58 at Djibouti.   Flown
back but WFU at Bordeaux.   Stored until scrapped 1961.

C/n 8   F-BAVI.   Built 1953.   To SAGETA (CofA 18 Mar
        54).   WFU at Orly May 59, later stored at
Bordeaux until scrapped ca 1964.

C/n 9   Aircraft built but not rolled out.   (Six more
        under construction but canc.)

BREGUET 763

The origin of the well-known (at least in French circles) Breguet Deux Ponts may be traced back to the pre-war Breguet 730 flying boat (qv), together with other aircraft built in prototype form only or uncompleted commercial developments.   Landplane double-deck derivatives were planned as early as 1935 under the designation of Br-760 and 761, later followed by the wartime Br-840, 842 and 843 projects.

In fact, only after a 1946 Air Ministry specification did the new 760 come into being, drawing much from the previous studies and using the slightly modified main-plane from the Br-730/731 flying-boat.

The aircraft was to be a combined freighter and passenger aircraft with a range of 2,000 km.   Work began immediately, but the maiden flight of the prototype did not take place until February 1949, fifteen production aircraft having already been ordered by the French Government in 1947.

Like the earlier SNCASO SO-30R prototypes, the Breguet 761-001 was powered by Gnôme-Rhône 14R power-plants, which were then widely used on large French aircraft.   But, as had also happened with the production SO-30P, American Pratt & Whitneys were preferred for subsequent aircraft, resulting in the new Br-761, with suffix letter S denoting 'série' - French for production.   In fact, the three Br-761Ss built were only pre-production machines, fitted with surplus 2,100 hp R-2800 B-31 engines. Starting in 1950, they were used for a long time on various trials and experiments including the dropping of heavy or outsized loads (much publicised trials being conducted with the first Br-761S on behalf of the DEFA), commercial evaluation as a freighter (the Air Algerie colour scheme being applied to the second aircraft, which was later painted in Silver City marks for operation in 1953 on the Hamburg-Berlin route) and the experimental ferrying by air of heavy loads such as armoured vehicles (761S-03).

Late in 1955 the three aircraft were taken over by the Armée de l'Air for use as military transports with ET 2/61, soon changed to 2/64 "Maine".   Codes were accordingly changed from the 61- to 64- prefix to become 64-PA, -PB and -PD respectively, corresponding to call-signs F-RAPA etc.   (F-RAPC was omitted as it had been allocated to Languedoc c/n 86, involved in a fatal crash at Dugny-Le Bourget in October 1952.)

The next version was the Breguet 763, named Provence after the sunny southern French province.   Performance was marginally better than on the earlier 761S thanks to the use of R-2800 CA-18 engines (also adopted on an SO-30P variant).   In 1949 the national airline, Air France, had evinced interest in the type but was now very reluctant to adopt the new airliner in service use.

This seems to have been a persistent attitude of the airline towards most French-designed airliners during the late forties and early fifties with, it must be admitted, generally good reasons.   In this particular case, however, Air France was later proved to be wrong.

Apparently Air France's reluctance to order the Provence was mostly concerned with payment, the airline battling to have the twelve aircraft subsidised by the State.   In what can only be described as an overt blackmail operation, extensive modifications of the Breguet transport were imposed on Breguet by Air France, and eventually the manufacturers were compelled to sell the aircraft at a loss, a serious blow for the company, from which they were never to recover.

The first Breguet 763 to come out from the assembly line at Villacoublay was not delivered until 1955 owing to successive certification trials under increasing loaded weight.   The first Provence delivered to Air France was actually the third production aircraft, put into service in March 1953.   Conforming with a then established French practice, a single block of registrations starting with F-BASN was allocated to the twelve aircraft (F-BASW being omitted, W being used only as a temporary prefix for aircraft on trials, eg the first aircraft was initially registered F-WASN and later changed to F-BASN on receiving the type certification). In Air France service, the Breguet 763 proved to be an excellent and reliable aircraft, being used mostly on the busiest domestic route from Paris to the Mediterranean and across the sea to Algeria.   It was a popular aircraft, usually referred to as the Deux-Ponts (French for double-decker), the official name of Provence being seldon used.   Accommodation was provided for 107 passengers (seating 59 on the upper deck) and could be raised to 112 or even up to 140. The lower deck could easily be converted into a capacious freight hold - an anticipation of today's 'quick change' aircraft.

After an uneventful career and with increased Air France operation of the Caravelle jet airliner, coupled with a slowdown of the trans-Mediterranean traffic following the French withdrawal from Algeria in 1962, six aircraft were modified to the so-called 'Universel' configuration (c/n 1, 2, 4, 8, 9 and 10 - but not in that order) in 1964-65. The modifications provided for heavy freight capacity with appropriate built-in handling equipment but still retained accommodation for some 30 passengers as necessary.   The Universel was a frequent visitor to the UK in those years.

At the same time, the six other Air France machines were transferred to the French Air Force.   With the termination of the Algerian War of Liberation in 1962, France had been compelled to remove its nuclear test range from the Sahara to French Polynesia.   There was a serious need for a military support fleet to fly supplies and passengers between the CEP test-range forward base (at Hao in the Tuamotu Archipelago) and the main staging centre in Tahiti.

Aircraft c/n 3, 5, 6, 7, 11 and 12 therefore entered service with the Armee de l'Air GAM 82 (with call-signs allocated in the range F-RBPN to F-RBPS) and saw extensive service on long overwater flights, until the last 763 was with-drawn from use in 1972.   Some aircraft were left derelict at Papeete-Faaa Airport in Tahiti.

Early military interest in the type had in fact been forthcoming in 1950 when the French Navy considered replacement of the ageing ASW Sunderlands and Wellingtons.   The original 760 prototype was modified as a flying mock-up of the Bréguet 764 ASW and maritime patrol version. This would no doubt have resulted in an excellent aircraft for the Navy, but escalating costs prevented the Br-764 from being ordered into production.   Aéronavale turned to the cheaper Hurel-Dubois HD-31 derivative until in 1953, when delivery of MDAP-funded Lockheed P2V-6s started, all hopes of developing the Br-764 were lost.

In 1951 the Armée de l'Air also evinced interest in a long-range version of the Deux Ponts, resulting in the Bréguet 765 Sahara - a heavy freighter with an outsize cargo capacity. The most conspicuous feature of this much-modified aircraft was the wing-tip tanks.

A large order was planned, later reduced to only twelve, but shortly after work had begun at the newly-erected Bréguet factory of Biarritz-Parme, the order was cancelled. Eight unfinished airframes were scrapped, and then it was found to be cheaper to complete the remaining four!   They finally entered service with the Air Force, c/n 501 through 504 being allocated (denoting the new 765 version as opposed to the ex-Air France Br-763 aircraft later added to the Armée de l'Air inventory, which took up a 3 prefix thus becoming 303, 305 etc).   After initial trials in 1958 they were delivered to ET II/64 'Maine' (call-signs being F-RAPE/RAPH).   These aircraft were phased out

in 1969, one being preserved in a dilapidated condition at Evreux.

While on the subject of preservation, it must be noted that a small number of Bréguet Deux Ponts were initially preserved, but some were unfortunately scrapped later.

Derivative projects of the basic design included an early Bristol Hercules 630-powered 763 for possible UK buyers and the Bréguet 766 and 767 with British turboprops.   There was also a 70 ton double decker with four Bristol Proteus turboprops, in fact bearing more relationship to the Br-978 than to the Br-760 series.   The Br-978 was the Bréguet answer to the SGACC requirement out of which the SE-210 Caravelle was born.

The Breguet 761/765 aircraft were good transports, and success could have been anticipated.   Lack of luck in the hard-sell airliner business, coupled with an unrealistic (and even irrational) industrial aeronautical policy, led to only a small number of aircraft being built.   Far from being the type which would have helped Bréguet to overcome growing difficulties if ordered in larger numbers, the sturdy but unlucky Deux-Ponts helped considerably to accelerate the manufacturer's decline - although Bréguet deserved praise nevertheless for producing this sound aircraft.

Specifications

1   Bréguet 761 prototype

Span: 41.66m
Length: 28.70m
Height: 9.53m
Wing area: 178.6 sq m
Empty weight: 22,050 kg
Loaded weight: 40,000 kg
Payload: 14,000 kg
Power plants: 4 x Gnôme-Rhône 14R-4/5 (1,590 hp each)
Maximum speed: 400 km/hr
Cruising speed: 320 km/hr
Ceiling: 6,000m
Range: 3,800 km

2   Bréguet 761S

Four-engined passenger or freight transport with fuselage divided into two decks.   Crew compartment in nose on upper deck.   Pre-production aircraft powered by four 2,100 hp Pratt & Whitney R-2800 B-31s with resulting increased all-up weight of 45,000 kg.

Upper deck capacity: 89.3 cu m
Lower deck capacity: 95.3 cu m
Floor area for each deck: 84 sq m
Passenger version: accommodation for 101

3   Bréguet 763 Provence

Span: 42.96m
Length: 28.95m
Height: 9.56m
Wing area: 185.4 sq m
Empty weight: 30,600 to 32,500 kg
Loaded weight: 48,000 to maximum 51,600 kg
Payload: 6,130 to 7,830 kg with 10,580 kg of fuel
Power plants: 4 x 1,927 hp Pratt & Whitney R-2800 CA-18
               (2,434 hp on take-off) with water injection
Cruising speed: 360 km/hr
Economical cruising speed: 330 km/hr
Maximum range: 4,100 km, or 2,700 km with 10,800 kg
               payload
Ceiling: 5,850m

Crew of three with accommodation for 59 to 112

passengers (59 on upper deck).

4 Breguet 765 Sahara

Span: 42.99m (less wing-tip tanks)
Length: 28.95m
Height: 10.22m
Wing area: 178.7 sq m
Volume of passenger or freight space on two decks:
          167 cu m
Empty weight: 30,000 kg
Loaded weight: 54,000 kg
Power plants: 4 x 2,400 hp Pratt & Whitney R-2800 CB-16
               (with water injection)
Maximum payload: 17,000 kg or 126-164 passengers
Cruising speed: 380 km/hr at 3,000m
Range: 4,000 km
Ceiling: 5,500m

Production List

C/n 01   761 F-WFAM. F/f 15 Feb 49. Naval conversion
         as Br-764 intended in 1951 but project canc
summer 1951. Scrapped at Villacoublay.

C/n 01   761S F-WASK. F/f 17 Feb 50. To French AF
         Nov 55 as "64-PA" (or initially "61-PA"). WFU
1961/63.

C/n 02   761S F-WASL. Built 1950. To F-BASL Breguet
         (CofA 22 Feb 52). Loaned to Air Algerie, del
29 Feb 52. Lsd to Silver City A/W Jul 53-Sep 53. To
French AF Nov 55 as "64-PB" (or initially "61-PB").
WFU 1961/63.

C/n 03   761S F-WASM. Built 1950/51. To French AF Nov
         55 as "64-PD" (or initially "61-PD"). WFU
1961/63.

C/n 1   763 F-WASN. Built 1953(?) To F-BASN Air
         France del 09 Aug 55. Cvtd to Universel
freighter 1965. WFU 03 Jun 71. Del 02 Jul 71 to Air
Afrique for ground training at Dakar-Yoff.

C/n 2   763 F-BASO. Built 1952. To Air France del
         26 Sep 53. Prior to this it was used by Air
France in a 800 hr operational testing programme 20 Aug
52-31 Dec 52. Cvtd to Universel freighter 1965. Del
21 Jun 70 to Salis Aviation, La Ferte Alais, for
preservation but scrapped early 1975.

C/n 3   763 F-BASP. To Air France del 27 Feb 53. To
         French AF 10 Dec 64 as "82-PN". WFU.

C/n 4   763 F-BASQ. To Air France del 31 Mar 53. Cvtd
         to Universel freighter 1965. Del Jul 71 to
Aero Club de Tarbes for preservation.

C/n 5   763 F-BASR. To Air France del 08 Jun 53. To
         French AF 21 Jan 65 as "82-PO". WFU at Papeete.

C/n 6   763 F-BASS. To Air France del 06 Jul 53. To
         French AF 10 Jun 64 as "82-PP". WFU 24 Aug 67.
Preserved by Aero Club de Fontenay Tresigny.

C/n 7   763 F-BAST. To Air France del 19 Sep 53. To
         French AF 05 Jun 64 as "82-PQ". WFU at Papeete.

C/n 8   763 F-BASU. To Air France del 16 Dec 53. Cvtd
         to Universel freighter 1964/65. WFU Sep 70.
Scrapped at Orly.

C/n 9   763 F-BASV. To Air France del 24 Mar 54. Cvtd
         to Universel freighter 1965. WFU Jan 69.
Scrapped at Toulouse by Aug 69.

C/n 10   763 F-BASX. To Air France del 03 Jun 54. Cvtd
         to Universel freighter 1965. Del Jul/Aug 71 to
Aero Club Air France at Toussus-le-Noble for preservation
                              but scrapped by Jun 79.

C/n 11   763 F-BASY. To Air France del 21 Jun 54. To
         French AF 10 Nov 64 as "82-PR". WFU at Papeete.

C/n 12   763 F-BASZ. To Air France del 30 Jun 54. To
         French AF 04 Jul 64 as "82-PS". WFU.

C/n 501 765 F-ZVME. F/f 06 Sep 58. To French AF Jan
         59 as "64-PE". WFU 18 Aug 69. Preserved by
Aero Club d'Evreux.

C/n 502 765 F-ZVMF. To French AF Feb 59 as "64-PF".
         WFU 27 Dec 69.

C/n 503 765. To French AF May 59 as "64-PG". WFU 30
         Dec 69.

C/n 504 765. To French AF Aug 59 as "64-PH". WFU 26
         Aug 69.

SE-161 LANGUEDOC

The SE-161 was actually a pre-war airliner,
since the original prototype had flown in 1939
as the SNCASO SO-161 derivative of the Marcel
Bloch MB-160. But, with the outbreak of World
War 2, flight tests of the prototype (F-ARTV)
were unable to be completed until January 1942.
A small production batch had been ordered by
the Vichy Government in December 1941, to be
built by the SNCASE in the still unoccupied
zone. With the German invasion of Southern
(Vichy) France late in 1942, the twenty aircraft under production for both Air France and
Lufthansa were more or less abandoned, as the
French were understandably not eager to deliver
the aircraft to the Germans, who had meanwhile
confiscated the prototype which had been used
for some time as a VIP transport by the Vichy
Government. During the occupation of France
the SO-161 was called Bordeaux.

Immediately after the liberation of France,
the Provisional Government authorised
production to be resumed, and the first SE-161
(SE- denoting transfer of the programme from
SNCASO to SNCASE) made its maiden flight in the
summer of 1945. With standard accommodation
for 33 passengers, 40 airliners were ordered by
Air France as there was an urgent need to
replace the elderly types still operated by the
airline (eight Dewoitine D-338s were still in
service in January 1946, five Bloch MB-221s in
May 1947). With the tradition of naming airliners after French provinces, the Languedoc, as
the SE-161 was now called, was first delivered
to Air France late in 1945; during 1946-47 the
expanding fleet of aircraft was introduced on
the European network and on routes to the three
major capitals in French North Africa (Casablanca,
Algiers and Tunis). All major European capitals
and many major towns were to see the Languedoc
airliner in the Air France colour scheme.

But the new aircraft suffered from many
teething problems. For instance, the
retractable two-wheel landing gear with
mechanical operation presented major difficulties.
The twin-fin and rudder endplates had to be
enlarged, and several other modifications had to

be incorporated as production was developing. Worst of all, the French Gnôme-Rhône 14N 68/69 power plants were a plague for the operator and had to be replaced by late 1946 with Pratt & Whitney R-1830s, thus giving birth to the re-designated SE-161/P7 version for the commercial Languedocs.

Only one export order was obtained, from the Polish national airline LOT, with four aircraft delivered in 1947 and a fifth in 1948. The whole batch was returned to France in January 1949 to be fitted with the American power plants.

From 1951 ex-Air France aircraft were trans-ferred to foreign operators, but only three more airlines were to use the airliner. The first was Air Liban, the large Lebanese operator in which Air France was the main shareholder (Air Liban was later to merge with MEA in 1963). Air Liban's first SE-161 was received by mid-1951, followed by another late in 1953, for a short-lived career. The second aircraft was written off at Beirut, and a third aircraft was obtained.

Five more Languedocs appeared in the Middle East, flown by the Egyptian airline Misrair (or Egypt Air, Misr being Arabic for Egypt). Three were delivered late in 1951, two more in 1952, to be used on the growing Middle East route network. Three aircraft were written off, the last one shortly after Misrair had ordered Viscounts to replace the Viking and SE-161 fleet.

The third and last foreign airline to use ex-Air France aircraft was also a Mediterranean operator, the Spanish carrier Aviaco, to whom nine aircraft were delivered (two in 1952, one in 1953, two in 1955 and four in 1956). Examples of the Spanish aircraft remained in service up to the early sixties.

Some Air France aircraft had also previously gone on lease to French North African operators such as Air Atlas (four aircraft) and Tunis Air (one, possibly two). Cie Air Transport also leased a Languedoc which was later delivered to Air Liban. From April 1951 Air France converted aircraft to a 44 seat all-economy layout for use on the routes to North Africa - and mention must also be made of a conversion carried out late in 1948, when five Languedocs had been modified as freighters.

Considerable (and costly) trouble having been experienced with the early operation of the type, compared with the proven reliability and serviceability rate of the DC-4 and Viscount, Air France was apparently eager to get rid of the SE-161. With no prospects of second-hand sales, an agreement was found with the French Government. From 1952 the fleet of 31 air-craft had dwindled to 13 late in 1954, and during the 1954-55 period ten ex-Air France aircraft were taken over by the SGACC for SAR operations. They were appropriately modified with a large ventral gondola, observation windows and a ventral search radar under a transparent fairing (similar to the one adopted on the French SAR Lancasters). The colourful blue and yellow scheme (also seen on the Lancasters) was applied to the Languedocs. The civil registrations were retained and, on SAR duties, call-signs were altered to the special F-SS.. prefix (for example F-BATP becoming F-SSTP).

Experience elsewhere has shown that an unsatisfactory airliner can make a more or less decent maritime patrol aircraft; this was the case with the Languedoc fleet used for about five years by EARS 99.

Three other Air France aircraft had also been used from 1954 by the French Air Force transport training centre, the CIET near Toulouse, before being delivered to Aviaco as mentioned.

The Armée de l'Air already had a large fleet of Languedocs, as the aircraft at the end of the production line had been earmarked for military use. With three aircraft retained by the DTI for experimental purposes, 16 Languedocs were delivered to the Air Force as the SE-161R (denoting Gnôme-Rhône 14R power plants), these aircraft having conspicuous four-bladed propellors. The aircraft were used between 1951 and 1955 by GT II/61 "Maine" at Le Bourget. They were allocated single letters, then 61-PX codes, the letters being the last of the call-sign in the F-RAP. block, but there was little use for this unsatisfactory variant introduced on the inventory in 1951. The SE-161Rs were phased out in two batches after limited service, nine on 12 December 1955 and five on 9 February 1956. The Armée de l'Air replacement for the type was to be the Bréguet 763 and 765 Sahara.

But the largest military operator of the Languedoc was the French Navy, with some 25 taken on strength over the years. First deliveries had taken place in 1949, and the Aéronavale version, known as the SE-161/1, was fitted with the original GR 14N 68/69 power plants. They were widely used from mid-1949 onwards as long-range transport aircraft with Escadrille 31S based at Paris-Orly (and later Le Bourget-Dugny). Fifteen were used by 31S, which then fulfilled the Navy's need for a long-haul communications fleet. In spite of technical problems, the Languedocs proved very useful, and even with the unreliable engines the main Moroccan base of Agadir was just eight hours away from Paris (with a stop-over in Casablanca). Aircraft of 31S had their codes and call-signs connected by the usual Aeronavale practice (eg SE-161/1 c/n 51 "31S-1" being F-YEBA, c/n 57 "31S-7" F-YEBG etc).

Two more Aeronavale units were also to receive the Languedoc: a flight of the experimental squadron, Escadrille 10S, had two aircraft at Cuers, near Toulon, during the late fifties, and the other aircraft went to Escadrille 56S at Agadir in Morocco and were used by the School for non-pilot aircrew. The Languedocs of 56S were appropriately modified as flying classrooms with the addition of a prominent radar-nose, together with a ventral "dustbin"-type radome. They included four unsold air-craft which had been put into storage by DTI at Nantes and were finally transferred to the Navy late in 1953 after refurbishing and necessary modifications. The Aéronavale Languedocs were all withdrawn from use by 1959.

The story of this unsuccessful four-engined airliner, with its production run of 100 air-craft - a high total for the then prevailing French standards - would not be complete without mention of the many DTI aircraft used for varied experiments.

These included flying test-beds c/n 1, 33 and 81 - while c/n 9 and 30 were used as development aircraft for the unsuccessful GR 14 power plants. Experimental flights were multifarious and included de-icing experiments with a Bréguet Atlantic wing section and a live airborne TVtransmission across the Mediterranean when General de Gaulle paid a visit to Algeria in 1958. Of particular interest is the use of four Languedocs (c/n 6, 31, 96 and 97) as mother-aircraft for experimental vehicles, the most publicised being those of the world's first piloted ramjet-driven prototypes. C/n 6 was ready for the trials from July 1946, but the experiments with the Leduc 010 did not start until 16 November 1946, leading six years later to the Leduc 021. The contribution of the unlucky airliner to French aeronautical development was therefore somewhat compensated for by this extensive role as company test-beds.

The last Languedoc, c/n 81, was withdrawn from use circa 1964-65. With too small a capacity for such a large aircraft, the SE-161 could not compete with its American and British counterparts in terms of load factors, and this was the principal reason behind its relative failure. It is important to state that, arising from the troubles experienced with the SE-161, Air France became increasingly reluctant to adopt French-designed airliners, and the SE-161 doubtless paved the way for some mistrust of native transport aircraft, from which it was hard to recover.

Specifications

Four-engined airliner with cantilever low-wing and twin-fin and rudder tail assembly. Crew of five (pilot, co-pilot/navigator, radio operator, flight engineer and steward). Standard cabin accommodation was for 33 passengers (eleven rows of three, two on starboard, one on port), but an alternative arrangement allowed for 24 seats. A 44 seat version was introduced by Air France in 1951.

SE-161/1 basic

Span: 29.39m
Length: 24.26m
Height: 5.14m
Wing area: 111.32 sq m
Weights: with 33 seat 'continental accommodation';
      figures in brackets relate to long-courrier -
      long-range - operation with 24 passengers only
      and 1,000 kg of freight included in both cases
Empty weight: 12,651 kg (same)
Loaded weight: 20,577 kg (21,001 kg)
Useful load: 3,970 kg (3,160 kg)
      As a freighter the useful load was 6,500 kg
      (Air France 1948 freighter conversions had
      5,300 kg payload with strengthened floor.
      Total capacity of the three freight
      compartments was 40 cu m.
Maximum speed: 440 km/hr at sea level, 430 km/hr at
      2,300m
Cruising speed: 375 km/hr at 2,300m (65% power)
Service ceiling: 7,200m
Maximum range: 3,200 km or 1,000 km under normal
      operational conditions (33 passengers)

Power plants and related figures

1   Bloch/SO-161-01

   4 x 900 hp Gnôme-Rhône 14N
   Maximum speed: 365 km/hr
   Loaded weight: 17,000-21,000 kg
   Range: 1,200-2,300 km

2   SE-161/1

   4 x 1,150 hp Gnôme-Rhône 14N 44/45
   Early commercial Air France aircraft
   Loaded weight: 21,100 kg (21,350 kg long-courrier)
   (23,000 freighter)

Range: 1,000 km (1,500 km) (800 km)
Cruising speed: 330 km/hr (335 km/hr) (325 km/hr)
Useful load: 33 passengers + 1,000 kg freight
      (24 passengers + 1,000 kg freight)
      (6,500 kg freight)

3   SE-161 delivered to Polish airline LOT

   4 x 1,150 hp Gnôme-Rhône 14N 54/55 when delivered,
   later changed to 14N 68/69 before final conversion
   to SE-161/P7 variant. The 14N 68/69s were fitted
   to the French Navy Languedocs.

4   SE-161R

   4 x Gnôme-Rhône 14R 24/28 rated at 1,220 hp (1,600 hp
   on take-off) for initial variant
   Loaded weight: 22,250 kg (22,600 kg long-courrier)
      (24,000 kg freighter)
   Cruising speed: 362 km/hr (370 km/hr) (350 km/hr
      with 6,185 kg freight)
   Aircraft delivered to the Armée de l'Air were fitted
with 1,325 hp (1,600 hp on take-off) Gnôme-Rhône 14R 28/29s (see Note).

5   SE-161/P7

   1946-47 Air France conversions
   P denotes Pratt & Whitney, 7 refers to seventh variant
   on drawing board.
   4 x 1,220 hp Pratt & Whitney R-1830 SIC-3-G

Note: French engine sub-types (eg GR 14R 28/29) denote
      clockwise or anti-clockwise operation and port or
      starboard location.
      - even numbers (eg 14R 28) = anti-clockwise/
                  starboard
      - odd numbers (eg 14R 29) = clockwise/port

Production List

C/n 01 F-ARTV. Prototype. F/f Sep 39. Used as VIP
      transport by Vichy Govt. Seized by Germany.
Force-landed and overturned at Chapelle Baloue 02 Apr 43.
One killed.

C/n 1 F-BATA. F/f 23 Aug 45 or 17 Sep 45. Used for
      various experiments including tests with TGAR
1008 jet engine mounted on top of fuselage Jan 51.

C/n 2 F-BATB. To Air France del 24 Oct 45. Written
      off 07 Apr 52 at Le Bourget.

C/n 3 F-BATC. To Air France del 29 May 46. Lsd to
      Air Atlas. Operated by CIET after 06 Jul 54.
To EC-ANP Aviaco del 26 Sep 56. Can 20 Sep 61.

C/n 4 F-BATD. To Air France del 11 Feb 46. Lsd to
      Air Atlas, then Tunis Air. Operated by CIET
after 06 Jul 54. To EC-ANQ Aviaco del 26 Sep 56. Canc
21 Jun 61.

C/n 5 F-BATE. With CEV/CATRE as F-BATE by mid-Apr 46,
      and del to Air France 13 Apr 46. Lsd to Tunis Air
15 Jun 53. WFU Sep 56.

C/n 6 F-BATF. Used for various experiments including
      launcher for Leduc 010 c/n 01 mounted on top of
fuselage (1946-49...).

C/n 7 F-BATG. To Air France del 04 May 46. Written
      off 14 Jun 48 at Coulommiers.

C/n 8 F-BATH. To Air France del 21 May 46. Dbr 10
      Feb 48 at Orly. Fuselage kept for ground
training until at least 1951.

C/n 9 F-BATI. To Air France del 29 May 46. Not
      modified with P&W R-1830 engines. Used instead
for development of Gnôme-Rhône version. Last flt Nov 50
then WFU.

Top   :   Armagnac F-BAVH of SAGETA at Le Bourget in May 1957
            (JMG Gradidge)
Centre :   Breguet 763 Deux Ponts F-BASV of Air France at
            Heathrow in March 1961 (JMG Gradidge)
Bottom :   Breguet 765 Sahara No.1, coded E of the Armee de
            l'Air at Dakar in 1960 (R Caratini via JMG Gradidge)

Top   :   SE.161 Languedoc F-BCUQ of Air France at Heathrow
            in October 1954 (JMG Gradidge)

Centre :   SE.(MB)161 Languedoc No.70 of Aeronavale at Le
            Bourget in June 1959 (JMG Gradidge)

Bottom :   SE.(MB)161 Languedoc No.80 of Aeronavale, coded
            56.S.10 at Dakar in 1960 (R Caratini via JMG Gradidge)

Top   : Latecoere 631 F-BDRA moored on Southampton Water
           in July 1947 (Flight)

Bottom : A Latecoere 631, showing clearly the retractable
           floats (E.C.P.Armees)

Caravelle III N420GE (c/n 42) with Douglas - Sud Aviation
titling (Douglas via J Wegg)

C/n 10 F-BATJ.   Built 1946.   With CEV/CATRE as F-BATJ
        by mid-Apr 46.   To Air France del 11 Mar 47.   To
EC-AGU Aviaco del 16 May 52.   Canc 20 Sep 61.

C/n 11 F-BATK.   To Air France del 22 Jun 46.   Written
        off 04 Feb 48 at Marignane.

C/n 12 F-BATL NTU.   Crash-landed at Cairo, repaired and
        retd to France.   Ferried from CEV Villacoublay to
Toulouse 31 May 48 (for storage?).   Fate not traced.

C/n 13 F-BATM.   Del 06 Jan 46 to War Office.   To Air
        France del 03 May 48.   Written off 23 Nov 48
at Toulouse.

C/n 14 F-BATN.   To Air France del 29 Jul 46.   Lsd to
        Air Transport 28 May 52-53.   To OD-ABU Air
Liban del 13 Oct 53.   Written off 06 Jan 54 at Beirut.

C/n 15 F-BATO NTU.   To CEV for de-icing tests 26 Feb 47-
        May 47.   To SP-LDD LOT del   47.   Fate
unknown.

C/n 16 F-BATP.   Built 1946.   To Air France del 15 Apr
        47.   Lsd to Air Atlas and to Cie Air Transport.
To SGACC (SAR service operated by French AF EARS 99) del
11 May 54.   WFU Jan 61 and SOC 02 Jan 61.

C/n 17 F-BATQ.   To Air France del 29 Nov 46.   Lsd to
        Air Transport May 52-03 Oct 52.   To SGACC/SAR,
operated by French AF EARS 99 del 17 Jun 54.   WFU Nov
56 and SOC 09 Dec 58.

C/n 18 F-BATR NTU.   To CEV Villacoublay   46 until
        final acceptance 30 Sep 47, before del to LOT as
SP-LDC   Oct 47.   Fate unknown.

C/n 19 F-BATS.   Built 1946.   To Air France del 15 Mar
        47.   To SU-AHZ Misrair del 30 Jan 52.   Written
off 24 Apr 54 at Damascus.

C/n 20 F-BATT.   To Air France del 30 Apr 47.   To
        OD-ABJ Air Liban "Beiteddine" del Jun 51.   Canc.

C/n 21 SP-LDA.   To LOT del 05 Jul 47.   Fate unknown.

C/n 22 SP-LDB.   To LOT del 05 Jul 47.   Fate unknown.

C/n 23 F-BATU.   To Air France del 01 Oct 47.   Written
        off 09 Apr 49 at Nice.

C/n 24 F-BATX.   To Air France del 12 Aug 47.   To
        SU-AIA Misrair del 12 Jun 52.   Reported to Air
Liban?

C/n 25 F-BATY.   To Air France del 15 Jan 47.   Written
        off 07 Oct 47 at Bone.

C/n 26 F-BATZ.   To Air France del 23 Oct 47.   Lsd to
        Air Atlas.   To OD-ABY Air Liban del 27 Jan 54.
To F-BAYY Air France del 27 Jun 54.   To EC-AKV Aviaco
del 29 Jul 55.   Written off 29 Sep 56 at Los Rodeos,
Tenerife.

C/n 27 F-BCUA.   To Air France del 07 Feb 47.   Lsd to
        Tunis Air 15 Jun 53-17 Oct 53.   To SGACC/SAR
operated by French AF EARS 99 del 17 Mar 55.   Dbr 20
Apr 55 at Le Bourget and SOC 08 Dec 55.

C/n 28 F-BCUB.   To Air France del 14 Nov 47.   Lsd to
        Air Atlas.   Operated by CIET after 03 Nov 54.
To EC-ANR Aviaco Sep 56.   Written off 04 Dec 58 in
Guadarrama Mts, Spain.

C/n 29 F-BCUC.   To Air France del 19 Dec 47.   Written
        off 26 Jan 48 at Romainville near Paris.

C/n 30 Military.   To CEV as SNECMA 14R prototype, for
        trials before del to CEAM.   Tested with new
engines at Villacoublay from Jun 49.   C-54 prop fitted
to no.3 engine Jul 49.   Ferried to CEAM Mont-de-Marsan
(Dept des Landes) 10 Aug 49.

C/n 31 F-BCUT.   Used for various experiments, including
        launcher for SNCAC NC-271 c/n 01 (1948-49),
SNCASO SO-M1 (1949), Leduc 010 c/n 02 (1950...) mounted
on top of fuselage.

C/n 32 F-BATO.   To Air France del 07 May 47.   Written
        off 29 Aug 48 at Le Bourget.

C/n 33 F-BCUU?   Used as engine test-bed for Arsenal 24H
        fitted in mid-1949.   F/f 07 Feb 50.   Last flt
Oct 50 and WFU.

C/n 34 F-BATV.   To Air France del 05 Sep 47.   To SGACC/
        SAR operated by French AF EARS 99 del 09 Jul 55.
WFU Mar 60 and SOC 17 Mar 60.

C/n 35 F-BCUE.   To Air France del 05 Jan 48.   To
        SU-AHG Misrair del 05 Nov 51.   Fate unknown.

C/n 36 F-BCUF.   To Air France del 21 Jan 48.   To SGACC/
        SAR operated by French AF EARS 99 del 16 Dec 54.
WFU Mar 60 and SOC 17 Mar 60.

C/n 37 F-BCUG.   To Air France del 16 Feb 48.   To SGACC/
        SAR operated by French AF EARS 99 del 17 Mar 55.
WFU Aug 57 and SOC 09 Dec 58.

C/n 38 F-BCUH.   To Air France del 23 Feb 48.   To
        EC-AHT Aviaco Jan 53 (reported in ferry marks
EC-WHT Apr 53 and del 26 Jul 53).   Canc 20 Sep 61.

C/n 39 F-BCUI.   To Air France del 11 Mar 48.   Written
        off 30 Jul 50 at Marignane.

C/n 40 F-BCUJ.   To Air France del 23 Mar 48.   To SGACC/
        SAR operated by French AF EARS 99 del 17 Dec 54.
WFU Jan 60 and SOC 19 Jan 60.

C/n 41 F-BCUK.   To Air France del 14 Apr 48.   To
        SU-AHH Misrair del 29 Nov 51.   Written off 24 Dec
51 at Teheran.

C/n 42 F-BCUL.   To Air France "Ciel de Geneve" del 04
        May 48.   Lsd to Tunis Air Oct 53-...   Canc Sep
56 but already WFU Oct 54 at Toulouse.

C/n 43 F-BCUM.   To CEV 26 Mar 48 before del to Air
        France 30 Mar 48.   Written off 03 Mar 52 at Nice.

C/n 44 SP-LDE.   To LOT del 1948.   Fate unknown.

C/n 45 F-BCUO.   To Air France del 26 Dec 47.   To SGACC/
        SAR operated by French AF EARS 99 del 17 Jun 54.
WFU Mar 60 and SOC 17 Mar 60.

C/n 46 F-BCUP.   To Air France del 16 Jun 48.   To SU-AHX
        Misrair del 26 Sep 51.   Written off 30 Jul 52 at
Almaza, Cairo.

C/n 47 F-BCUQ.   To Air France del 22 Oct 48.   Operated
        by CIET after 08 Oct 54.   To EC-ANS Aviaco 08
Oct 56.   Canc 20 Sep 61.

C/n 48 F-BCUR.   To Air France del 14 Jun 48.   To SGACC/
        SAR operated by French AF EARS 99 del 09 Jul 55.
WFU Mar 60 and SOC 17 Mar 60.

C/n 49 F-BCUS.   To Air France del 22 Jul 48.   To EC-AMH
        Aviaco Sep 55.   Canc 20 Apr 61.

C/n 50 French Navy.

C/n 51 French Navy del 12 Jul 50.

C/n 52 French Navy.

C/n 53 French Navy.

C/n 54 French Navy.

C/n 55 French Navy.

C/n 56 French Navy.

C/n 57   French Navy.

C/n 58   French Navy.

C/n 59   French Navy.

C/n 60   F-BATR.   To Air France del 30 Mar 48.   To
         EC-AGV Aviaco del 06 Jun 52.   Canc 1962.

C/n 61   F-BCUN.   To Air France del 29 Jun 48.   To
         SGACC/SAR operated by French AF EARS 99 del 26
May 55.   Lost in Mediterranean 01 Dec 56.

C/n 62   French Navy.

C/n 63   French Navy.

C/n 64   French Navy.

C/n 65   French Navy.

C/n 66   F-WCUD.   To DTI.

C/n 67   French Navy?

C/n 68   French Navy.

C/n 69   To DTI (special missile launcher?). CEV trials
                                              1949/50.
C/n 70   To DTI, then French Navy.

C/n 71   To DTI, then French Navy.

C/n 72   French Navy?

C/n 73   French Navy.

C/n 74   To DTI, then French Navy.

C/n 75   To DTI (special missile launcher?).

C/n 76   To DTI, then French Navy.

C/n 77   French Navy.

C/n 78   French Navy.

C/n 79   French Navy.

C/n 80   French Navy.

C/n 81   To DTI.   Used by CGTM as engine test-bed (Palas,
         Bastan..).   Mostly coded "F" (F-ZWRF).   Last
Languedoc to fly in 1964.   WFU at Cazaux 1964-65.

C/n 82   To French AF del Sep 50.   To CEAM Sep 50-Jun 51
         coded "BZ", later GT II/61.   Mostly coded "61-
PA".   WFU Dec 55 and SOC 12 Dec 55.

C/n 83   To DTI.   Used as engine test-bed for SNECMA
         Atar, mounted on   .   Mostly coded "P"?
Ferried Toulouse-Blagnac to Brétigny 10 Sep 51.

C/n 84   To French AF del Jun 51.   GT II/61, mostly coded
         "61-PQ".   WFU Dec 55 and SOC 12 Dec 55.

C/n 85   To French AF del Oct 51.   GT II/61, mostly coded
         "61-PD" and "61-PB".   WFU and SOC 12 Dec 55.

C/n 86   To French AF del   .   GT II/61, mostly coded
         "61-PC".   Written off 23 Oct 52 at Dugny near
Le Bourget.

C/n 87   To French AF del   .   GT II/61, mostly coded
         "61-PQ" and "61-PD".   WFU Feb 56 and SOC 09 Feb
56.

C/n 88   To French AF del   .   GT II/61, mostly coded
         "61-PE".   WFU Feb 56 and SOC 09 Feb 56.

C/n 89   To French AF del Sep 52.   GT II/61, mostly coded
         "61-PF".   WFU Dec 55 and SOC 12 Dec 55.

C/n 90   To French AF del Oct 52.   GT II/61, mostly coded
         "61-PH" and "61-PG".   WFU Dec 55 and SOC 12 Dec
55.

C/n 91   To DTI coded F-ZLAX.   To French AF del Feb 53.
         GT II/61, mostly coded "61-PI" and "61-PH".   WFU
Feb 56 and SOC 09 Feb 56.

C/n 92   To French AF del Dec 52.   GT II/61, mostly coded
         "61-PJ" and "61-PI".   WFU Dec 55 and SOC 12 Dec
55.

C/n 93   To French AF del Sep 52.   GT II/61, mostly coded
         "61-PG" and "61-PJ".   WFU Dec 55 and SOC 12 Dec
55.

C/n 94   To French AF del May 53.   GT II/61, mostly coded
         "61-PK".   WFU Dec 55 and SOC 12 Dec 55.

C/n 95   To French AF del Apr 53.   GT II/61, mostly coded
         "61-PL".   WFU Feb 56 and SOC 09 Feb 56.

C/n 96   To DTI.   Mostly coded F-ZLAW.   Used as launcher
         for Leduc 021 c/n 02 in 1954...

C/n 97   To DTI.   Mostly coded F-ZLAV.   Used as launcher
         for Leduc 021 c/n 01 in 1953...

C/n 98   To French AF del Apr 53.   GT II/61, mostly coded
         "61-PM".   WFU Feb 56 and SOC 09 Feb 56.

C/n 99   To French AF del May 53.   GT II/61, mostly coded
         "61-PN".   WFU Dec 55 and SOC 12 Dec 55.

C/n 100 To French AF del   .   GT II/61, mostly coded
         "61-PO".   Dbr in ground accident.

NOTE: The Languedocs used for search and rescue duties
         flew with a civil registration, although operated
by the French AF EARS 99 Squadron.   When on an SAR
mission, a special call-sign applied, beginning with F-SS..
but retaining the last two letters of the civil
registration;   for example on an SAR mission F-BCUA used
the call-sign F-SSUA.   The only exception was F-BCUO,
which used the call-sign F-SSUD.

LATECOERE 631

The Latécoère 631 was a large six-engined
commercial flying boat designed to meet the
1936 Ministère de l'Air specification for a
transatlantic flying-boat.   Two more contenders in this ambitious programme were the
SNCASE SE-200 (qv) and the Potez-CAMS 161.

A prototype was ordered in 1938, and the
Société Industrielle d'Aviation Latécoère

(SIAL) began construction of the components.
The design team was led by Messrs Moine and
Jarry.   For obvious reasons (tooling and construction beginning in 1939), completion at
Toulouse was not reached until 1942.   With no
appropriate surface of water in the area, and
with the Biscarosse Latécoère seaplane facilities
on the Atlantic coast - part of the then German-
occupied territories - only the Mediterranean

shores were left to undertake flight-testing.
This meant a long 562 km tranfer by road on
several specially designed trailers, an extra-
ordinary feat in itself considering the size
of the flying-boat.   Final assembly was then
undertaken at Marignane, and the Laté (short
for Latécoère) 631 prototype was flown from
the Berre Lake only a few days before the
German invasion of Southern France following
Allied landings in French North Africa.

The Laté 631 was powered by Six Wright two-
row radial engines (1,290 hp each), the planned
French Gnôme-Rhône 18L power plants not being
available.   Like its SE-200 contemporary, the
Laté 631 was seized by the German authorities
and transferred to Germany, only to be sunk by
Allied bombs.

Early in 1943 the French Secrétariat à l'Air
had issued the letter E to Latécoère for use as
initial letter to aircraft names, and the Late
631 was named Eole after the Greek god of
winds.   This was, however, never to be used.

Meanwhile two more Laté 631s had been
ordered, the second airframe being completed in
March 1943 also at Toulouse.   The Germans were
anxious to confiscate the Laté, and when told
by the French that road transport to the sea
was no longer feasible, they suggested fitting
a temporary gear for initial take-off from
Toulouse.   But they were eventually
convinced that the aircraft should be dismantled and hidden from Allied bombers on
different locations in southwest France.

Following the Liberation, it was once more
assembled at Biscarosse this time and first
flown in March 1945.   Power plants were Wright
Cyclones.   After a trip to French West Africa
in August of that same year the flying-boat
left France in October on a demonstration tour
to South America, having been named 'Lionel de
Marmier' after the Armée de l'Air pilot
recently lost over the Mediterranean who had
been General de Gaulle's own pilot and had
pioneered Free French air military transport
during the war.   When in Brazil on 31 October,
in-flight disaster was avoided after the loss
of a propellor blade on an inboard engine, but
with two fatalities on board.   Emergency
repairs were made on location, and the 'Lionel
de Marmier' was flown back to France.

In the meantime production had got under
way.   The total government order had now risen
to eleven, and following completion of boats 2,
3 and 4 at Toulouse the next batch was to be
built by SNCASO at Le Havre and St Nazaire (c/n
5, 6, 8 and 10 to be erected at St Nazaire and
7, 9 and 11 at Le Havre).   SNCAN was also
involved in the programme with an order for six
hulls.

Air France was the first operator of the
Laté, using three production boats powered by
1,900 hp Wright R-2600 engines.   The French
West Indies route was inaugurated in July 1947,
the French terminal being at Biscarosse, still
a long way from Paris.   Luxury accommodation
was provided for 46 passengers in small cabins
with collapsible sleeping berths for night
flying.   There was no galley for the French
cuisine;   instead, a real kitchen and two
dining-rooms.   Total crew complement reached
14 with the catering staff!

Unfortunately the story of this flying
'passenger ship' was soon to be marked with
tragedy.   Disaster first struck in February
1948;   while on delivery from Le Havre to

Biscarosse the seventh Laté 631, F-BDRD, crashed
in the Channel owing to the lack of proper de-icing
equipment, with the loss of its crew.   Six
months later an Air France machine, F-BDRC,
disappeared in unexplained circumstances during
a transatlantic flight with a heavy toll in
lives.   Air France was faced with ceasing
operations after hardly more than a year of
successful crossings.   The type was withdrawn
from use pending the results of the accident
investigation.   By then, it was clear that the
future of the Laté 631 no longer lay in
passenger service.

Operations were however resumed, this time
with a newly-formed State-backed company, SEMAF
(Société d'Exploitation du Matériel Aérien
Français), for cargo flights between France and
French colonies in Africa.

Unfortunately while on tests in March 1950
the third Laté 631 crashed into the Atlantic off
Cap Ferrat, southwest of Bordeaux.   Lengthy and
costly investigation did not reveal any evidence
from the salvaged parts, and this latest fatal
crash brought SEMAF activities to an end.   The
following year, however, after the discovery of
more wreckage by a trawler in August 1950,
tentative conclusions to explain the crash were
given.   The primary cause was found to be the
power plants, with ensuing but unrelated failure
in aileron controls;   wind tunnel tests with
outer wings added support to this theory.

A last chance was then given to the huge
flying-boat.   Trials were conducted with the
eighth Laté, including a trip to French Indochina in April 1952.   A new operator, France-
Hydro, was incorporated to carry on with Laté
cargo operations.   After suitable conversion
F-BDRE was used for more than a year to air-
ferry heavy loads of cotton from Lake Léré in
Northern Cameroons, and by mid-1954 the French
Government was planning to transfer the
surviving flying-boats to France-Hydro (Hydro
being colloquial French for 'hydravion' or
flying-boat).

With cargo conversion of the remaining Latés
under way, all plans were cancelled after the
crash of F-BDRE in September 1955, once more
with fatalities.   The flying-boats were put in
storage at Biscarosse, and the last blow came
in 1956 when, following an exceptional snowfall,
the hangar there collapsed, destroying several
Laté 631s.

Thus came to an end the story of this huge
and elegant flying-boat.   Derivatives of the
Laté 631 had been envisaged, including the four-
engined Laté 634 and the six-engined Laté 636,
both with an all-up weight of about 82 metric
tons.   These two projects soon fell by the wayside.

In conclusion, it can be said that it was
unfortunate perhaps that at the time of its
entry into Air France service six-engined flying-
boats like the Laté 631 belonged to a bygone
pre-war era.   Recent experience was showing
that the new generation of landplanes such as
the Constellation was safe and reliable on the
transatlantic run.   It was also more convenient
to take off from Orly on the outskirts of Paris
rather than travel by train all the way to
Biscarosse finally to board a flying-boat.   All
things considered, this was an unrealistic
approach to modern air travel.

Last but not least, any possible success of
the Laté 631 as a freighter suitable under
specific geographical conditions, especially

overseas, was hampered from the start by a succession of fatal accidents. The loss of F-BANU had been a hard blow; the probable cause of the accident was supposed to be the fatigue failure of the aileron control couplings, resulting from the simultaneous occurrence of several vibratory phenomena. The reducing gear was found to be one source for critical resonance, unknown to the crew until it was too late.

The famous pre-war Latécoère name disappeared from the French aeronautical scene with the Laté 631 and the company then only developed missiles, including a most unusual experimental postal rocket, and various missiles built in quantities for the French Navy (Engin Postal 110, Latécoère 231/232/233 Malafon, 258 Malface and Masalca missile respectively). In 1961 a twin Astazou high-wing executive project was announced as the Latécoère 870. Another project was the 853, a twin Astazou cargo aircraft, followed by the 863 which, powered by four Bastan IV turbo-props, would have carried 38 passengers. None left the drawing board stage.

Specification

High-wing six-engined all-metal (except controls) transport flying-boat. Wing floats retractable under outboard engine nacelles.

Fortysix passengers in Air France service (accommodation including sleeping-berths). In cargo configuration, maximum payload was 25 metric tons or 169.5 metric metres volume available in holds.

Span: 57.43m
Length: 43.46m
Height: 5.66m
Draught at maximum weight: 1.78m
Empty weight: 32,000 kg
Maximum all-up weight: 75,000 kg
Cruising speed: 320 km/hr
Maximum speed: 395 km/hr at 1,850m
Range: 6,000 km
Typical payloads: 8 metric tons on 6,000 km flight
                  14 metric tons on 4,000 km flight
                  25 metric tons on 2,000 km flight

Power plants

1 1938 prototype: 6 x Gnôme-Rhône P-18 (1,500 hp), not fitted

2 Prototype: 6 x Wright Double Row Cyclone (1,290 hp)

3 Air France c/n 2, 3, 4: 6 x Wright Cyclone GR-2600-A5B (1,600 hp each)

4 France Hydro c/n 8: 6 x Wright R-2600-20 (1,900 hp)

5 Later modification: R-2600-C14BB Cyclone 14

Production List

C/n 01 F-BAHG. F/f 04 Nov 42, Marignane. Captured by German forces after invasion of Vichy France and flown to Friedrichshafen, where sunk by Allied Mosquitoes 17 Apr 44. TT 40 hrs.

C/n 2 F-BANT built Toulouse 1943. Dismantled Mar-Apr 44 and hidden until late 1944. F/f 06 Mar 45 Biscarosse, SGACC (CofA 09 Oct 45). Named 'Lionel de Marmier". To Air France del Oct 47. Canc Sep 49. Cargo conversion for France-Hydro 1954 but WFU. Seized at Biscarosse Sep 57 and scrapped.

C/n 3 F-BANU. F/f Nov 46. To Air France as 'Henri Guillaumet' del 02 Jun 47. Later to SEMAF, cargo conversion. Re-regd F-WANU and lost on test flight 28 Mar 50 off Cap Ferrat. Ten occupants killed.

C/n 4 F-BDRA. F/f Jul 47, del Air France 18 Aug 47. Canc Oct 50. Intended for transfer to France-Hydro 1954 but WFU and scrapped.

C/n 5 F-BDRB. F/f Sep 47. For France-Hydro del 1954. NTU and WFU. Damaged by hangar collapse Feb 56. Seized at Biscarosse Sep 57 and scrapped.

C/n 6 F-BDRC. F/f 15 Nov 47. To SGACC. To Air France del 11 Jun 48. Lost 01 Aug 48 on trans-atlantic crossing between Fort de France and Port Etienne. Twelve crew and 40 passengers lost.

C/n 7 F-BDRD. F/f 26 Jan 48. Crashed near St Marcouf (Cherbourg peninsula) 21 Feb 48 during del flight to Biscarosse. Seven crew and 12 passengers killed.

C/n 8 F-BDRE. F/f 28 Nov 48. To RF-SEMAF (CofA 04 Aug 49). WFU Mar 50. Flown again 23 Feb 52 after overhaul and cargo conversion (R-2600-20 engines). To France-Hydro Mar 52. Crashed in bad weather 10 Sep 55 Sambolabo area, French Cameroons.

C/n 9 F-WDRF. F/f Nov 48. Intended for transfer to France-Hydro 1954 but NTU and WFU. Seized at Biscarosse Sep 57 and scrapped.

C/n 10 F-WDRG. F/f 1949. Intended for transfer to France-Hydro 1954 but NTU and WFU. Damaged in hangar collapse Feb 56. Seized at Biscarosse Sep 57 and scrapped.

C/n 11 (F-WDRH). Not flown. Del also intended to France-Hydro 1954. Damaged in hangar collapse Feb 56. Seized at Biscarosse Sep 57 and scrapped.

NOTE 1 C/n 2 to 4 were manufactured at Toulouse, 5 to 11 at Le Havre (of which 5, 6 and 8 were assembled at St Nazaire and 10 intended) and 7 was assembled at Le Havre with 9 and 11 intended.

NOTE 2 Three were ordered by FAMA on 20 Feb 46 and allotted c/n 02, 03 and 04. Order canc Jul 46. On a further allocation Air France was to receive c/n 2, 3 and 4, and 5, 6 and 7 were intended for Mexico in Jul 47.

SE.210 CARAVELLE

The historical background and development of the Caravelle begins on the following page, the technical details and specifications commence on page 42, and these are followed by an airline operators section starting on page 48 and the production list on page 57 et seq.

Scenario

With six years of war finally over in Europe in May 1945 the thoughts of the major aircraft manufacturers - at that time effectively only those in the USA and UK - turned towards peaceful applications of new found technology. In Britain the Brabazon Committee had already drawn up a number of civil specifications to cover the needs of both its domestic and Empire air routes while in the USA there was a fine line of unwanted military transports waiting to be developed for airline customers. In Continental Europe only France was in a position immediately following the cessation of hostilities to begin rebuilding her aviation industry.

Government aid was forthcoming to many concerns including the Société Nationale de Construction Aéronautique Sud-Est, SNCASE. Formed on 21 December 1936 in accordance with the Law of Nationalisation of Aeronautical Industries, SNCASE had absorbed the factories formerly owned by Lioré-et-Olivier, Romano and SPCA; expanding in 1941 with the acquisition of the Societe Nationale de Construction Aéronautique du Midi which brought into the fold the former Dewoitine plants. As described in an earlier chapter SNCASE lost no time in applying itself to the civil market with the resurrection of the Bloch 161 four-engined transport, originally flown in 1939. Redesignated SE.161 and named the Languedoc it was produced in reasonable numbers for Air France, L'Armée de l'Air and a few other customers. Considerable design effort was being expended in 1946 on the design of a new long range transport, the SE2010 Armagnac. Although flown in the same year, 1949, as the DC-6A, it suffered development problems which delayed airline deliveries until 1953 and only a handful were completed.

By the turn of the decade it was obvious that if the French aviation industry was to regain some of its pre-war eminence a radical change in design philosophy was required in the civil field. It was no longer viable to try and compete with the long range airliners that were being produced by the USA en masse. And the efforts de Havilland were making in Great Britain with their revolutionary jet airliner, the Comet could not go unnoticed. Rather naturally, thoughts turned to a medium range aircraft-one that would particularly satisfy Air France's requirements on the prime North African routes.

Evolution

In order to steer the industry in the right direction the official civil aviation agency in France, the SGACC (Secrétariat Générale à l' Aviation Commerciale et Civile) began to draw up an outline requirement for a medium-haul airliner (Moyen-corrier) capable of carrying a 7-tonne payload over ranges of 1000-1200 miles at speeds in excess of 330 kts. The key SGACC meeting was held on 12 October 1951 and the resultant specification issued to the industry for bids on November 6. To provide the necessary performance the committee favoured the use of the indigenous SNECMA Atar jet engine, although the possible use of a yet-to-be-developed turbofan jet was also noted.

Preliminary design studies were submitted by a number of companies and ranged through the whole spectrum of designs of that era. Bréguet offered the Br-978A with four turboprops and another design jointly with SNCAN utilising three Atars, the 978T. Dassault also proposed a turboprop while SNCASO put forward the SO.60 with two Rolls-Royce RA.7 Avons assisted by two Turboméca Marborés. Using their experience with high aspect

ratio wing design Hurel-Dubois presented the HD.45, powered by two Atars. The engines were to be hung on the high-lift bracing struts of the shoulder mounted wing. The SNCASE design team under the leadership of Pierre Satre submitted a "Tri-Atar" design in a whole range of studies numbered X-200 upwards. The X-210 featured two Atars mounted in nacelles on the rear fuselage with the third buried in the T-tail structure. Another proposal made use of turbofans, but it was admitted that this was rather an ambitious project for 1951. On 28 March 1952 the SGACC met with the CMC (Comité du Matériel Civil) with the result of the elimination of the Dassault, Bréguet and Bréguet/SNCAN proposals. Of the remainder, the X-210 was considered the most promising. By May it had become apparent that with the prospects of increased power, the Avon offered the greater potential and a better service background than the Atar. With the Avon now projected at 9,000lb thrust (versus the 6,000lb Atar) only two would be required and SNCASE was instructed to modify the X-210 accordingly.

Satre's team was convinced that the rear engined layout had many advantages over burying the engines in the wing structure (as British designers generally favoured) or hanging them from wing-pods as was the vogue in the USA. This feature was therefore retained, although with preliminary wind-tunnel tests on the hazards of the T-tail, this was deleted in favour of mounting the horizontal tailplane midway up the fin. The result was an aerodynamically clean wing and the prospect of a very quiet passenger cabin. Other points were the reduced fire hazard in the event of a wheels-up landing - the engines being well removed from the wing fuel cells, easy handling characteristics in a one-engine out situation and ease of engine access for maintenance The main passenger access was from the integral stairs mounted underneath the tail, previously only used on the Martin 2-0-2 and some Convair 240s. This finalised design was resubmitted to the SGACC in July and two months later was confirmed as the winner.

A contract authorising construction of two flying and two static test airframes was signed by the Secrétariat d'Etat à l'Air on 3 January 1953. The designation SE.210 was applied to the design and the name "Caravelle" bestowed, after the French name for fleet vessels of the 15th and 16th Centuries. In order to speed up development and get the prototypes into the air as quickly as possible SNCASE purchased two complete Comet nose sections from Hatfield. These were spliced onto the Caravelle fuselage and the Comet cockpit layout was retained. No doubt at the time there was also a thought that customers for the Comet would look to France for their medium range route requirements and the similarity in cockpit design would be a big advantage in training crews.

Flight trials

The first prototype was rolled out of the SNCASE Toulouse factory on 21 April 1955. It differed little from the original project except for taking advantage of the 10,000lb thrust Avon that had become available. Just over a month later, Pierre Nadot and Andre Moynet were at the controls when F-WHHH made its first flight of 41 minutes on May 27. Twenty flights had been completed by the time 01 made its public debut at the Salon du Bourget in June, making a number of flypasts but not being statically shown due to the pressures of the test programme. An intensive series of tests were carried out from Toulouse, Istres and Brétigny. By October over 100 hours had been logged and high speed dives had been made to Mach 0.86. Single-engine take-off tests indicated that the then-designed maximum weight of 40,000 kg would be comfortably exceeded. It was decided that production aircraft would have the 11,400lb RA.29/1 Avon offering a 41,000 kg gross weight and a maximum range of 1500 miles with a full 16,000lb payload. Passenger configurations would range from 65-90 in five-abreast seating. Hot trials were conducted in the French Sahara in 1957, with the cold weather counterparts in Sweden in February 1958.

By November 1955 Air France plans to order 12 + 12 options were public, although the contract was not signed until February 3 of the following year.

The initial test programme for 01 was completed on 19 March 1956 and the Caravelle received its C of A (CdN or Certificat de Navigabilité) from the French authorities on May 23. The second prototype had flown earlier that month, on the 6th with Nadot and Leopold Galy. Essentially it was similar to 01 except for the deletion of the wing leading edge slats and the provision for a passenger cabin. Following its return to Toulouse for refurbishing and the removal of some test equipment, 01 was handed over to Air France for a series of operational evaluation flights. Between 23 May and 2 October 1956, F-BHHH made 209 flights totalling 505 hours, principally on the Paris-Algiers route and usually carrying cargo in the cabin. Crews were provided by the CEV at Brétigny although Air France crews were carried as observers. Notable among the trial flights were two round trips between Orly and Casablanca on 28 August, all being made on a single engine.

The first of the static airframes known as "EF" (for Cellule d'Essais de Fatigue) was delivered to the Etablissement Aéronautique de Toulouse in 1957. Tests were undertaken at CEAT Toulouse (Dept de la Haute-Garonne) from the autumn of 1957, and water tank testing, initially to 30,000 hrs equivalent flight time and later to destruction, commenced in late May 1958. The second airframe "ES" (Cellule d' Essais Statiques) followed to the CEAT in June 1958. A wooden structural mock-up was also built to study equipment accommodations.

Sales Tours

With cabin installed, 02 made a number of demonstration flights to European capitals in 1956 including Rome, Lisbon, Brussels, Amsterdam, Cologne-Bonn, Stockholm, Copenhagen and Helsinki. However, by the spring of 1957, 02 could be spared from the test programme to embark on a major sales tour of the Americas. Accordingly in March F-BHHI was converted into a 52-seat configuration, 13 rows of four abreast in a blue and grey scheme cabin and revised external paint applied incorporating the emblem of Sud Aviation which had been formed on March 1 with the amalgamation of Sud-Est (itself just established on 28 August 1956) with Sud-Ouest Aviation. To emphasise the low cabin noise during flight speakers were installed to soothe prospective buyers with the music of Mozart and Vivaldi.

On 18 April 1957 F-BHHI left Paris for Casablanca and Dakar piloted by Nadot and two Air France pilots, Lionel Casse and Andre

Lesieur. The remainder of the flight crew was from Air France and also on board was Sud President, Georges Hereil. The Caravelle crossed the South Atlantic to Recife and then continued to Rio-Sao Paulo - Porto Alegre-Buenos Aires-Montivedeo-Belem-Caracas before heading north to Miami and then New York-Idlewild. It thus became the first commercial jet to land at Idlewild as the 707 and Comet were banned at the time due to noise regulations. The extensive US tour covered Washington, Baltimore, Atlanta, Houston, Dallas, Kansas City, St.Louis, Denver, Los Angeles, San Diego, San Francisco, Seattle and Chicago before returning to France on June 25 by way of Montreal, Toronto and Gander. During the 28,000 mile trip over 3,000 officials had been carried on nearly 100 flights with no major maintenance problems or delays.

Although no firm orders were immediately announced as a result there must have been some consolation when three days later 02 brought SAS President Henning Throne-Holst to Paris for the signing of a contract for 6 Caravelles with 19 options.

Into Service

Largely as a result of slippages in the delivery of Avons, scheduled services did not start until 1959. The Caravelle was awarded its full transportation certificate on 2 April 1958 followed by FAA certification six days later, and the first production Series I was flown on 18 May 1958. Besides having the RA.29/1 Avon 522 engines it differed from the prototypes in that the fuselage was lengthened by 1.5m and a long extension of the dorsal fin along the top of the fuselage housed communications antennae, this latter feature being subsequently retro-fitted to 01 and 02. Gross weight had now risen to 43,500 kg.

Air France's first aircraft (F-BHRB) was handed over at Toulouse on 19 March 1959 and the first scheduled service was flown on 6 May from Orly to Istanbul via Rome and Athens. However, SAS had become the first operator one week earlier with an inaugural flight from Copenhagen to Beirut on 26 April. These two carriers remained the sole operators until Varig introduced the type on its New York run in December. Air Algérie initiated scheduled service on 12 January 1960 and was soon followed by Finnair, Royal Air Maroc, Swissair and Alitalia.

Production

Right from the start of the Caravelle programme in 1956 production was planned to be sub-contracted to various concerns with final assembly at Sud's St.Martin-du-Touch plant near Toulouse. Noting Vickers' unhappy experience in underestimating production requirements for the Viscount, Sud arranged with SNCASO and SNCAN for them to build 50% of the fuselage. Arrangements were also made with Fokker and Fiat and some German companies for certain sub-assemblies. If large US orders came in, Republic at Farmingdale was interested in opening up a line capable of producing up to seven aircraft per month. In the event this was never required but the early arrangements made by Sud in Europe held when changes were made to the whole structure of the French industry. Sud factories were responsible for the basic wing box structure (at Bouguenais); the rear wing section and leading edge (at St.Nazaire); the flaps (at Rochefort); the forward fuselage (at Marignane) and the cylindrical fuselage sections (at Toulouse). The rear fuselage was contracted to Latécoère at Toulouse, engine nacelles and tail to FIAT's Turin plant, control surfaces to Forges de la Boissière and the landing gear became the responsibility of Hispano Suiza. After the initial flurry of orders the production line was kept open by authorisations for additional aircraft to be built on speculation by the French Government. This system remained until Caravelle production ended in 1972.

Following twenty Series Is were twelve 1As, fitted with the Avon RA.29/1 Mk.526, there being no external differences. Initially Caravelle model designations were linked to the stage of Avon engine development and therefore the Avon RA.29/3 was the powerplant of the Series III. This engine was fitted with noise suppressors giving a slightly longer nacelle, the only visible difference between the I and III. The increased thrust of the RA.29/3 coupled with the results of static testing of the "ES" airframe authorised an increase in gross weight to 46,000 kg and improved the cruising speed to 430 kts. All earlier production aircraft except one Series I were modified to III standard between 1960 and 1961.

With an extra compressor stage the Avon RA.29/6 developed 12,500lb thrust and allowed the gross weight to climb to 48,000kg and the cruising speed to 445 kts. Following a 352-hour test flight programme with the re-engined III prototype, the first Caravelle VI entered service with Sabena on 18 February 1961.

Four Alitalia IIIs were subsequently converted to VI standard. Early production aircraft were priced around \$1-\$3M apiece although spares and support raised this figure. Complicated financing arrangements and offsets always cloud the selling price of an aircraft but individual contracts are mentioned where known in the operator section.

First Year in Service

For pioneering medium haul jets on their routes the early operators of the Caravelle, Air France and SAS, were rewarded with a remarkably trouble-free aircraft introduction period. Over 13,000 hrs were accumulated by the end of 1959 by both companies and by the end of 1960 total Caravelle hours by all operators had topped 76,000. Engine failures were 0.37 per 1000 hours by the end of 1959 compared to 2.9 per thousand with the best pistons. Both Air France and SAS recorded load factors between 60 - 85% on internal European and Middle East routes and average daily utilisation was around 7.5 hours.

SAS cost comparisons between the Caravelle and DC-7C found that Caravelle variable operating costs per round trip and per seat mile were 20 - 25% lower and on a ton-mile basis 17% lower. Direct operating costs were estimated to be 2.1 ¢ per seat-nm or \$1.47 per aircraft-nm (1960 prices). In the first 6 months SAS averaged 47 technical delays (in excess of 30 minutes) per 1000 departures compared to 77 with the DC-7C.

Pilot reaction was extremely favourable on the flying characteristics and passengers were equally happy with the unusually quiet cabin. Cockpit visibility was considered to be in need of some improvement and as no thrust reversers were fitted the braking system was backed up by a braking parachute, something SAS found welcome on wet and icy runways.

Modifications to the Caravelle in service were low, totalling around 200 compared to 3000 for the DC-6/6B and 1000 for the DC-7C and CV-440.

SAS was the unfortunate victim of the first fatal Caravelle accident which occurred during an approach to Ankara on 19 January 1960. The Board of Enquiry could not reach a decision on the cause of the accident, but a number of recommendations were made including the modifying of the drum-type altimeter and revised procedures for a high altitude descent. The second fatal accident also occurred during approach, involving an Air France aircraft at Rabat. This was attributed to poor visibility and lack of vigilance on the part of the crew.

US Breakthrough

In early 1959 Sud had made renewed efforts to interest US operators. The Electra had entered service in January that year with American Airlines and Eastern, so the main drive was directed at the major trunk operators not yet committed to re-equipping their medium haul fleets. TWA had shown interest since the visit of 02 to Culver City in 1957 for a demonstration to Howard Hughes but was struggling with financial problems and was pre-occupied with the introduction of the CV-880. United were therefore the most likely target although it had already stated that it preferred more than two engines to Boeing during early discussions on the 727 design.

To assist in the promotion to US operators, Sud went looking for a sales agent in the USA. Republic had an agreement with Sud as noted earlier in the production review but it was felt by Sud that assistance was required from a more civilian orientated company. As Boeing had more than enough to keep busy with on the 707 family and the projected 727, discussions were held with Convair in San Diego and Douglas in Long Beach. By the end of the year United was showing strong interest in the Caravelle as an interim aircraft to compete with the turboprops of its competitors until the 727 became available. A batch of 20-30 was planned and United particularly wanted to trade in a number of DC-6s and DC-7s as part payment so the drift towards Douglas as an agent increased. In February 1960 it was announced that Douglas would represent Sud for sales and post delivery support in virtually the whole of the Western Hemisphere with an option to manufacture should sales warrant that step. Sud was to continue with sales operations in Continental Europe and in French-speaking parts of the world. The agreement clinched United's order for 20 Series VI-Rs with 20 options and the contract was signed on 25 February for the first deliveries in 1961. Cost was reported to be \$65M.

Actually Douglas already had a Caravelle competitor on paper at least - the DC-9 - which was abandoned for the duration of the Sud agreement. At that time Douglas was in severe financial troubles and facing increased production headaches with the DC-8 so Donald Douglas Sr was anxious to get that programme flowing smoothly before embarking on a decision for production of a completely new type. There were serious hopes that enough orders for the Caravelle would be secured to open up a production line at the disused Santa Monica facility. In reality the agreement with Sud was allowed to dissolve at the end of 1961 due to the lack of sales and a certain amount of French unhappiness with a US company representing

Sud. Douglas then rather belatedly proceeded with its model 2086, later finalised as the DC-9 and becoming another Caravelle competitor.

The model for United, the VI-R (R standing for reverse thrust) was specifically developed for US operators. Besides an engine change to the RA.29/6 Avon 532R (later the improved turbine bladed 533R was standard) several other new features were introduced. The cockpit was completely reworked and slightly bulged at the top to provide more space and the cockpit windows were considerably enlarged to give better visibility. Additional wing spoilers and better brakes improved landing performance and the dorsal spine was deleted, although other VI-Rs retained it.

New Engines

As early as 1956 Sud was thinking of alternative power plants to the Avon. Initial thoughts were of a Conway version capable of carrying a 16,000lb payload 2500 miles and having a gross weight of 47,500 kg. The Pratt and Whitney J75 was also considered at this time, having been eliminated earlier in favour of the Avon due to weight. By early 1957 the J75 had been shelved. A US engine was still desirable for future market prospects therefore interest centred around the General Electric CJ-805 rated at 12,000lb thrust.

A serious feasibility study was initiated in March 1959 for the model now designated Series VII to be powered by two CJ-805-23C engines fitted with thrust reversers. Airline deliveries were proposed for early 1962 and the fuselage would accomodate 68-90 passengers in a 112ft 6ins length. The redesigned cockpit of the VI-R would be used and maximum weight would be 52,000 kg. Performance was significantly increased, the GE engines offering a maximum cruise of 480 kts (Mach 0.80) at 25,000 ft with a maximum range of 2100 miles. With the Douglas/Sud pact signed, Douglas was to undertake design and manufacture of the engine nacelles.

In December 1959 GE announced its purchase of a single Series III Caravelle to be modified as a VII testbed. The aircraft, registered N420GE and named "Santa Maria" after the caravel of Christopher Columbus, was handed over to GE in Toulouse on 18 July 1960 and then left for an extensive US sales tour. Jointly handled by Sud, Douglas and General Electric, the tour visited New York, Montreal, Washington, Miami, Atlanta, St.Louis, Tulsa, Dallas, Kansas City, Chicago, Minneapolis, Denver, Seattle, Vancouver, San Francisco and Los Angeles. There, a side trip was made to Long Beach for a temporary re-paint in Douglas-Sud titles for publicity photos before being delivered to GE's facility at Edwards AFB for fitting of the CJ-805s.

The first flight of the CJ-805 powered aircraft was made on 29 December 1960 and an intensive test programme was conducted. The aircraft was then flown to Paris for the Salon in May and followed this up with a sales tour to 16 cities in 11 countries of Europe and the Middle East totalling 64 demonstration flights. The Series VII was certificated in May and strong interest was shown by a number of companies including National Airlines, South African, Air India, American and KLM.

At the time of its US demonstration tour, Douglas announced another re-engined Caravelle for the US market, the Series VIII. This was projected with RB.141-11As and had virtually an identical specification to the VII. This model along with the XIV (also proposed with the RB.141-11A for the US) were dropped the following year due to lack of airline interest.

TWA and the 10

In March 1961 Douglas announced yet another new model, the Series X (10) with JT8Ds. The VII designation was still being applied to the stretched CJ-805 version at this time and TWA was showing considerable interest for about 20. The commonality in engines with the CV-880 was attractive and so were the early delivery dates offering a good lead over 727 buyers. In fact so confident was Sud of an order that in May work was started on five VIIs on a speculative basis. Keen to get TWA's order, Sud offered TWA the VI-R on a lease basis until the VIIs were ready. However due mostly to Howard Hughes' indecision, time slipped by and the CJ-805 became more competetive as regard to delivery deadlines. About July 1961 the designation 10A was adopted for the stretched CJ-805 model and in September TWA ordered 20 with 15 options for a reported cost of $100M and first deliveries by 1 January 1963. To meet such an early deadline Sud planned to modify a couple of aircraft on the line based on an American Airlines order once nearly signed but dropped in favour of the 727.

But all was not well with TWA's bank account and by December the delivery schedule had already slipped 30 days as a result of financing problems. The production decision continued to slip into 1962 and with mounting bank pressures to "buy American", TWA ordered 10 727s in March. The option on the 10As dragged on well into that year before being cancelled, largely due to the early delivery and extremely favourable terms offered.

Sud had earmarked an airframe(c/n 63) for its use as a VII/10A prototype and this was finally completed to the latter standard, flying in August 1962. The name "Caravelle Horizon" was adopted and this was the first model to feature significant production changes. A revised leading edge wing chord at the root was adopted, the window line was raised 4", an anti-shock acorn smoothed the fin/tailplane junction and double-slotted flaps were introduced. An APU was planned as a standard feature for production aircraft. The Horizon was displayed at the following years' Le Bourget Salon, but with TWA out of the running little interest was shown.

The Super 10s

With the outlook for the 10A poor, attention was turned to re-engining the improved airframe of the series to suit other customers. The JT8D version first projected in 1961 became known as the 10B or Horizon B and an Avon 533A model the 10C. Air France was particularly interested in the latter for standardisation purposes but Sud was keener to work on the 10B for its US potential and interest from KLM. It was actually Finnair who opened the 10B order book in December 1962 with an order for six 10B3s with the JT8D-1. Finnair preferred the term "Super B" for its fleet and this name was adopted for all the 10B3 series, later often just referred to as "Super Caravelle". This latter name had initially been used on an SST design up to mid-62 - eventually appearing as the Concorde. In turn the name Horizon was given to the four-seat touring aircraft, the GY.50, built by Sud as the GY.80.

A prototype of the 10B3 was first flown in Finnair colours on 3 March 1964 and acted as a development and demonstration aircraft until sold to Finnair two years later. Finnair commenced service with the new aircraft in August 1964. After certification of the 10A in July 1963 the prototype c/n63 was little used and eventually broken up for spares, the nose section being used to complete one of the Finnair 10B3s.

Further Offers

With the JT8D proving itself an outstanding engine and therefore a popular one, Sud offered the 10B1N and 10B1R, both re-engined versions of the VI-R airframe with some changes in structural strength and revised layout, greatly improving the lower baggage hold capacities. No orders were received for the 10B1N, but 20 of the 10B1R model were sold, the first flying in January 1965. Simplification of the designation among operators led to this version being known simply as 10R. The improved JT8D-7s brought the gross weight to 54,000 kg.

A demand for a mixed cargo/passenger version particularly from some African operators led to Sud developing the 11R. Derived from the 10R, it featured a 0.93 m plug in the forward fuselage to adjust the centre of gravity and to allow a large cargo door to be installed. Cabin floor strength was considerably improved and a moveable bulkhead enabled a number of mixed traffic configurations to be offered. In an all-passenger cabin the maximum load was 99. First flight was made in April 1967, with the first two deliveries going to Air Afrique. The only other customers were Air Congo and TransEurope, taking two aircraft each.

Final Stretch

On 1 January 1970 the French Government merged Sud, Nord Aviation and SEREB to form the Société Nationale Industrielle Aérospatiale (SNIAS). The new name did not reflect a very healthy Caravelle line. The Super had resulted in relatively few orders, with the balance of the production over the past three years being the 10R with a handful of older models going to established customers. The US market had been all but abandoned since 1963 with the loss of the TWA order the same year as United dropped further options in favour of the 727. Ironically, Boeing had speeded up development of the 727 to combat Douglas plans for a production line in California. There had also been a number of short-medium haul jets that had gained orders the Caravelle might have secured.

To keep the Toulouse factory open, Sud had talked to Sterling Airways in Denmark for some time about a further stretched model which would be ideally suited to that company's IT operations and carrying around 140 passengers. The Series 12 was launched by Sterling in April 1969 with an order for four plus four options. The fuselage was stretched by the addition of two plugs of 2.0m and 1.21m respectively fore and aft of the wing. Five extra seat rows were installed and power came from the JT8D-9 rated at 15,000lbs. With a general structural strengthing and revised

undercarriage the gross weight went to 56,000 kg. The first example was flown on 29 October 1970 with the first delivery to Sterling in March the following year. SNIAS hope to get further orders to increase authorised production by 25 from 275 to 300 airframes with the 12. Sterling, however, remained the only customer, and based on their options a dozen 12s were completed, the later examples receiving clearance for a maximum weight of 58,000 kg. Sterling ordered the 727-200 in 1972 and dropped some options for the Caravelle 12, the unsold aircraft going to Air Inter on a lease basis.

Retrospect and Future

Success is often relative and the case of the Caravelle is no exception. From a strictly financial point of view the aircraft undoubtedly lost money. Early break-even figures varied between 80-150 and by 1961 this had risen to 225. With the development costs of the later series this figure must have risen considerably higher than the final total of 280 aircraft (of which there were several unsold prototypes and returned aircraft). However, the type provided the industry in France with work for 20 years along with the associated spin-offs. It was the pioneer of the rear-engine grouping concept, quickly adopted by the UK, USA and USSR, although it can today be argued from an engineering view-point that this arrangement is not so beneficial as was previously supposed. It proved that the jet could make money for its operators on short-medium haul stage lengths, and, despite accidents in service, performed the job of transporting passengers in relative comfort and safety. Indeed it continues to do so and the total hours flown on the type are now approaching 7.5million, with 5 million landings and several aircraft now beyond 40,000 hours.

With the majority of the original customers disposing of their Caravelle fleets, the type has found a ready market in second-hand sales. Air France continues to be the major user of the type and some of the early models have flown well over 30,000 hrs and are in need of withdrawal in the near future. This immediate requirement was to have been met by the Boeing 737. However the proposed Air France lease of twelve of the type was abandoned due to opposition from the pilot's union operating with a two-man cockpit crew. Caravelles, particularly the non-Avon variety, are likely to be in service with charter operators in Europe and elsewhere well into the eighties, but a factor involved in this projection is the possible flood of early 727s and DC-9s likely to be on the second-hand market shortly.

The Caravelle was used to certificate the Lear-Siegler autoland system down to CAT IIIA in the mid-sixties. The first entirely automatic landing anywhere in the world in commercial service was made by an Air Inter aircraft on 9 January 1969 on a scheduled flight from Lyons to Paris. SNECMA has used one Caravelle as an engine test-bed since 1972. In July 1973 it was flown with the SNECMA M53 and currently is doing work with the CFM56. Numerous flight tests have been conducted with the Caravelle by the CEV, the latest evaluating the Thompson TC-125 civil HUD (Head Up Display) system.

TECHNICAL DESCRIPTION

Structure

The fuselage is a circular cross-section monocoque structure with a maximum diameter of 3.20m (10'6"), built up in a conventional manner with circular rings and frames skinned to form a complete annular section. The forward fuselage and rear fuselage were assembled separately and interior installations and systems added before the two-piece three-spar wing was mated on the fuselage centreline. The vertical tail section was then added followed by the horizontal tail and engine nacelles. The wing box was made in halves and joined on the centreline by pin joints, each joint being further braced by vertical and diagonal tubes. Landing gear, flap and aileron loads were taken into the box through forged heavy ribs. The window shape is inverted heart and was largely dictated by considerations of avoiding stress-raisers.

Flying Controls and Flaps

Two-piece ailerons on each wing are operated hydraulically by duplicated actuators with electric stand-by power and slotted Fowler trailing-edge flaps. Air brakes on the upper and lower wing surfaces are situated ahead of the flaps. There are three-section spoilers on the trailing edge of each wing on the Series VI-R. The Super A and B Series have an increased operating range of flaps from 35° to 45° with double-slotted type. The aircraft has an hydraulically-powered rudder and elevators, using duplicated actuators and electric stand-by power.

Engines

The RA.29 Avon is an axial-flow turbojet with sixteen-stages (the RA.29/6 has an additional "OO" compressor stage, making 17 stages in all). Thrust reversers and fitted on the Mk.532R and 533R. The combustion chamber is annular with 8 flame tubes of Nimonic 75. Both the R.A.29/3 and 29/6 incorporate a two-position variable-area silencing nozzle. Starting is by electric motor mounted in a bullet in the air intake.

The Pratt and Whitney JT8D is a two-spool axial flow turbofan, with a two stage fan, four stage intermediate pressure compressor, seven stage high pressure compressor, single stage high pressure turbine and three stage low pressure turbine. The combuster arrangement is annular with nine flame tubes. There is no water injection, but Sud-designed cascade thrust reversers are standard. The engine is housed in a SNECMA-designed nacelle.

Undercarriage

The main units are equipped with a four-wheel bogie retracting sideways into the fuselage and are manufactured by Hispano-Suiza. The twin nose-wheel retracts forward. Maxaret anti-skid brakes are fixed on the main wheels.

Systems

The air-conditioning system utilises two turbo-compressors, driven by engine-bleed air, and includes a cold air unit. Pressure differential is 8.25lb/sq.in. (0.58kg/cm^2). The hydraulic system, with a pressure of 2,500lb/sq.in (175kg/cm^2) is used for landing gear actuation, nose-wheel steering, brakes, flying controls and air brakes. An electrical system includes two 30V DC engine-driven generators and inverters for 115V 400 c/s AC. The 10B3 and 12 have an APU for engine starting, air-conditioning of flight deck and cabin on ground and flight up to maximum cruise altitude, and operation of a third 40K VA alternator both on the ground and in the air. There is a Lear L-102B autopilot on Series I/III/VI, later models have provision for Aérospatiale/Lear automatic landing system for operation in Category IIIA weather conditions.

The basic fuel system is four integral tanks in the wings with a total capacity of 19000 litres (4180 imp.gals). The 10B3 and 12 offer the optional installation of an additional centre tank, giving a maximum fuel capacity of 22,000 litres (4840 imp.gals).

It is necessary to comment on the presentation of the dimensions, weights, loadings and performance specifications. All measurements are given in Metric units, the Caravelle was a European aircraft, with the English equivalents. An exception is the engine power rating, given in pounds. As with any aircraft and in particular an airliner, single measurements for weights, speed, range, payload and take-off/landing performance are not enough. They must be qualified by load, route length, air temperature, runway elevation, prevailing weather conditions and the economic policies of the operator. Therefore the quoted figures are merely representative.

DIMENSIONS

	I/III/VI/10B1R/11R	1OB3 "SUPER"	12
Wing Span	34.30m (112ft 6in)	34.30m (112ft 6in)	34.30m (112ft 6in)
Overall Length	32.01m (105ft 0in)	33.01m (108ft $3\frac{1}{2}$in)	36.24m (118ft $10\frac{1}{2}$in)
Overall Height	8.72m (28ft 7in)	8.72m (28ft 7in)	9.01m (29ft 7in)
Wing Area (gross)	146.7m^2 (1,579ft^2)	146.7m^2 (1,579ft^2)	146.7m^2 (1,579ft^2)
Tailplane Span	10.60m (34ft 9in)	12.00m (39ft 4in)	12.00m (39ft 4in)
Cabin Length (excluding cockpit)	22.5m (73ft $8\frac{1}{2}$in)	23.45m (76ft $11\frac{1}{2}$in)	26.40m (86ft 7in)
Cabin Height (max)	2.00m (6ft 7in)	2.00m (6ft 7in)	2.00m (6ft 7in)
Cabin Width (max)	3.00m (9ft $9\frac{1}{2}$in)	3.00m (9ft $9\frac{1}{2}$in)	3.00m (9ft $9\frac{1}{2}$in)
Cabin Floor Area	60.00m^2 (646ft^2)	62.7m^2 (675ft^2)	71.28m^2 (767ft^2)
Cabin Volume	110.m^3 (3885cu ft)	125m^3 (4414cu ft)	142m^3 (5015 cu ft)
Cabin Hold Volume	5.70m^3 (201 cu ft)	4.33m^3 (153cu ft)	5.70m^3 (201cu ft)
Underfloor Hold Volume	10.60m^3 (374cu ft)	12m^3 (423cu ft)	16.50m^3 (582cu ft)
Track	5.21m (17ft 10in)	5.21m (17ft 0in)	5.21m (17ft 0in)
Wheelbase	11.79m (38ft 7in)	12.50m (41ft 0in)	14.80m (48ft $6\frac{1}{2}$in)
Max passengers	99	104	140

Notes

1. Prototype length was 31.50m (103ft 4in), track was 5.28m (17ft 4in), wheelbase was 11.33m (37ft 2in), otherwise as I/III
2. Underfloor Hold Volume on I/III was 9.60m^3 (339cu ft), Underfloor Volume on 10B1R was 12m^3 (423cu ft)
3. Length of 11R 32.71 m (107ft 4in), cabin length 23.37m (76ft $8\frac{1}{2}$in), cabin floor area 62.25m^2 (670ft^2), cabin volume 124m^3 (4379cu ft). Otherwise as 1OR

WEIGHTS AND LOADINGS

	I	III	VI-N	VI-R
Manufacturers Empty Weight	23,400kg (51,600lb)	24,185kg (53,320lb)	24,915kg (54,922lb)	26,280kg (57,935lb)
Basic Operating Weight		27,210kg (59,985lb)	27,330kg (60,250lb)	28,655kg (63,175lb)
Max payload	8,370kg (18,454lb)	8,400kg (18,520lb)	7,900kg (17,415lb)	8,200kg (18,080lb)
Max take-off weight	43,500kg (95,900lb)	46,000kg (D) (101,413lb)	48,000 kg (105,822lb)	50,000kg (110,230lb)
Max landing weight	41,430kg (91,340lb)	43,800kg (96,560lb)	45,700kg (100,750lb)	47,620kg (104,990lb)
Max Zero-Fuel Weight	35,000kg (77,162lb)	35,500kg (78,260lb)	35,500kg (78,260lb)	37,000kg (81,570lb)
Max Wing Loading		313kg/m^2 (64.2 lb/sq ft)	326kg/m^2 (66.8lb/sq ft)	341kg/m^2 (69.8lb/sq ft)

	I	III	VI-N	VI-R
Max Power Loading		4.44 kg/kg st	4.33 kg/kg st	4.37 kg/kg st
	10B1R	10B3	11R	12
Manufacturer's Empty Weight	26,725kg (58,920lb)	27,623kg (60,897lb)	28,841kg (63,585lb)	29,500kg (65,050lb)
Basic Operating Weight	29,075kg (64,100lb)	30,055kg (66,260lb)		31,800kg (70,100lb)
Max Payload	9,400kg (20,720lb)	9,100kg (20,060lb)	9,095kg (20,050lb)	13,200kg (29,100lb)
Max take-off weight	52,000kg (A) (114,640lb)	56,000kg (B) (123,460lb)	52,000kg (A) (114,640lb)	58,000kg (C) (127,870lb)
Max landing weight	49,500kg (109,130lb)	49,500kg (109,130lb)	49,500kg (109,130lb)	49,500kg (109,130lb)
Max Zero-Fuel Weight	38,500kg (84,880lb)	39,500kg (87,082lb)	40,000kg (88,185lb)	45,000kg (99,200lb)
Max Wing Loading	354kg/m^2 (72.6lb/sq ft)	368kg/m^2 (75.4lb/sq ft)	354kg/m^2 (72.6lb/sq ft)	381kg/m^2 (78.0lb/sq ft)
Max Power Loading	4.10 kg/kg st	4.25kg/kg st	4.10kg/kg st	4.25kg/kg st

Notes

A) Standard undercarriage; with strengthened u/c increased to 54,000 kg (JT8D-7)
B) This version certificated 3/70 (JT8D-7) original model was 52,000kg (JT8D-1) second increase to 54,000kg certificated 1968. (JT8D-7)
C) Originally 56,000kg, increase to 58,000kg certificated 1971
D) Some Series III aircraft were modified during service to enable an increase in max take-off weight to 48,000kg. Those aircraft so modified are noted in the main production list where known.

<u>PERFORMANCE</u>

	I	III	VI-N	VI-R
Max Cruise at 25,000ft (7,620m)	400kts (740Km/h) (at AUW 39,000kg)	434Kts (805Km/h) (at AUW 41,000kg)	456Kts (845Km/h) (at AUW 41,730kg)	456Kts (845Km/h) (at AUW 43,000kg)
T-O balanced field length: ISA at S/L (at max take-off weight)	1,800m (5,900ft)	1,830m (6,000ft)	1,950m (6,400ft)	2,073m (6,800ft)
Landing Distance at Max Ldg Wt.	1,035m (3,400ft)	1,800m (5,900ft)	1,965m (6,450ft)	1,720m (5,650ft)

	I	III	VI-N	VI-R
Range with max. fuel with payload, reserves		995nm (1,845Km) 7,620kg (16,800lb)	1,350nm (2,500Km) 7,620kg (16,800lb)	1,380nm (2,560km) 7,620kg (16,800lb)
Range with max payload Reserves/ (300m/260nm/480Km div.)	810nm (1,500Km)	915 nm (1,700Km)	1,270nm (2,350Km)	1,240nm (2,300Km)

PERFORMANCE

	10B1R	10B3 (JT8D-1)	11R	12 (JT8D-9)
Max Cruise at 25,000ft (7,620m)	432Kts (800Km/h) (at AUW 47,250kg)	445Kts (825Km/h) (at AUW 52,000kg)	432Kts (800Km/h) (at AUW 52,000kg)	445Kts (825Km/h) (at AUW 50,000kg)
T-O balanced field length: ISA at S/L (at max take-off weight)	1,950m (6,400ft)	2,090m (6,850ft)	1,950m (6,400ft)	2,460m (8,070ft)
Landing Distance at Max. Ldg Wt.	1,620m (5,315ft)	1,580m (5,180ft)	1,550m (5,085ft)	1,520m (4,985ft)
Range with max fuel with payload, reserves	1,780nm (3,295Km) 7,620kg (16,800lb)	1,758nm (3,260Km) 7,620kg (16,800lb)		1,710nm (3,170Km) 11,240kg (24,780lb)
Range with max. payload Reserves/ (300m/260nm/480Km div.)	1,565nm (2,900Km)	1,435nm (2,655Km)	1,511nm (2,800Km)	1,370nm (2,540Km) 13,200kg (29,100lb) (3,000kg) (6,600lb)

MODEL REFERENCE LIST

MODEL	ENGINES	STATIC THRUST (LBS)	MAX TOW (KGS)	FIRST FLIGHT	REMARKS	NUMBER BUILT
Prototypes	RA.26 Mk.521	10,000	41,000	27 May 55	Original projection with 9000lb.st RA.16 (Max TOW 38,400kg)	2
I	RA.29/1 Mk.522	10,760	43,500	18 May 58	All converted to Series III standard except c/n 14	20
IA	RA.29/1 Mk.526	10,760	43,500	11 Feb 60	All converted to Series III	12
II					Designation not used	
III	RA.29/3 Mk.527	11,400	46,000	30 Dec 59	c/n 42 to Series VII. 5 aircraft converted to Series VI-N	78
IV					Designation not used	
V					Designation not used	
VI-N	RA.29/6 Mk.531	12,200	48,000	10 Sep 60	Original designation VI changed to VI-N with introduction of VI-R	53
VI-R	RA.29/6 Mk.532R RA.29/6 Mk.533R	12,600	50,000	6 Feb 61	Avon 533R has improved turbine blades	56
VII	GE CJ805-23C	16,100	52,000	29 Dec 60	Initially applied to stretched fuselage this later re-designated 10A. Construction of 5 begun on speculation, not completed. One prototype converted from Series III	-
VIII	RB.141-11-A	15,000	52,000	-	3ft 4in stretch from VI-N Douglas proposal for USA. 78 mixed-class passenger configuration	-
IX					No details available	-
10A	GE CJ805-23C	16,100	52,000	31 Aug 62	Intended for TWA (originally designated VII) Also known as 'Super A' & 'Horizon A'	1
10B1.N	JT8D-1	14,000	48,000	-	Basic VI airframe. No thrust reversers	-
10B1.R	JT8D-1 (R) JT8D-7 (R) - 7A	14,000	52,000 54,000	18 Jan 65	Also known as 10R Strengthened u/c	20
10B.2					No details available	
10B.3	JT8D-1 (R) JT8D-7 (R) JT8D-9 (R)	14,000 14,000 14,500	52,000 54,000 56,000	3 Mar 64 24 Jan 68 18 Apr 70	Also known as 'Horizon B' &'Super B' Strengthened u/c Strengthened u/c	22
10BF					F = Fast; 1,000lbs extra fuel. No other data available	-
10C	RA.29/6 Mk.533R	12,600	-	-	10B Airframe-Proposed for Air France	-
10D	RB.174					-
11R	JT8D-7R	14,000	52,000 54,000	21 Apr 67 9 Oct 68	Mixed cargo/passenger configuration Strengthened u/c	6

12	JT8D-9R	14,500	56,000	29 Oct 70	Stretched Super B	12
			58,000	22 Feb 72	Strengthened u/c	
13					Designation not used	
					TOTAL PRODUCTION	282
14	RB.141-11-A	15,000	47,200	–	Douglas wing design M.O.84 cruise 95 passengers	
15/16/17/18/19					Details not available	
20	"advanced" RR Conway	22,000	72,575		'Air-bus' for 130 pax, cruise M.O.82	
21/22/23					Details not available	
24	RB.183-1 Spey (derated)	8,850	26,000		'Baby Caravelle' studied in 1964	

NOTES:

1. Roman numerals were used for Caravelle designations until the appearance of the 10A in 1961 Further designations adopted Arabic numerals. In this monograph the Roman numerals have been retained, although the early models are now frequently "Arabicised" in official SNIAS documents and airline records.

2. It should be noted that in 1959 the Caravelle VII was referred to by UK periodicals as being powered by the RB.141-3 (14340lb). However by 1961 the VII was the CJ805 version. It is not clear whether Sud ever officially applied VII to a RB.141-3 projection.

   Similarly, in 1963 the 10B1R was referred to as a basic VI-R airframe fitted with CJ805-23Cs without thrust reversers.

   In 1961 the Model 15 was known as the "Baby Caravelle" but was presumably a different model from the 24 which was studied 3 years later. No official list of Caravelle design studies/designations has been released by Sud.

INTRODUCTION

This section lists the airline users of the Caravelle, the individual aircraft operated and brief details of the fleet service history. Included are operators that announced orders but later cancelled, and those who announced a strong interest in acquiring the type.

The aircraft are listed by the model type and registration. Leased aircraft are shown marked with an asterisk (*). Complete details of order and contract signing dates are not available but are mentioned where known.

AER LINGUS, Dublin

By early 1962 Aer Lingus had completed its evaluation of the new jets and had planned to order two Caravelles. Lack of Government approval forced a postponement of the order and the company had to wait until 1965 before it introduced the BAC 1-11 on its short/medium haul routes.

AEROFLOT, Moscow

As early as the spring of 1956, the USSR declared its interest in the Caravelle for Aeroflot. After a number of visits by trade delegations to Toulouse, it was believed that three Caravelles would be acquired, no doubt primarily for comparison with the Soviet Union's own jet transport efforts. Interest apparently was dropped after 1961.

AEROLINEAS ARGENTINAS, Buenos Aires

VI-N: I-DAXT*/LV-HGX/HGY/HGZ/III

Aerolineas Argentinas' order for 3 aircraft was announced in Sep 61 by Douglas and they were used on domestic and regional routes from Mar 62. A fourth aircraft was acquired to replace one aircraft lost in 1963. The three survivors were transferred to the Air Force in Jul 73.

AEROLINEAS TAO, Bogota

In 1972 it was reported that the company would acquire Caravelles from Alitalia when that operators' fleet was phased out. Evidently TAO's financial footing was not sound as it ceased operations in 1975.

AEROMARITIME, Paris

(Compagnie Aéromaritime d'Affrétement SA)

Formed on 11 Jan 67 as a subsidiary of UTA to operate charters, Aéromaritime started operations later in the year using Caravelles and other types wet-leased from UTA.

AEROTAL COLOMBIA, Bogota

(Taxi Aereo El Llanero SA)

VI-R: HK-1778/1779/1780/1709-X/PP-PDZ (last two for spares use only)

Aerotal acquired the LAN-Chile Caravelle fleet in late 1975 to operate scheduled flights in Colombia. The company initiated jet services to Arauca and Leticia but these routes were subsequently taken over by TAME DC-4s and Avianca B727s. Aerotal now operates services to Barranquia, Cali, Pasto, San Andres, Bucaramanga and Cucuta with Caravelles.

AEROTOUR SA, Paris

III: F-BUFH

VI-N: F-BVPU/F-BYAT/F-BYAU/F-BYAI/F-BYCY

VI-R: F-BVPZ

Aérotour commenced IT and charter flights in May 1976 with two ex Catair machines. Two ex Alitalia aircraft have subsequently been added, and three ex JAT aircraft were purchased in 1978

AIR AFRIQUE, Abidjan

(Société Aérienne Africaine Multinationale)

VI-N: OO-SRB*/TU-TXR*
VI-R: PH-TRS*
10B1R: TU-TCN/TXQ
11R: TU-TCO/TCY

The first customer for the mixed cargo/passenger 11-R model, Air Afrique started Caravelle services in Sept 1967 throughout its West African network, replacing DC-6Bs. Two Caravelles continue in service to 15 points in West Africa as well as to Las Palmas.

AIR ALGERIE, Algiers

(Compagnie Générale de Transports Aériens)

I/III: F-OBNG/OBNH/OBNI/OBNJ/OBNK/OBNL
       (7T-VAG/VAI/VAK/VAL)
VI-N: OO-SRF* F-BLCZ*/HB-ICZ*/7T-VAE

Entirely owned by French interests in the fifties (as Algeria was then still linked with Metropolitan France) Air Algérie & Air Transport were combined 23 May 53 into CGTA-AA, which became Air Algérie in about 1960. It was a natural and early customer, placing its first order on 13 Mar 58 for two model Is with one option (taken up 18 Jun 58) to replace Constellations and DC-4s. Scheduled services started on 12 Jan 60 between Paris and Algiers intitially in 24 first class/ 50 economy configuration. Gradually replaced by B727s and B737s, the Caravelle was displaced to domestic routes and made its last flight with Air Algérie on 12 Jun 76 between Ouargla and Algiers.

AIR BURUNDI, Bujumbura

III: 9U-BTA

Formed in 1971 as Société de Transports Aériens du Burundi (STAB), the airline took its present title in Jun 75 with the acquisition of an ex Air France Caravelle, donated by the French Government. The only service flown is a once-weekly round-trip to Nairobi via Kigali. Otherwise its use is limited to cargo charters and flights for the Government.

AIR CAMBODGE - See ROYAL AIR CAMBODGE

AIR CENTRAFRIQUE, Bangui

III: TL-AAI

A single aircraft was acquired in Dec 1970 when the intention was announced to withdraw from the Air Afrique consortium. When this plan was dropped the aircraft was used by President (now Emperor) Jean-Bedel Bokassa, retaining airline titles for a time.

AIR CHARTER INTERNATIONAL, Paris

(Société Aérienne Française d'Affrétement until 1 Jan 70)

III : F-BJTE*/BJTF*/BJTG/BJTH/BJTI/BJTJ/BJTL*/BJTO

A wholly owned subsidiary of Air France formed to operate charter and IT services, the airline initially bought time on Air France aircraft, later leasing its own aircraft from the parent company.

AIR CONGO - see AIR ZAIRE

AIR FRANCE, Paris

(Compagnie Nationale Air France)

III: F-BHRA/RB/RC/RD/RE/RF/RG/RH/RI/RJ/RK/RL/RM/RN/
     RO/RP/RQ/RR/RS/RT/RU/RV/RX/RY/RZ. F-BJTA/TB/TC/
     TE/TF/TG/TH/TI/TJ/TL/TM/TN/TO/TP/TQ/TR/TS.
     F-BKGZ/BLKF/BOHA/BOHB/BOHC/BSGZ/BNKA*

Air France's involvement has already been covered in the general history text. Suffice to say it was the largest operator of the Caravelle (with a maximum of 43 aircraft in service 1968-69). Air France's financial interests in Air Inter, Air Algérie, MEA, Tunis Air and Royal Air Maroc guaranteed orders from these airlines also. Twenty-three aircraft continued in service in early 1979 and these were due to be phased out as they

Top   : Caravelle III HB-ICX (c/n 38) of Swissair at
         Schiphol in June 1967 (J Bossenbroek)

Bottom : Caravelle III F-BHRA (c/n 1) of Air France at
         Schiphol (NGJ Roozen)

Top   : Caravelle VI-N OO-SRI (c/n 175) of Sabena landing
           at Frankfurt-Rhein-Main in April 1969 (Peter J Marson)

Bottom : Caravelle VI-N I-DABW (c/n 150) of Alitalia at
           London-Gatwick in September 1974 (J Wegg)

Top   : Caravelle III PH-TRR (c/n 48) at Schiphol (J Wegg)
Bottom : Caravelle 10B 3 EC-CUM (c/n 212) of TAE at Palma
         in March 1976 (A Camarasa)

Top   :   Caravelle 10B 3 OH-LSG (c/n 169) of Finnair at
             Malaga in March 1975 (J Wegg)

Bottom :   Caravelle III F-BSRD (c/n 38) of Catair at Le
             Bourget in May 1972, with a Euralair aircraft in
             the background (Peter J Marson)

come up for major overhauls. Ten are scheduled to be withdrawn in 1979, ten in 1980 and the last three in 1981. Air France placed a modest deposit on options to lease 13 Boeing 737s in February 1978 to fill its need for a 100-seat Caravelle replacement. A dispute with the company's pilots and the main French pilots' union SNPL over the three-man crew issue in the 737s forced the carrier to give up its options in April the same year. The lack of capacity will be met partially by an order for five Boeing 727s, but these are 150-seat aircraft.

AIR FRET, Nimes

VI-R: F-BTON*

Air Fret chartered a model VI-R in 1972 for a total of 250 hours from Sovatour, although it appears there was a close connection between these companies.

AIR GABON, Libreville

VI-R: TR-LWD

On its withdrawal from the Air Afrique consortium in May 1977, Air Gabon took over the Government Caravelle for regional services pending delivery of Boeing 737s. The first service was flown Libreville-Lagos-Lomé on 3 June 77.

AIR INTER, Paris

(Lignes Aériennes Intérieures)

III: F-BNKA/KB/KC/KD/KE/KF/KG/KH/KI/KJ/KK/KL
　　F-BHRQ/RR/RS/RZ. F-BSRR/RY

VI-N: OO-SRF*/SRI*
VI-R: F-BRGX*
12: F-BTOA/OB/OC/OD/OE. F-BVPY*

Initially Air Inter chartered Air France Caravelles, flying a Paris-Marseille service starting on 6 Mar 64. The build-up of its own fleet began in 1966 and Air France and other aircraft have been leased as demand required. The use of an automatic landing system was very important for the airline, as this was the only way to compete properly with fast railway services in winter time, France being a comparatively small country (by US standards at least).(For full details see previous section). The first Caravelle 12 service was flown on 8 Nov 72 between Paris and Bordeaux in a 128 seat economy class configuration. The Caravelle continues to be the mainstay of the fleet and flies to 23 cities on the route network.

AIR JUGOSLAVIA - see JAT

AIR LIBAN - Beirut

III: OD-ADY/ADZ

A 30% interest was held in this airline by Air France and Caravelle service was started Beirut - Paris on 20 Dec 60 with a chartered Air France aircraft. Two aircraft were operated on lease until the company was merged with MEA under a joint development programme in Mar 63.

AIR MALI, - Bamako

III: F-BRUJ*

One aircraft was leased from Transunion in early 1971 for regional routes in West Africa.

AIR-SEA SERVICE, Basle

Formed in May 74 the company started scheduled service to Luxembourg and in Mar 77 announced its intention to purchase one Caravelle. To date no further action has been taken.

AIR TOURING - see SOVATOUR

AIR VIETNAM, Saigon

III: XV-NJA

VI-R: B-2503*

Air Vietnam operated one aircraft from August 64 to Sept 68 on its Far East routes, until withdrawn in favour of the Boeing 727. To meet wartime demands in 1975 one aircraft was leased from FEAT

AIR ZAIRE, Kinshasa

11-R: 9Q-CLC/CLD

Air Zaire was the name adopted in Oct 71 by the independent Zaire for its national airline which had up to then been called Air Congo. Air Congo had introduced the 11-R in Nov 67 and these two aircraft continued to be operated for cargo charters until sold to UTA for the French Air Force.

ALIA, ROYAL JORDANIAN AIRLINES, Amman

10B1R: JY-ACS/ACT/ADG

Impatient to move into jet operations, Alia ordered two Caravelles in May 65 instead of waiting for 1967 delivery of DC-9s (the then favoured equipment). Services started in August 65 to Rome from Amman and Jerusalem with Paris and London added in June and August 1966 respectively. A third aircraft was added to the fleet in 1968. The Caravelles were all subsequently sold with the introduction of Boeing 727s.

ALISARDA S.p.A., Olbia

VI-R: OY-SAL*

This Sardinia operator chartered a Sterling Caravelle for two short periods during the summer of 1977.

ALITALIA, Rome

(Linee Aeree Italiane S.p.A.)

VI-N: I-DABA/BE/BF/BG/BI/BL/BM/BP/BR/BS/BT/BU/BV/BW/BZ
　　I-DAXA/XE/XI/XO (del as III) DAXT/DAXU.

Alitalia placed its initial order for four IIIs on 15 Oct 59 and they entered service to London on 23 May 60, initially operating from Ciampino airport and replacing DC-6Bs. Fiumicino was opened in time for that summers' Olympic Games and the Alitalia fleet was adorned with special markings commemorating the event. A peak of 21 Caravelles was reached in the mid-sixties. Many aircraft was operated by Alitalia's charter subsidiary SAM until the type was withdrawn from service on 28 Feb 77, having been replaced by DC-9s and later the Boeing 727.

AMERICAN AIRLINES INC, New York

Sud came very close to finalising an order for the Caravelle VI-R from American in mid-1961. In fact, production of one or two aircraft was started in anticipation, but the order was still-born in favour of the Boeing 727. The airframe assemblies were then assigned for conversion and completion to 10As for TWA which in turn cancelled their order. It has not yet been possible to identity the airframes but they were eventually completed as VI-Rs.

ATI, Naples

(Aero Transporti Italiani, S.p.A)

VI-N: I-DABT*

Formed Dec 63 as an Alitalia subsidiary to operate domestic services, ATI frequently operated Alitalia aircraft, but only one Caravelle has been reported carrying their titles.

AUA - AUSTRIAN AIRLINES, Vienna

VI-R: OE-LCA/LCE/LCI/LCO/LCU

AUA's initial Caravelle order dated 29 Oct 62 was for two aircraft and the first service was flown to Frankfurt on 1 Apr 63 in a 12 first class/ 68 economy class configuration. Replacing Viscounts on AUA's European routes, the Caravelles were in turn replaced by DC-9s and the last scheduled flight was made on 26 Jul 72. The Caravelles were frequently operated for AUA's charter subsidiary, Austrian Airtransport without any change of livery.

AVENSA, Caracas

(Aerovias Venezolanas SA)

III: YV-C-AVI

50

One Caravelle was purchased in Aug 64 to supplement
CV-580s on regional routes, entering service that
November. The aircraft was written-off in a landing
accident at Barquisimento in Aug 73.

AVIACO, Madrid

VI-R: EC-ARI/ARK/ATX/AXU/AYD*
10B1R: EC-BIB/BIC/BID/BIE/BIF/CAE

Traditionally Aviaco has operated ex Iberia types
on their withdrawal from that company's fleet. The
Caravelles were used extensively on European charters
and domestic scheduled services in a 94 seat, economy
class configuration from May 72. Prior to this,
Caravelles had been chartered from Sabena. In fact
Aviaco ordered four 10B1Rs in 1968 but none were
delivered.

AVIACTION, Düsseldorf

This short-lived German charter company was
planning to acquire Caravelles either to supplement
or replace its F-28s in 1975. In the event,
liquidation occurred at the end of 1973.

BIAS - BELGIAN INTERNATIONAL AIR SERVICES SA, Brussels

VI-R: OO-CVA

This charter operator purchased one ex United
VI-R with one option in March 71 to replace a DC-6B.
It flew European charter flights from 1 Apr 71 but was
sold when BIAS ceased operations the next year.

CAAC (General Administration of Civil Aviation of China), Peking

Following demonstrations at Hong Kong in 1961
it is understood that CAAC was extremely interested
in acquiring Caravelles. However a ban on the
export of sophisticated jet engines to China
effectively frustrated the efforts of the Sud
sales team.

CARIBBEAN INTERNATIONAL AIRWAYS, Caracas

Little is known of this company which
advertised one Caravelle available for charters in
1976. It was presumably a title front for another
airline or aircraft broker.

CATAIR, Paris

(Compagnie d'Affrétements et de Transports Aériens)

III: F-BHRX* F-BSRD/BUFH
VI-N: F-BVPU/BYCA/BYCB/BYCD OO-SRE*/SRF*
VI-R: F-BRGX* F-BTON/BUFC/BUFF
12: F-BVPY*

Catair's first Caravelle arrived in May 71
to supplement its Super Constellations on
European charter and IT services. Since then a
variety of models has been operated. Catair's
financial situation deteriorated in 1978 and the
remaining aircraft were reported by SNIAS
to be owned by the concern "ALBATROS". The
company finally ceased operations during 1978
and its Caravelles were disposed of to other
French operators.

CHINA AIRLINES, Taipei

III: B-1850/1852/1854/1856

Two ex Swissair Caravelles were acquired for domestic
and charter services in 1971, being joined by two
more via Catair in 1973/74. One was destroyed by sabotage
and another returned to SNIAS. The only scheduled
route flown by 1977 was Taipei-Chiayi, and the last
aircraft (B-1850) is scheduled to be withdrawn in
1978.

CRUZEIRO DO SUL, Rio de Janeiro

(Servicos Aereos Cruzeiro do Sul SA)

VI-R: PP-CJA/CJB/CJC/CJD/PDV/PDX/PDZ

Besides four aircraft acquired 1962-63, Cruzeiro
took over three ex Panair do Brazil machines after that
company's demise. The five survivors were withdrawn
in 1975 in favour of Boeing 727s and 737s and put
into storage at Porto Alegre. To date only one
has been sold, as a spares aircraft, to Aerotal.

CTA, Geneve

(Compagnie de Transports Aériens)

10B1R: HB-ICN/ICO/ICQ

CTA was formed on 1 Nov 1978 to succeed SATA
which ceased operations on 31 Oct 78. The company
is wholly owned by Swissair and took over SATA's
three remaining Caravelles for European charter and
IT work. The first services were flown on 2 Nov 78.

EGYPTAIR, Cairo

VI-R: OY-SAH*/SAL*

Due to shortage of equipment caused by accidents
and operating problems with its Soviet-built fleet,
Egyptair leased two Caravelles from Sterling for
domestic services during the summer of 1975.

EURALAIR, Paris

VI-R: F-BSEL/BTDL/BUFC

This small air-taxi and executive charter company
purchased two ex AUA aircraft in 1971 to expand into
passenger and cargo charters in January 1972.

EUROPE AERO SERVICE (EAS), Perpignan

VI-N: F-BYCD/BYCA/BX00/BYCB/GBMI/GBMJ/GBMK

VI-R: F-BUFF/(Spares use only)

EAS acquired 3 Catair aircraft and one Sabena
aircraft in the spring of 1978 for charter work within
Europe, and subsequently acquired further aircraft
from International Air/Nevada Air Tours in late 1978,
after these had been stored at Lawton, OK.

FAR EASTERN AIR TRANSPORT CORP, Taipei

VI-R: B-2501/2503/2505

Three ex Iberia aircraft were acquired for
domestic and regional routes in 1973, two being
delivered via an Irish broker. One was leased to Air
Vietnam in 1975 and present scheduled routes include
Taipei-Hualien-Kaohsiung.

FILIPINAS ORIENT AIRLINES, Manila

VI-R: PI-C969*/C970*

Two aircraft were leased from Sterling from
Aug 72 for charter operations, which commenced in
October. PAL took over FPO by Government decree in
1973 but continued to operate under its own name until
late 73 when its operating licence was withdrawn.

FINNAIR, Helsinki

(Aero O/Y)

IA/III: OH-LEA/LEB/LEC/LED
VI-R: OH-LER*
10B3: OH-LSA/SB/SC/SD/SE/SF/SG/SH/SI*/SK*

Finnair placed its first order for three IAs
on 18 Jan 58 and put them into service on 1 Apr 60
to Stockholm and Frankfurt, initially in a 16 first
class/52 economy class configuration, replacing CV-440s.
The four Series IIIs were traded in to Sud for six
10B3s in Dec 62, marking the first order for that
type and the first example of a new jet airliner being
financed in part by trading in used jet transports.
The 10B3 entered service on 16 Aug 64 and enabled
expansion of Finnair's European network to take place.
Finnair Caravelles were operated on behalf of KLM
(1964), Lufthansa (1962-64 and 1967-68) and for Kar-Air
on IT and charter flights. The type has now been
displaced from most European routes on to the domestic
network, by the DC-9-51

IBERIA, Madrid

(Lineas Aereas de Espana)

VI-R: EC-ARI/ARJ/ARK/ARL/ATV/ATX/AVY/AVZ/AXU/AYD/AYE/BBR/BIA
10B1R: EC-BDC/BDD/BIB/BIC/BID/BIE/BIF/BRJ*
11R: EC-BRX*/BRY*

Iberia's first order for four VI-Rs and four options came in Oct 60 after lengthy negotiations with Sud. They went into service on 1 May 62 from Madrid-Zürich, replacing CV-440s. The fleet was increased to 20 VI-Rs and 10Rs by 1967 and additionally two 11Rs were operated in pool with TransEuropa from 1969. The Caravelle fleet was gradually displaced by the DC-9 and Boeing 727 on European routes and several were transferred to Aviaco for charter and domestic services.

INDIAN AIRLINES, Delhi

VI-N: VT-DPN/DPO/DPP/DSB/DUH/DUI/DVI/DVJ/DWN/ECG/ECH*/ECI*

Three VI-Ns were initially ordered in May 63, the sale being aided considerably by a French long-term loan of $60m. Another factor in the choice of the Caravelle over the preferred Boeing 727 was the ease of engine overhaul by Air India in Bombay. First service was flown Bombay-Delhi on 1 Feb 64 and the Caravelles displaced the Viscount, which in turn went to the regional routes to replace over 40 DC-3s. Indian's experience with the Caravelle was not a particularly happy one as five aircraft were written-off in service. In 1973 three aircraft were leased from SNIAS due to an equipment shortage, one of these eventually being purchased.

INEX-ADRIA, Ljubljana

(Inex-Adria Aviopromet)

III: YU-AJE*

This Jugoslav charter and IT airline leased a single aircraft from the manufacturers for the summer of 1972. A total of 1,188 hours was accumulated in service before returning to SNIAS in Nov 72.

ITAVIA, Rome

(Aerolinee Itavia S.p.A.)

VI-R: PH-TRX*

This domestic operator considered leasing Alitalia Caravelles in 1971 for charter work but in the event did not follow through its plans. A single aircraft was wet-leased from Transavia-Holland in May 75 to cover DC-9 overhauls.

JAT, Belgrade

(Jugoslovenski Aerotransport)

III: YU-AJG*
VI-N: YU-AHA/HB/HD/HE/HF/HG/HK*

JAT ordered three VI-Ns in Feb 62 to replace DC-6Bs and CV-440s on its routes to Cairo, Athens, London, Frankfurt, Paris and Rome. Eventually displaced by the DC-9, the Caravelle continued in service on domestic routes until 1976. Air Jugoslavia, JAT's wholly-owned charter subsidiary used JAT Caravelles extensively without any change of titles. JAT also acted as agent for one Caravelle delivered to the Jugoslav Air Force/ Government.

KAR-AIR O/Y, Helsinki

Finnair holds a 35% interest in this charter company and aircraft have been leased from Finnair as required. The Caravelles are used frequently on European IT flights generally without change of titling, although OH-LSA did once appear briefly with a small Kar-Air sticker, and OH-LSB was painted in their full colours during 1978.

KINGDOM OF LIBYA AIRLINES, Tripoli

III: F-BJTI*
VI-R: 5A-DAA/DAB/DAE

Formed in 1964, operations started in Aug 65 with Caravelles on the Tripoli-Paris route. Air France provided technical assistance and flight crews and the initial two aircraft were ordered originally by VASP. A fourth aircraft was ordered by KLA for Jun 68 but was never

delivered. The country became a Republic on 1 Sept 69 with a successful military coup against King Idris, and the airline was renamed LIBYAN ARAB AIRLINES. The Caravelle fleet was replaced by a fleet of Boeing 727s.

KLM - ROYAL DUTCH AIRLINES, Amsterdam

KLM showed considerable interest in the 10B model during 1961 to replace its Electra fleet. Heavy financial losses in 1962, however, caused a medium-range jet re-equipment programme to be abandoned until the purchase of DC-9s in 1965. KLM's Amsterdam-Madrid service was operated by Finnair 10B3s from 1964 until the DC-9 entered service in 1966.

LAN - CHILE, Santiago

(Linea Aerea Nacional de Chile)

VI-R: CC-CCO/CCP/CCQ

Three VI-Rs were ordered in July 1963 primarily for shorter international routes. Entering service in Apr 64 they were eventually replaced by the Boeing 727 but remained on domestic routes until 1975 when they were disposed of to Aerotal.

LEBANESE INTERNATIONAL AIRWAYS, Beirut

A single Sabena Caravelle was leased during 1965 (identity not yet confirmed) and used on LIA's Middle East routes, primarily to Teheran.

LUFTHANSA, Köln

(Deutsche Lufthansa AG)

III : F-BHRN*/BHRQ*/BLKF*

To cover shortages of equipment, Lufthansa signed an agreement with Air France for that company to operate Hamburg-Düsseldorf-Paris, Frankfurt-Vienna and Frankfurt-Paris (two daily flights) from 1 Apr 64 to 31 Mar 65. Two Caravelles were used at any one time carrying Lufthansa stickers on the forward fuselage. Although only 3 aircraft have been recorded with these titles, it is probable that other aircraft were also used.

Prior to the agreement with Air France covering the Paris-Frankfurt route, this had been flown with Finnair Caravelles (the first service on 1 Apr 62 with OH-LED). Finnair also flew Frankfurt- Nice - Palma and Frankfurt-Palma until 1 Oct 64. A new agreement with Finnair commenced on 1 Apr 67 in which a Super B was used between Frankfurt and Zürich until 11 May 67 and later between Frankfurt -Hannover and Düsseldorf. The charter agreement ended on 31 Mar 68.

LTU , Düsseldorf

 (Luftransport - Unternehmen GmbH & Co KG)

III: D-ABAF/ABAM*
10B1R: D-ABAF(2)*/ABAP/ABAV/ABAW/ANYL

LTU became the first European independent operator to order the Caravelle when it purchased an ex Finnair aircraft in Aug 64, putting it into service to Palma in Feb 65. A second III was introduced in 1966 but returned to Sud with the delivery of the 10B1Rs. The latter flew the first revenue service on 25 Dec 67 to Las Palmas. Normal charter configuration was 89 economy class or 12 first class/74 economy class for the winter service but Turkish "Gastarbeiter" (ie foreign workers) flights were made with 83 economy class configuration, the extra cabin space being used for baggage. The 10B1Rs of LTU were delivered with the JT8D-7 engines but later modified to JT8D-7As.

LUXAIR, Luxembourg

(Société Anonyme Luxembourgeoise de Navigation Aérienne)

VI-R: LX-LGE*/LGF/LGG

Luxair announced in 1969 that it would lease a Caravelle from Sud for two years. Originally ordered by KLA the aircraft had been stored at Toulouse since 1968. LX-LGE entered service on 22 Mar 70 to Palma in an 88 economy class configuration. Two AUA VI-Rs were acquired in 1972 (initially leased then purchased) to provide expansion on the popular holiday destinations and Monastir and Ibiza were added to the route network that year. The Caravelles were replaced by two Boeing 737s in 1978, and sold to SAN Ecuador.

MIDDLE EAST AIRLINES, Beirut

(Middle East Airlines-Air Liban SAL)

III: OD-AEM*   CN-CCY*/F-BNGE*
VI-N: OD-AEE/AEF/AEO

MEA's delivery of two aircraft in 1963 (ordered Aug 62) coincided with a joint operation agreement with Air Liban, which was essentially a merger. The Israeli attack on Beirut destroyed two Caravelles and to cover the shortage of equipment aircraft were leased from Air France and Royal Air Maroc. In addition, capacity was chartered on Alia Caravelles on the Beirut-Amman route.

MIDWEST AIR CHARTER, Elyria, Ohio

VI-R: N902MW/N903MW/N904MW/N905MW

Midwest acquired the four remaining Sterling VI-Rs in 1978 on behalf of Airborne Freight Corp (q.v) with the first being delivered at the end of July and entering service in September. All Caravelles were stripped of furnishings and operated scheduled small-packet freight services within the USA.

MINERVE SA, Paris

(Compagnie Française de Transports Aériens)

VI-N: F-BRGU*/GATZ
VI-R: F-BUZC/GAPA

Formed in June 1975, Minerve commenced charter and IT flights from Le Bourget in November 75 following delivery of an ex Sterling Caravelle.

MISRAIR (United Arab Airlines), Cairo

Misrair considered the Caravelle to be its first choice of new jet equipment over the DH Comet in 1958. No doubt due to British influence the company eventually chose the Comet.

In 1963 UAA operated a Vienna-Cairo service in pool with SAS and AUA, SAS Caravelles were used with Scandinavian flight crews and cabin crews from all three airlines.

NATIONAL AIRLINES INC. Miami

In late 1960 National was showing interest in the Series VII to replace the Electra on such routes as Boston-Washington to Jacksonville-Tampa. By 1962 the study was concentrating on the 10A or 10B but in the event the expected order did not materialise, the Caravelle again losing to the Boeing 727.

PANAIR, Düsseldorf

III: F-BRIM*

Founded as Pan Europa this charter company was re-named Panair in the spring of 1969. A Caravelle was sub-leased from Transunion until the delivery of BAC One-Elevens when the airline was then again re-named Paninternational

PANAIR DO BRAZIL, Rio de Janeiro

VI-R: PP-PDU/PDV/PDX/PDZ

Panair ordered four VI-Rs in Oct 61 with two options and the type was introduced on domestic routes in Sep 62. One aircraft was lost but the remaining three were in service when Panair's operating authority was suspended on 10 Feb 65 by President Castelo Branco because of a "truly chaotic financial situation". Cruzeiro subsequently took over the Caravelles and most of Panair's domestic route network (in conjuntion with VASP and VARIG.)

PSA, San Diego

(Pacific Southwest Airlines Inc)

This California intra-state airline announced its intention to purchase two Caravelles plus one option in Jun 57. Nothing further came of this interest.

ROYAL AIR CAMBODGE, Phnom-Penh

III: XU-JTA/JTB

Air France had a financial interest in RAC and provided a single Caravelle for regional routes in 1969. This was destroyed by a rocket attack on Phnom-Penh in Jan 71 and by then the airline had dropped the "Royal" prefix. A second Caravelle was acquired in Apr 73 and was reported derelict at Bangkok in late 1977.

ROYAL AIR LAO, Vientiane

III: XW-PNH

Another airline in the Far East with an Air France interest, RAL used a Caravelle on regional routes during 1974/75. Escaping the Communist take-over of the country, the aircraft was flown to Bangkok and eventually re-entered service with Air France.

ROYAL AIR MAROC, Casablanca

(Compagnie Nationale de Transports Aériens)

IA/III: CN-CCT/CCV/CCX/CCY/CCZ F-BJTF*/BJTI*/BNGE*/
         BSRR*
VI-N: OO-SRD*

With strong French interest and management RAM became an early Caravelle customer ordering one and one option on 19 Jul 58. The initial service was on 20 May 60 between Paris and Casablanca with a 20 first class/55 economy class configuration. Further deliveries replaced RAM's four Constellations on its routes to Paris, Marseille, Nice and Frankfurt. The Caravelles enabled further route expansion to European capitals to take place including Milan, Rome, Brussels, London, Copenhagen and Amsterdam. The fleet was withdrawn from service in 1977.

SABENA, Brussels

(Société Anonyme Belge d'Exploitation de Navigation Aérienne)

III: F-BHRP*
VI-N: OO-SRA/SRB/SRC/SRD/SRE/SRF/SRG/SRH/SRI/SRK

The first customer for the Series VI, Sabena signed a contract for four aircraft and four options on 14 Dec 59. The first service was flown on 18 Feb 61 to Nice and the Caravelle replaced the DC-6B and DC-7C on its longer stage lengths to Ankara, Beirut, Casablanca, Cairo, Teheran, Tel Aviv, Tripoli and Benghazi while the CV-440 was replaced on the prime European runs. With the delivery of Boeing 727s some Caravelles were turned over to Sabena's subsidiary Sobelair for charter work but the fleet was not finally replaced until the advent of Boeing 737s.

SAETA, Quito

(Sociedad Anonima Ecuatoriana de Tranportes Aéreos)

VI-N: HC-BAD/BAE/BAI/BDS/I-DABE/BW

Operating services in Ecuador, SAETA purchased six Alitalia Caravelles to replace Viscounts on its main routes from Quito to Cuenca and Guayaquil.

SAM, Rome

(Societa Aerea Mediterranea S.p.A.)

VI-N: I-DABG/BI/BL/BM/BP/BT/BV/BW/BZ/XA
VI-R: PH-TRU*

Formed in Dec 59 by Alitalia, SAM has always flown charters and IT flights within Europe using ex Alitalia aircraft as required. Alitalia Caravelles were flown in SAM titles but short-term usage often occurred with no change of livery.

SAN, Quito

(Servicios Aéreos Nacionales SA)

VI-R: HC-BAJ/BAT   CS-TCC

SAN purchased the TAP fleet of three Caravelles and like SAETA used them from Quito to Cuenca and Guayaquil. One aircraft remains unused at Quito, as a spares source. Two additional aircraft were acquired from Luxair in 1978.

SAS, Stockholm

(Scandinavian Airlines System)

I/III: LN-KLH/KLI/KLN/KLP/KLR
OY-KRA/KRB/KRC/KRD/KRE/KRF/KRG
SE-DAA/DAB/DAC/DAD/DAE/DAF/DAG/DAH/DAI

With Air France, SAS was a pioneer user of the Caravelle following its order for six and 19 options on 28 Jan 57. The company had the distinction of becoming the first operator of the Caravelle, putting the type into scheduled service on 26 Apr 1959 from Copenhagen to Beirut. This was the result of SAS leasing O2 for crew training in March and April that year. The Caravelle replaced the DC-6B and DC-7C on the Middle East routes and the CV-440 and DC-6B in Europe. Initially, operations were centred at Bromma but in 1962 all international flights moved to Arlanda. SAS also ordered Caravelles for Swissair and operated those leased to Thai International. Despite a large fleet of DC-9s SAS kept the Caravelle in service as long as economically possible, making its last flight on 27 Sep 74 with OY-KRC from Göteborg to Arlanda.

SATA, Geneva

(Société Anonyme de Transport Aérien)

VI-R: HB-ICP*
10B1R: HB-ICK/ICN/ICO/ICQ

Starting services in Mar 70 with a new Caravelle from SNIAS, SATA increased its fleet (and European charter business) with aircraft from Alia, Sterling and UTA. One aircraft was lost in an accident in 1977. The company ceased operations on 31 Oct 78 and was succeeded by the wholly-owned Swissair concern CTA.

SOBELAIR, Brussels

(Société Belge de Transports Aériens SA)

III: OO-SBQ*
VI-N: OO-SRB/SRC/SRD/SRI

As Sabena's charter subsidiary, Sobelair acquired three ex Sabena aircraft in Apr 71 for European IT work, supplementing these with an ex Swissair aircraft that had been purchased by TEN-BEL Touring. Other Caravelles were also utilised as required from the parent company without change of titles.

SOVATOUR, Paris

(Société de Valorisation du Touristique)

VI-R: F-BTON

This IT operator owns Air Touring and used one Caravelle for the summer season in 1972. Air Fret was used for maintenance and also chartered the aircraft for 250 hours.

SPECIAL AIR TRANSPORT, Düsseldorf

10B1R: D-ABAP/ABAV/ABAW

A subsidiary of Kurfiss Aviation, Special AT received their first Caravelle from LTU at the end of August 1978. The others will be delivered as LTU receives further TriStars.

STARLINE, Naples

VI-R: OY-SBW

Sterling, with Italian business interests, proposed the formation of this charter company in 1976. Two Caravelles, OY-SAJ/SAL were earmarked for service in 1977 but the airline did not develop as planned. OY-SBW was used from 19 Sep 77 to 9 Mar 78, but it is believed that its intended re-registration as I-STAE was not taken up.

STERLING AIRWAYS A/S, Copenhagen

VI-R: OY-SAH/SAJ/SAK/SAL/SAM/SAN/SAO/SAP/SAR/SBV/SBW/SBY/
SBZ
10B1R: OY-SAY*/SAZ*
10B3: OY-STA/STB/STC/STD/STE/STF/STG/STH/STI/STK/STL/STM
12:    OY-SAA/SAB/SAC/SAD/SAE/SAF/SAG

After Air France, Sterling had the largest fleet of Caravelles and easily had the most utilisation, with a peak of 12.49 hours average per aircraft per day during 1967. Sterling's requirements for increased fuel capacity and take-off weight led to improvements in the Super B series culminating in the 12. The original order was for only one 10B3 (dated 22 Jun 64) but this was gradually increased to a total of 12 of this model. The first order for the Series 12 came on 23 Apr 69 and committed Sud to a go-ahead. On 6 Nov 71 Sterling purchased 13 VI-Rs from United although some were then leased to other operators. Initial flight dates were Apr 65 (10B3). 19 Mar 71 (12) and 24 Mar 72 (VI-R). Charter flights to North America commenced on 23 Jun 70 (to Omaha) using OY-STK/L/M with extra fuel capacity and the increased take-off weight of 56,000kg. The 10B3 is operated with 109 seats, the VI-R with 99 and the 12 with 131. Sterling has also had an interest in operations in the Philippines with FPO, Transasian and Sterling Philippines.

SWISSAIR, Zürich

(Schweizerische Luftverkehr AG)

III: HB-ICR*/ICS/ICT/ICU/ICV/ICW/ICX/ICY/ICZ

Under the joint development agreement signed with SAS, Swissair's initial Caravelles were ordered by that company. The first scheduled service was flown to London on 21 May 60 replacing CV-440s.

SYRIAN ARAB AIRLINES, Damascus

(later Syrianair)

10B3: YK-AFA/AFB/AFC/AFD

Syrian Arab ordered two new "Super Caravelles" in 1965, introducing them to Nicosia, Athens and Rome on 1 Dec 65. Later two more "Supers" were acquired from Sterling and the four aircraft continue in service on both international and domestic routes.

TAA (Trans-Australia Airlines), Melbourne

In 1957 TAA announced that it would resist political pressure to buy British turboprops and instead order Caravelles or Electras. Sud undertook the redesign of the Caravelle in order to meet TAA's requirement for a non-stop Adelaide to Perth sector. The Australian government refused TAA's application for import in May 1958 and later that year TAA ordered the Electra as a substitute.

TAC - COLOMBIA, Bogota

(Aerovias del Cesar Ltda)

VI-R: HK-1810/1811/1812   EC-ARI (spares only)

TAC was formed in 1968 as Transportes Aereas de Cesar to operate taxi services and adopted its present title in 1975. The following year three Caravelles were purchased from Aviaco for scheduled services within Colombia to Monteria, Valledupar, Barranquilla, Bucaramanga and Medellin.

TAE, Palma de Mallorca

(Trabajos Aéreos y Enlaces SA)

10B3: EC-CMS/CUM

TAE operated two DC-7Cs on European charter flights from Apr 67 until 1970 when operations were suspended. It resumed activities in Apr 73 and has leased two Caravelles from Sterling until the end of 1979.

TAP, Lisbon

(Transportes Aereos Portugueses SARL)

VI-R: CS-TCA/TCB/TCC

In 1960 TAP reached a pool agreement with Air France to use the latter's Caravelles on the Lisbon-Paris route (the DC-4s and L.1049Gs of the Portugeuse carrier hardly being competitive).

TAP returned to the route with its own Caravelles in July 1962 (at the same time replacing BEA Comets that had been operated to London under a similar pool arrangement). With the introduction of the Boeing 727 five years later the Caravelle was gradually relegated from the prime European sectors and the fleet was eventually withdrawn and sold to SAN in 1975.

TAR, Nice

(Transports Aériens Réunis)

VI-N: OO-SRE*

One Sabena aircraft was intended to be purchased in November 1974. However the sale fell through and the aircraft never operated, although OO-SRE was painted in TAR titles, for a short period.

THAI INTERNATIONAL AIRWAYS, Bangkok

III: HS-TGF/TGG/TGH/TGI/TGK/TGL
      SE-DAA/DAB/DAD/DAF OY-KRF   LN-KLI
      (all leased from SAS)

Formed on 24 Aug 59 by SAS, Thai leased a number of Caravelles. The first service was flown on 2 Jan 64 with HS-TGF, followed by HS-TGG on 29 Mar 64. The Caravelles took over from CV-990s on a varied route network throughout the Far East. Stagelengths varied from Bangkok-Hong Kong (1,064 miles) to Kuala Lumpur - Singapore (194 miles) and the Caravelles regularly achieved a utilisation rate of over 9½ hours per day. Thai provided the first regular jet service to Katmandu, Nepal with the Caravelle in Dec 69.

TOP AIR, Düsseldorf (?)

Top Air was a German charter operator that intended to purchase three Caravelles in late 1972. No transactions were finalised and Top Air only operated a Merlin and Queen Air on air-taxi flights.

TRANSASIAN AIRLINES CORP, Manila

VI-R: RP-C970

Formed by Sterling, Transasian was to have operated two ex Sterling Caravelles on charters. One aircraft was painted in their colours but no services were operated.

TRANSAVIA HOLLAND BV, Amsterdam

III: PH-TRM*/TRN*/TRO/TRP/TRR
VI-R: PH-TRH/TRS/TRU/TRX/TRY/TVT*/TVZ* OY-SAK*/SBW*
VI-N: PH-TVV/TVW

Transavia initially leased two Caravelles from SNIAS and introduced the type on 18 Feb 69 on a charter to Lisbon. Three Swissair IIIs were purchased in late 69 and five ex UAL VI-Rs followed in 1970. Additional aircraft were leased from Sterling and Alitalia, the latter company's two aircraft later being purchased. The last revenue flight of a Transavia Caravelle was made on 1 Aug 76 to Reus, by which time the type had carried 2.6 million passengers in 83,100 commercial flying hours.

TRANS EUROPA, Madrid

(Compania de Aviacion SA)

10B1R: EC-BRJ/CIZ/CPI/CYI/DCN
11R: EC-BRX/BRY

TECA ordered two 11Rs in Mar 69 and operated them in pool with Iberia and Aviaco. A 10B1R was acquired under the same joint agreement in early 1970 followed by second hand purchases of 10B1Rs.

TRANSPOMAIR, Brussels

This Belgian operator announced an order for one ex UAL VI-R in late 1970 for services to start in Apr 71. However the deal failed to materialise and the airline changed its name to Pomair and acquired DC-6Bs.

TRANSUNION, Paris

III: F-BRIM*/BRUJ*
12:   OY-SAA*

Transunion leased Caravelles from Sud in 1969 and 1970. Services were suspended on 1 Sep 71 but plans for the revival of the company were laid in 1974. In the event the proposed lease of a Sterling 12 fell through but OY-SAA did operate charters for Transunion in the summer of 1974.

TRANS WORLD AIRLINES INC. New York

TWA's order for 20 model 10A's and 15 options has been covered in the general history. Suffice it to say the initial delivery was originally planned for 1 Jan 63 with the balance by Jul 63. Configuration was to have been between 68 first class and 85 economy class.

TUNIS AIR, Tunis

(Société Tunisienne de l'Air)

III: F-BRUJ*/BUFM*/OCPJ* PH-TRP* TS-IKM/ITU/MAC/TAR
VI-N: OO-SRE*

With French financial interests, Tunis Air introduced the Caravelle in Sep 61. The fleet was gradually increased to four but a number of aircraft were leased on occasions. The type was withdrawn from the fleet in 1977.

UNITED AIRLINES, Chicago

VI-R: N1001U to N1020U

United's significant order for the Caravelle on 25 Feb 60 has been covered in the general history. The Caravelle was used on central and eastern US routes but overhauls were carried out in San Francisco. Throughout its life with UAL, the configuration remained at 64 all first class. In actual fact, there were several allocations of airframes for the United order. At one time c/n 62 was slated to be N1002U and the remaining c/ns adjusted, with N1020U c/n 104. These allocations were all cancelled by the FAA on 17 Mar 61 due to the number of times United had changed the allocations for its fleet. Evidently, such was the uncertainty at this stage, the changes were costing the FAA too much time and money in file maintenance!

United in fact reserved the registrations N1001U through to N1050U. They also requested (but never followed up to reserve) the registrations N1051U to N1100U.

UTA, Paris

(Union de Transports Aériens)

10B1R: F-BNRA/BNRB
10B3:   F-BOEE*

UTA ordered two Caravelles for its routes across the South Pacific, introducing the type on the Noumea-Sydney service in Dec 66. The first Caravelle service had been flown in Feb 66 to West Africa.

VARIG, Rio de Janeiro

(Viacao Aérea Rio Grandense)

I/III: PP-VJC/VJD/VJI

Varig ordered two Caravelles on 16 Oct 57 and became the third operator of the type on 19 Dec 59, with the inauguration of jet services to New York from Buenos Aires via Montevideo, Sao Paulo, Rio, Belem, Port of Spain and Nassau. Seating configuration was 68 first class for the short-lived service as the Boeing 707 took over in June 60. The Caravelles then operated on domestic and South American routes. Two model VI-Rs were ordered but never delivered to Varig.

VASP, Sao Paulo

(Viacao Aérea Sao Paulo SA)

VASP ordered four model VI-Rs in Sept 62 but the order was cancelled the following year.

VIASA, Caracas

(Venezolana Internacional de Aviacion SA)

III: YV-C-AVI

VIASA leased Avensa's Caravelle for a period, operating it jointly with its subsidiary PIASA.

Other civil operators

AEROSERVICE CORP - see Litton Industries

AIRBORNE FREIGHT CORP, Seattle

This air freight forwarder purchased the remaining four Sterling VI-Rs in 1978 to enter the expanding overnight small packet service in the USA. Midwest Air Charter (q.v.) is operating the Caravelles on its behalf.

ATLANTA SKYLARKS, Atlanta

VI-R: N555SL/N777VV

An air travel club, the Skylarks purchased two ex Transavia Caravelles through their associate Independent Air Inc in 1976. One aircraft, N777VV was delivered to a Dr Alex Gazaui in May 77 for conversion to a luxury interior standard for charters to entertainers. However payments fell in arrears almost immediately and the aircraft was impounded at Burbank, CA. pending legal action. It was then sold to Ronald J.Clark in October 77. N555SL was used as a back-up to the Skylarks' Boeing 720 until sold in Feb 78 to Kearney & Trecker Corp.

CENTRAL AFRICAN REPUBLIC,

III: TL-AAI
10B3: TL-ABB

The Central African Republic obtained a Series III from Air France in late 1970, initially intended for use by Air Centrafrique, which had proposed a break with the Air Afrique consortium. In the event the airline stayed with the group and the Caravelle was used by President Bokassa. In 1975 an ex Sterling 10B3 was acquired for presidential use, being repainted in Central African Empire titles in early 1977 in line with the country's new status as proclaimed by the self-styled Emperor Bokassa.

FRANCE.

III: F-BJTI

The DGAC (section of La Formation Aéronautique) obtained an ex Air France aircraft in Sept 73 probably for use on test or calibration of navaids work

GABON,

VI-R: TR-LWD

The Gabon Government obtained an ex UAL/Sterling VI-R via UTA in 1976. It was transferred temporarily to Air Gabon on 1 Jun 77.

GARRETT CORPORATION,

VI-R: N210G

The Garrett Corp ordered a single aircraft in Dec 62 for use as an executive aircraft and travelling showcase. It was intended that N210G be flown "green" to Los Angeles for fitting of a customized interior by AiResearch Aviation Service (a division of Garrett). At this time Garrett held the option to buy all 20 of United's fleet plus exclusive sales rights to all used Caravelles that Sud might own. However, with the death of Cliff Garrett in Jul 1963 the corporation decided to abandon its planned introduction of large jets into the executive fleets.

The aircraft was in fact, purchased by Garrett on 28 Jul 63 and the acceptance flight flown on 28 Aug 63 but N210G remained at Toulouse in storage until resold to Iberia in 1965.

GENERAL ELECTRIC CO

III/VII: N420GE

General Electric announced its order for a single aircraft on 7 Jan 60 and the early story of this aircraft has been outlined in the general history. After appearing at the Le Bourget Salon in May 1961 followed by a sales tour of Europe and the Middle East, N420GE returned to the USA. It was disposed of to Sud, leaving Edwards AFB, CA. on 12 Dec 61 for a non-stop flight to Idlewild, NY covering the 2530 miles in 4 hours 40 minutes at an average ground-speed of 540 mph; arriving at Toulouse via Gander on 14 Dec 61.

GOODYEAR AEROSPACE - See Litton Industries

N.V. Huygen & CO

III: HB-ICU (OO-SBQ)

Mr van Huygen was the owner of TEN-BEL Touring and purchased an ex Swissair Caravelle in Oct 70 for charter flights between Tenerife and Belgium. Due to difficulties securing landing permits from Spanish authorities, HB-ICU was operated by Sobelair (being re-registered OO-SBQ) on behalf of TEN-BEL. Although eventually operated on other charters by Sobelair, Huygen remained the owner until the aircraft was sold in 1974 to Catair.

INDEPENDENT AIR - See Atlanta Skylarks.

The three aircraft were registered to Nevada Air Tours in Mar 78 (q.v.)

INTERNATIONAL AIR INC

VI-N: N45SB/46SB/49SB

A US aircraft broker, International Air purchased the three ex Argentine Air Force Caravelles in 1975 via Fokker. They were all flown to Lawton, OK. for a long-term storage in July of that year.

KEARNEY & TRECKER Corp

The Atlanta Skylarks' N555SL was sold to this company in Feb 78, but it was not confirmed if this aircraft would be used as an executive jet.

LASA

(Engenharia e Prospeccões SA)

VI-R: PT-DUW

A Cruzeiro subsidiary, LASA leased the Litton Caravelle for aerial survey work both in Brazil and Indonesia from May 71 to Mar 73.

LITTON INDUSTRIES

VI-R: N1001U

Litton purchased an ex United Caravelle in Jan 71 and registered it to world-wide mapping and surveying subsidiary, Aero Service Corporation. Fitted with Goodyear Electronic Mapping System (GEMS) radar, the Caravelle undertook extensive aerial surveys of Brazil while leased to LASA. During this time aerial survey work was also carried out for Venezuela, Indonesia and Australia. On its return to the US in 1973, N1001U was re-registered to another Litton subsidiary, Western Geophysical Co. Goodyear Aerospace Co. (Arizona Division) is responsible for the operation and maintenance of the Caravelles radar equipment.

MAURITANIA

VI-R: 5T-CJW (Air Mauritanie use)/MAL(military use)/ RIM (Government use)

The Mauritanian Government obtained an ex Sterling Caravelle via UTA in May 74. In less than a year this single aircraft had managed to wear three different registrations although utilisation is believed to be low.

NEVADA AIR TOURS

VI-N: N45SB/46SB/49SB

The three Independent Air Caravelles were re-registered to this concern in Mar 78. This was only a paper transfer of title and the three aircraft remained in store at Lawton until sold to EAS in late 1978.

OTAL

This Angolan carrier has been reported using a Caravelle VI from Luanda but no further details are known.

PUSHPAKA

Two Indian Airlines aircraft VT-DUH/DUI were transferred to this concern in 1978. It is not yet known whether this is an operating company or broker.

RON CLARK ENTERPRISES INC.

VI-R: N777VV

Ronald J.Clark provides a specialised leasing service, primarily with Viscounts, and catering particularly to rock musicians etc from Burbank CA. Operating entities are Go Inc (not currently used), Go Transportation Inc and Go Leasing Inc. An ex Atlanta Skylarks aircraft was purchased in October 1977 and awaits conversion to a luxury interior standard at Burbank for use as an executive aircraft.

RWANDA

III: 9XR-CH

Presumably with financial assistance from France, Rwanda acquired a single aircraft from SNIAS for government use in Mar 74.

SENEGAL

III: 6V-AAR (6V-ACP ntu)

To replace a Constellation (also 6V-AAR), the Government of Senegal obtained an ex Air France aircraft in Dec 71.

SOGERMA

III: 193/F-ZACF

SOGERMA (Société Girondine d' Entretien et de Réparation de Matériel Aéronautique), a part of the Aérospatiale group, obtained a Caravelle from SNIAS for use as an engine test-bed in Dec 71. The first venture tested the SNECMA M53 (first flight 18 Jul 73) and work continued with the CFM56 (first flight 17 Mar 77).

TCHAD

VI-R: TT-AAM

Tchad joined the list of African governments operating a Caravelle in their executive fleet by obtaining an ex Transavia aircraft via UTA in Apr 76.

TEN-BEL TOURING - See N.V. Huygen and Co.

WESTERN GEOPHYSICAL - See Litton Industries

Military operators

ARGENTINE

(Fuerza Aerea Argentina)

VI-N: T-91/T-92/T-93 (all 1 Brigade)

FRANCE

(Armée de l'Air)

III: 141/F-RAFG 158/F-RAFA*   IOR: 201/F-RAFH   (GLAM)
IIR: 240/F-RBPR 251/F-RBPS   264/F-RBPT   (ETOM.82)

The GLAM Caravelles are used for government officials, 158/F-RAFA being used for General de Gaulle's tour of South America , while the three 11Rs are more of a regular military air transport variant for use in the Pacific Nuclear Test Range (CEP) replacing earlier prop-driven transports there for shuttles between Tahiti and the test range.

(Centre d' Essais en Vol)

III: 116/F-ZACE

The CEV is the official French test and research establishment(similar to the British RAE). Aircraft are flown with military-style roundels, but codes and call-signs are all in the F-Z range.

JUGOSLAVIA

(JRV)

VI-N: 7601/74101

SWEDEN

(Flygvapen)

III: 85172"81"/85210"82" (F13)

57

CARAVELLE PRODUCTION SEQUENCE

*The French expression for sequence or line number is "numéro de rang" (N/R)

N/R	c/n	N/R	c/n	N/R	c/n	N/R	c/n	N/R	c/n	N/R	c/n
1	1	48	46	95	99	142	165	189	198	236	238
2	2	49	49	96	100	143	144	190	195	237	246
3	3	50	50	97	101	144	145	191	192	238	243
4	4	51	51	98	102	145	153	192	196	239	248
5	5	52	55	99	103	146	148	193	194	240	242
6	6	53	54	100	104	147	157	194	184	241	249
7	7	54	56	101	111	148	152	195	199	242	244
8	8	55	58	102	81	149	151	196	171	243	245
9	9	56	59	103	114	150	156	197	200	244	236
10	10	57	60	104	107	151	154	198	209	245	239
11	11	58	61	105	82	152	168	199	203	246	254
12	12	59	62	106	112	153	159	200	204	247	234
13	13	60	64	107	113	154	138	201	201	248	247
14	14	61	65	108	116	155	158	202	210	249	252
15	15	62	66	109	108	156	162	203	188	250	251
16	16	63	67	110	109	157	155	204	202	251	250
17	17	64	69	111	121	158	128	205	207	252	237
18	18	65	71	112	122	159	136	206	186	253	253
19	19	66	72	113	85	160	130	207	208	254	241
20	20	67	57	114	106	161	173	208	205	255	261
21	23	68	74	115	110	162	169	209	190	256	256
22	24	69	76	116	123	163	163	210	206	257	255
23	21	70	52	117	118	164	140	211	212	258	257
24	25	71	77	118	115	165	170	212	211	259	260
25	26	72	53	119	124	166	166	213	213	260	263
26	22	73	73	120	120	167	178	214	214	261	259
27	29	74	86	121	127	168	177	215	216	262	264
28	30	75	68	122	105	169	132	216	217	263	258
29	28	76	87	123	119	170	164	217	220	264	262
30	31	77	79	124	117	171	181	218	222	265	265
31	35	78	75	125	63	172	182	219	224	266	266
32	27	79	88	126	125	173	160	220	215	267	267
33	32	80	89	127	126	174	185	221	221	268	268
34	33	81	83	128	131	175	187	222	223	269	269
35	34	82	70	129	137	176	189	223	218	270	270
36	36	83	84	130	147	177	180	224	225	271	271
37	37	84	90	131	143	178	174	225	228	272	272
38	39	85	91	132	129	179	134	226	219	273	273
39	40	86	92	133	149	180	183	227	230	274	274
40	38	87	78	134	133	181	175	228	226	275	275
41	41	88	93	135	139	182	172	229	232	276	276
42	43	89	94	136	141	183	176	230	240	277	277
43	44	90	80	137	135	184	197	231	231	278	278
44	42	91	95	138	161	185	191	232	227	279	279
45	47	92	96	139	142	186	193	233	229	280	280
46	48	93	97	140	146	187	179	234	233		
47	45	94	98	141	150	188	167	235	235		

Introduction

The aircraft are listed in constructor's number order with subsequent production data presented as follows:

1. Following the c/n, the sequence number or "numéro de rang" of each Caravelle appears in brackets

2. Series designation. Throughout, Roman numerals are retained for Caravelles up to the Series 10, although Arabic numerals are now often applied retroactively to the earlier models. The Series 10 introduced the practice of modifying the basic structure to accomodate higher take-off weights without any change of model number. The Series 10B3 "Super" for example, was produced with three different maximum weights, 52000kg, 54,000kg and 56,000kgs. Where known, models with the higher weights are so designated, eg. Series 10B3/54T (indicating 54 Tonnes a.u.w.).

3. French temporary test registration, if known. All export Caravelles were flown at Toulouse prior to acceptance by the customer with these registrations. In addition, some French registered aircraft were initially flown as F-W... prior to being awarded their full CofA. The majority of these "bloc-navettes" allocations remain unknown.

4. First flight date.

5. Initial permanent registration.

6. Initial customer. Airline names are abbreviated using a form that should be readily understood. Full names and other details can be found in the "Airline Operators" section.

7. Individual aircraft name if allocated. In some cases names were not applied until after delivery. In the case of an interval of several years elapsing before receiving a name, this is shown by "later named...".

8. Delivery date for the first owner is the official Sud delivery date, representing the date handed over to the customer in Toulouse. Arrival date at customer's headquarters is given (where known) only if it differs significantly. For subsequent owners, the date of the actual transaction is given where possible. Failing this, or sometimes in addition to this, the delivery date or date of registration is shown.

<u>01</u>   Prototype. r/o 21 Apr 55: F-WHHH f/f 27 May 55. Re-regd F-BHHH on CofA issue 23 May 56 to SGACC. Painted in SAS colours 1957, "Finn Viking". Last flight 16 Apr 66 and preserved at Orly, Paris since 1 Jun 66.

<u>02</u>   Prototype F-WHHI f/f 6 May 56. Re-regd F-BHHI with SGACC 19 Mar 57. Cold weather trials at Lulea Jan 59. Lsd by SAS for crew training 1 Mar 59 - Apr 59. Withdrawn 1968 and to Air France's Centre d'Instruction Vilgenis. Nose to Musée de L'Air, Le Bourget, Paris Jan 76, remainder of aircraft scrapped.

<u>1</u> (1) Series I F-WHRA f/f 18 May 58. F-BHRA Air France "Alsace", del 3 Apr 59. Cvtd to Series III 1960. Wfu 19 Dec 75 and ferried by road Feb 76 to Centre d'Instruction Vilgenis.

<u>2</u> (2) Series I F-WHRB f/f 14 Mar 59. F-BHRB Air France "Lorraine", del 19 Mar 59. Cvtd to Series III 1961. Wfs 25 Dec 77 and partially broken up 16 Jan 78 (TT 33,094hrs.Still derelict at Orly Nov 78.

<u>3</u> (3) Series I   f/f 23 Mar 59. LN-KLH SAS "Finn Viking", del 10 Apr 59. Cvtd to Series III Oct 60. Wfs 29 Aug 74 and flown to Oslo/ Gardemoen 27 Sep 74 for storage for Norwegian air museum (TT 31822hrs).

<u>4</u> (4) Series I   f/f 15 Apr 59 SE-DAA SAS "Eskil Viking", del 25 Apr 59. Cvtd to Series III Oct 60. Lsd to Thai International (as-'DAA) 21 Jun 70 - 1 Oct 70. Wfs 28 Jul 74 and used at Stockholm-Arlanda for fire practice (TT 31710 hrs)

<u>5</u> (5) Series I   f/f 9 May. F-BHRC Air France "Anjou", del 15 May 59. Cvtd to Series III 1961. Sold 28 Dec 71 to Senegalese Govt. (6V-ACP ntu) 6V-AAR(2) "Fleche des Almadies".

<u>6</u> (6) Series I   f/f 24 Apr 59. OY-KRA SAS "Vagn Viking", del 5 May 59. Cvtd to Series III 1960. Damaged at Stockholm-Arlanda 25 Jan 74 and wfu. Broken up Aug 74. (TT 30878 hrs).

<u>7</u> (7) Series I   f/f 6 Jun 59 LN-KLI SAS "Einar Viking", del 17 June 59. Cvtd to Series III Mar 60. Damaged by fire in hangar while on overhaul at Stockholm-Arlanda 21 Feb 68 and repaired with new cockpit section by 23 Jun 68. Lsd Thai International 17 Oct 68-27 Jan 69. Wfs at Arlanda 30 Jun 74 and broke-up Sep 74. (TT 31425 hrs).

<u>8</u> (8) Series I   f/f 19 Jun 59 F-BHRD Air France "Guyenne", del 30 Jun 59. Cvtd to Series III 1961.

<u>9</u> (9) Series I   f/f 23 Jul 59 F-BHRE Air France "Artois", del 31 Jul 59. Cvtd to Series III 1961.

10 (10)Series I F-WJAP f/f 31 Aug 59 PP-VJC Varig "Brasilia", del 16 Sep 59. Cvtd to Series III 1961. Retd to Sud May 64 and lsd Air Vietnam as XV-NJA, del 17 Aug 64. Retd to Sud for Air France 25 Sep 68 and re-regd F-BNGE but never entered service. Lsd MEA 9 Jan 69 - 31 Mar 69, and Royal Air Maroc - 69. Sold 28 Dec 70 to the Central African Republic as TL-AAI and painted as "Air Centrafrique" but never used on airline services. Repainted in Republique Centrafricaine titles "Jean Bedel Bokassa", later in Empire Centrafricain titles. Wfu and stored at Le Bourget, Paris since Aug 77.

11(11) Series I   f/f 4 Sep 59 SE-DAB SAS "Ingemar Viking", del 18 Oct 59. Cvtd to Series III 1960. Lsd Thai International 11 Jul 69 - 7 Apr 70. Wfu 21 May 74 and scrapped Stockholm-Arlanda Sep 74 (TT 31890 hrs)

12(12) Series I   f/f 8 Oct 59 F-BHRF Air France "Auvergne", del 17 Oct 59. Cvtd to Series III 1961.

13(13) Series I   f/f 24 Oct 59 F-BHRG Air France "Berry", del 6 Nov 59. Cvtd to Series III 1961. Wfu Aug 78.

14(14) Series I   f/f 8 Nov 59. OY-KRB SAS "Orm Viking", del 18 Nov 59. Crashed on approach to Ankara-Esenboga 19 Jan 60, 42 killed. (TT 473 hrs).

15(15) Series I   f/f 26 Nov 59 PP-VJD Varig del 11 Dec 59. Cvtd to Series III 1961. Landed short and dbf at Brasilia, 27 Sep 61, no fatalities (TT 4054 hrs).

16(16) Series I later IA   f/f 2 Dec 59 F-BHRH Air France "Bourgogne", del 11 Dec 59. Cvtd to Series III 1961.

17(17) Series I   f/f 10 Dec 59 F-BHRI Air France "Bretagne", del 22 Dec 59. Cvtd to Series III 1961.

18(18) Series I F-WBNG f/f 21 Dec 59 F-OBNG Air Algerie del 26 Dec 59. Cvtd to Series III 1961. Re-regd 7T-VAG 7 Jan 64 and later named "Ovarsenis". Wfu Jun 76 and stored at Algiers.

19(19) Series III F-WJAQ f/f 30 Dec 59. Cvtd to Series VIN and f/f 10 Sep 60. Re-regd F-BJAQ 3 Mar 61 to Sud Aviation for tropical trials at Johannesburg. Re-regd F-WJAK for test flights prior delivery to Aerolineas Argentinas. Sold to Aerolineas Argentinas "Aldebaran". del 15 Jan 62 as LV-PRR and re-regd LV-HGX. Wfs 17 Apr 73 and transferred to T-91 F.A. Argentina Jul 73. Sold via Fokker to International Air Inc (a broker) as N45SB and del to Lawton, OK. 24 Jun 74,for long-term storage on behalf of Fokker N.V. Sold to

International Air 12 May 75 and re-regd to Nevada Air Tours 28 Feb 78. Sold to EAS as F-GBMI and del 6 Jan 79.

<u>20</u> (20) Series I F-WJAK f/f 8 Jan 60 F-OBNH Air Algerie, del 14 Jan 60. Cvtd to Series III 1961. Sold to Varig 15 Dec 61 as PP-VJI. Stored by Lockheed Air Services at New York-JFK from 18 Dec 63. Sold via International Aviation 13 Aug 64 to Avensa as YV-C-AVI. Lsd to VIASA/PIASA - 69. Wing struck ground on landing at Barquisimento 20 Aug 73 and wfu (TT 25656 hrs).

<u>21</u> (23) Series IA F-WJAK f/f 11 Feb 60 OH-LEA Finnair "Sinilintu", del 18 Feb 60. Cvtd to Series III 1961. Sold to LTU Aug 64 and del to Sud 20 Sep 64 for refurbishing. Del to LTU 5 Feb 65 D-ABAF "Nordrhein-Westfalen". Retd to Sud 8 Aug 68 as F-WLGA. Lsd Transavia as PH-TRM 11 Feb 69 - 13 May 70. Stored Toulouse until regd F-BSRR 2 Mar 71 with SNIAS. Lsd Air Inter 19 Mar 71, and believed also lsd to Royal Air Maroc.

<u>22</u> (26) Series IA f/f 23 Mar 60 OH-LEB Finnair "Sinisiipi", del 30 Mar 60. Cvtd to Series III 1962. To Sud 31 Aug 64 as F-BJTR, to Air France, CofA 28 May 65 "Vercors", later "Principaute de Monaco".

<u>23</u> (21) Series I later IA F-WHRJ f/f 18 Jan 60 F-BHRJ Air France "Champagne", del 9 Feb 60. Cvtd to Series III 1961. Lsd MEA 20 Jan 64 as OD-AEM. Crashed into Persian Gulf 17 Apr 64 during approach to Dhahran during sandstorm, 49 killed. (TT 8771 hrs).

<u>24</u> (22) Series I F-WJAM f/f 30 Jan 60, cvtd to IA prior del. (LN-KLJ NTU) LN-KLP SAS "Trond Viking", del 8 Mar 60. Cvtd to Series III Aug 60. Wfs 27 Aug 74 and ferried Stockholm-Arlanda to Malmo-Sturup 30 Aug 74 for LFV (Swedish CAA) as ground instruction airframe. (TT29837 hrs).

<u>25</u> (24) Series IA f/f 25 Feb 60 SE-DAC SAS "Arne Viking", del 11 Mar 60. Cvtd to Series III 1960. Lsd Thai Intl 29 Sep 66 as HS-TGI "Chiraprapa". Crashed on approach to Hong Kong 29 Jun 67 during typhoon "Anita", 27 killed (TT 17360 hrs).

<u>26</u> (25) Series IA F-WHRK f/f 29 Feb 60 F-BHRK Air France "Corse", del 18 Mar 60. Cvtd to Series III 1961. Wfu 79 and used for crew training. Derelict by May 79 at Orly.

<u>27</u> (32) Series IA f/f 29 Apr 60 OH-LEC Finnair "Sininuoli", del 4 May 60. Cvtd to Series III 1961. To Sud 1 Oct 64 as F-BJTS, to Air France Cof A 15 Jun 65 "Principaute de Monaco", later "Vercors".

<u>28</u> (29) Series IA f/f 10 Mar 60 F-OBNI Air Algerie, del 19 Mar 60. Damaged in mid-air collision with Stampe SV.4 F-BDEV during approach to Orly 19 May 60 and repaired as Series III, flying again 11 Feb 61. Re-regd 7T-VAI 14 Jan 64. Damaged on landing at Algiers/Dar-el-Beida 23 Sep 73 and declared written-off.

<u>29</u> (27) Series IA f/f 23 Mar 60 OY-KRC SAS "Faste Viking", del 2 Apr 60. Cvtd to Series III 1960, later modified to Series III/48T. Lsd Thai International 5 Oct 65 - 3 Apr 70 as HS-TGH "Srisoonthon". Wfu 27 Sep 74 and handed over to LFV Fire Service, Stockholm-Arlanda Oct 75. (TT 34284 hrs).

<u>30</u> (28) Series IA f/f 7 Apr 60 LN-KLR SAS "Hall Viking", del 15 Apr 60. Cvtd to Series III 1960, later modified to Series III/48T Lsd Thai International 3 Aug 67 - 10 May 70 as HS-TGL "Srichulalak". Wfu 9 Jun 74 and scrapped Stockholm-Arlanda Oct 74. (TT 31658).

<u>31</u> (30) Series IA f/f 14 Apr 60 F-BHRL Air France "Dauphine", del 26 Apr 60. Cvtd to Series III 1961.

<u>32</u> (33) Series IA F-WJAL f/f 4 May 60 CN-CCV Royal Air Maroc del 11 May 60. Cvtd to Series III 1961. Crashed Berrechid 1 Apr 70 on approach to Casablanca-Nouasseur, 62 killed. (TT22779 hrs).

<u>33</u> (34) Series III F-WJAM f/f 25 Apr 60 HB-ICW Swissair (lsd from SAS until purchased 30 Mar 65) "Solothurn", del 30 Apr 60. Damaged at Basle 26 Sep 61. re-entered service 23 Dec 61. Sold Transavia as PH-TRO 15 Oct 69, later named "Provincie Gelderland". Wfu 22 Nov 75 and sold for scrap Dec 75. Broken up Mar 76 - nose to Aviodome Museum at Amsterdam-Schiphol. (TT 31997 hrs).

<u>34</u> (35) Series IA f/f 20 May 60 SE-DAD SAS "Torolf Viking", del 25 May 60. Cvtd to Series III 1960. Lsd Thai International 2 Dec 64 - 18 Feb 65 (as DAD). Lsd again as HS-TGK "Thepamart" 27 Nov 67 until damaged in landing at Bangkok-Don-Muang 9 Jul 69 and declared written off (TT 23150 hrs).

<u>35</u> (31) Series III F-WJAO f/f 21 Apr 60. I-DAXA Alitalia "Altair", del 29 Apr 60. Cvtd to Series VI-N 1962. Lsd to SAM 10 Jun 68 but later back to Alitalia. Wfs 17 Feb 77. Sold to SAETA as HC-BAD - 76. WFU at Quito.

<u>36</u> (36) Series III f/f 13 May 60. I-DAXF Alitalia "Aldebaran", del 19 May 60. Cvtd to Series VI-N 1962. Wfs 19 Apr 73. Lsd to Transavia as PH-TVW and del 28 Apr 73, purchased by them 28 Sep 73, "Provincie Overijssel". Wfs 10 Jun 74, and sold to Luchthaven Schiphol 18 Apr 75 for use by fire service. Named "Florianus" and used as emergency evacuation procedures trainer. (TT 29062 hrs).

<u>37</u> (37) Series III f/f 27 May 60. F-BHRM Air France possibly as "Flandré" initially but "Quercy", by 1961, del 3 Jun 60. Due to be wfs Apr 79.

<u>38</u> (40) Series III F-WJAM f/f 17 Jun 60. HB-ICX Swissair (lsd from SAS until purchased 30 Mar 65) "Chur", del 24 Jun 60. Sold to Catair and after block overhaul Nov 70 rolled out with Catair blue scheme, no titles. Operated for Swissair until Mar 71, re-regd F-BSRD and del to Catair 27 Apr 71. Sold to China Airlines as B-1854 and del 17 Jan 73. By Jul 77 had logged TT 40,518 hrs (37,289 landings) which prompted SNIAS interest in the aircraft. After 17 years of operation, research could be undertaken on the airframe, which was acquired by SNIAS and returned to Toulouse, Aug 77. Re-regd F-WJAL and transferred by road early 1978 from St Martin du Touch to CEAT Toulouse for a detailed "post-mortem examination" (research will take one or two years on both airframe and systems, to be conducted by CEAT and Direction Etudes Avions et Support Exploitation of Aérospatiale).

<u>39</u> (38) Series III f/f 11 Jun 60. F-BHRN Air France "Gascogne", del 18 Jun 60. Operated Lufthansa services 64-65. Wfs 1978 and broken up Orly Aug 78.

<u>40</u> (39) Series III f/f 1 Jun 60. I-DAXI Alitalia "Antares", del 10 Jun 60. Cvtd to Series VI-N 1962. Wfs 31 Mar 75. Sold to SAETA as HC-BAE and del ex Rome 16 Jun 75.

<u>41</u> (41) Series III f/f 24 Jun 60. F-BHRO Air France "Ile de France", del 1 Jul 60. Wfu Angers-Arille Jun 78 after sale to private buyer. Currently open for public tours.

<u>42</u> (44) Series III F-WJAM f/f 15 Jul 60. N420GE General Electric Co. "Santa Maria", del 18 Jul 60. Cvtd to Series VII with CJ-805-23C engines and f/f 29 Dec 60 at Edwards AFB, CA. Retd to Sud 14 Dec 61 and re-regd F-BJAO 5 Apr 62 for tour of Far East and Australasia Apr - May 62 (still named "Santa Maria"). Re-cvtd as series III for Air France and f/f Jun 63 as F-WLKF. Del as F-BLKF "Angoumois" 31 Jul 63. Lsd to Lufthansa - 64. Later modified as Series III/48T.

<u>43</u> (42) Series III f/f 1 Jul 60. HB-ICY Swissair (lsd from SAS until purchased 30 Mar 65) "Lausanne", del 8 Jul 60. Overran runway at Zürich 1 Jan 62 without serious damage. Sold to Transavia as PH-TRP and del 11 Nov 69. Later named "Provincie Noord Brabant". Lsd to Tunis Air 2 Nov 72 - 14 Jan 73. Used for Osmonds pop group tour of Europe Oct-Nov 73. Wfs - 75 and sold for scrap Dec 75

60

Tail section del to Delft Technical University 23 Feb 76.
(TT 31711 hrs).

44 (43)   Series III   f/f 12 Jul 60. I-DAXO
           Alitalia "Deneb", del 18 Jul 60. Cvtd to Series
VI-N 1962. Wfs 19 Apr 73. Lsd to Transavia as PH-TVV
and del 25 May 73. Purchased by them 28 Sep 73,
"Provincie Zuid-Holland". Used for Osmonds pop group
tour May 75. Wfs 31 Aug 75 and scrapped at Amsterdam-
Schiphol Jun 76 (TT 30022 hrs).

45 (47)   Series III   f/f 4 Aug 60. F-BHRP
           Air France "Languedoc", del 12 Aug 60. Lsd
to Sabena 1 Apr 63.- 31 Mar 64. Due to be wfs Jun 79.

46 (48)   Series III   f/f 19 Aug 60. F-BHRQ
           Air France "Limousin", del 26 Aug 60. Operated
Lufthansa services - 64. Sold to AirInter 31 Mar 72.

47 (45)   Series III   f/f 26 Jul 60. OY-KRD
           SAS "Ulf Viking", del 30 Jul 60. Lsd to
Swissair 6 Jan 62 - 16 Jan 62. Wfs 21 Aug 74 at
Copenhagen - Kastrup and stored for Danish Air
Museum. (TT 29845 hrs).

48 (46)   Series III   f/f 8 Aug 60. HB-ICZ Swissair
           (lsd from SAS until purchased 30 Mar 65)
"Bellinzona", del 13 Aug 60. Wfs 28 Jan 69 then lsd
to Air Algerie 20 Jun 69 - 20 Oct 69 but actually not
del to Algiers until 3 Jul 69. Lsd to them again 15
Jan 70 - 15 Mar 70. Sold to Transavia as PH-TRR
and del 27 Mar 70. Named "Feyenoord EEN" during
Europa-Cup Jun 70 then renamed "Provincie Drenthe".
Used for David Cassidy European tour Mar 73.
Wfs 28 Feb 75 and sold for scrap Dec 75. (TT30559 hrs).

49 (49)   Series III   f/f 14 Sep 60. OY-KRE
           SAS "Knud Viking", del 21 Sep 60., later
modified to Series III/48T. Lsd Thai International.
HS.TGG 26 Mar 64 - 10 Oct 70 "Thepsatri". On return
to SAS re-named "Gaut Viking". Wfs 18 Sep 74 and sold
to LFV Fire Section at Stockholm-Arlanda Oct 75.
(TT 34249 hrs). Scrapped Oct 77.

50 (50)   Series III   f/f 17 Sep 60. F-BHRR
           Air France "Lyonnais", del 23 Sep 60.
Sold to Air Inter 31 Mar 73..

51 (51)   Series III   f/f 8 Sep 60. F-OBNJ
           Air Algérie, del 12 Sept 60. Lsd to Air
Liban as OD-ADZ Jan 62 - Jun 63. Regd to Air France
3 Aug 63 as F-BLCZ but lsd to Air Algérie. Re-regd
7T-VAE to Air Algérie 16 Sep 67, later named
"Djurdjura". Damaged at Algiers/Dar-El-Beida
11 Nov 73 when veered off runway on landing.
Returned to service 18 Nov 73 until wfs Jun 76.
Stored at Algiers. Reportedly with Algerian AF.

52 (70)   Series III   f/f 8 Mar 61. F-BHRZ
           Air France "Flandre", del 16 Mar 61. Lsd
Air Inter 1 Apr 71 until sold to them 1 Jul 72.

53 (72)   Series III   f/f 29 Mar 61. F-BJTA
           Air France "Comte de Nice", del 7 Apr 61.
Used on Caribbean services from about Nov 67 -
72 with name "Antilles"., later renamed "Champagne".
Sold to Air Cambodge 10 Apr 73 as XU-JTB "Siemreap
Anghor". Wfu at Bangkok and believed still
derelict there.

54 (53)   Series III   f/f 13 Oct 60. F-BHRS
           Air France "Normandie", del 26 Oct 60.
Sold to Air Inter 20 Mar 74.

55 (52)   Series III   f/f 4 Oct 60. F-BHRT
           Air France "Picardie", del 14 Oct 60.

56 (54)   Series III   f/f 20 Oct 60. SE-DAE
           SAS "Alrik Viking", del 28 Oct 60.
Later modified to Series III/48T. Lsd Thai
International as HS-TGF "Suranaree" 26 Dec 63 -
2 Oct 70 although used by SAS in Feb 70 (as HS-TGF).
On return to SAS renamed "Amund Viking". Wfs
9 Sep 74 and sold to LFV Fire Service at Stockholm-
Arlanda Dec 75. Scrapped Oct 77.(TT 33415 hrs).

57 (67)   Series III   f/f 4 Feb 61. CN-CCX
           Royal Air Maroc, del 25 Feb 61, later named
"Nador". Wfu at Casablanca-Anfar on CofA expiry
17 May 76.

58 (55)   Series III   f/f 3 Nov 60. F-BHRU
           Air France "Poitou", del 10 Nov 60.

59 (56)   Series III   f/f 15 Nov 60. F-BHRV
           Air France "Provence", del 22 Nov 60.

60 (57)   Series III   f/f 18 Nov 60. F-BHRX
           Air France "Savoie", del 29 Nov 60. Lsd
to Catai Jun 75 -   -75.

61 (58)   Series III   f/f 29 Nov 60. F-BHRY
           Air France "Touraine", del 15 Dec 60.
Damaged at Marseilles-Marignane in wheels-up landing
9 Sep 69. Returned to service.

62 (59)   Series VI-R prototype. F-WJAP f/f 6 Feb 61
           in United Airlines colour scheme. Reg
N2001U allocated but ntu for US demonstration tour.
Re-regd F-BJAP to Sud 22 Aug 61. Sold to Cruzeiro and
reverted to F-WJAP for pre-delivery tests. Regd
PP-CJC and del 30 Jul 63, arriving Rio 8 Aug 63. Wfu
and stored at Porto Alegre since CofA expiry 30 Sep 75.

63 (125) Series VII prototype-not completed, cvtd
           to Series 10A prototype F-WJAO f/f 31 Aug 62.
Wfu and cocooned at Toulouse until scrapped in 1969.
Nose section used in construction of c/n 259.

64 (60)   Series VI-N. F-WJAK f/f 14 Jan 61. OO-SRA
           Sabena, del 20 Jan 61. Wfs 8 Aug 74 and
stored at Brussels until donated to Musee de l'Air,
Brussels Jan 77. (TT 24,244 hrs).

65 (61)   Series VI-N. F-WJAL f/f 27 Jan 61. OO-SRB
           Sabena, del 1 Feb 61. To Sobelair 31 Mar 71.
Lsd Air Afrique 6 Jan 75 - 24 Mar 75. Retd to Sobelair.

66 (62)   Series VI-N F-WJAM f/f 10 Feb 61. OO-SRC
           Sabena, del 16 Feb 61. To Sobelair 31 Mar 71.
Sold to Catair 5 Apr 76 as F-BYCA, later operated
on behalf of Catair by ALBATROS. Sold to Europe Air
Service Apr 78.

67 (63)   Series VI-N. F-WJAK f/f 20 Mar 61. OO-SRE
           Sabena, del 24 Mar 61. Overran runway at
Köln-Bonn 10 Jan 65. Lsd Tunis Air 27 Jun 74 -
30 Sep 74. Proposed sale to TAR Nov 74 fell through.
Wfs - 75 and lsd to Catair Apr 76 until sold to
them 30 May 76 as F-BYCD. Later opeated by ALBATROS
for Catair. To .Europe Air Service Mar 78.

68 (75)   Series III   f/f 17 May 61. F-BJTB
           Air France "Bearn", del 30 May 61. Crashed
on approach to Rabat 12 Sept 61 in poor visibility,
all 78 on board killed. (TT 688 hrs).

69 (64)   Series VI-N. F-WJAN f/f 23 Feb 61. OO-SRD
           Sabena, del 28 Feb 61. To Sobelair
31 Mar 71. Lsd Royal Air Maroc Dec 73 until written-off
22 Dec 73 on Hadj flight to Tangier. Crashed on
Mt Mellaline in the Rif range, 20 km from Tetuan in
poor visibility and high winds, 106 killed. (TT 23202 hrs).

70 (82)   Series VI-N. F-WJAK f/f 3 Jul 61 then
           re-regd F-BJAU. OO-SRG Sabena, del 12 Jul 61.
Retd to SNIAS 19 Sep 73 and lsd to Indian AL as
VT-ECG named "Vijayadoot" and del 5 Nov 73. Purchased
by Indian AL   75. Wfs on CofA expiry 25 Oct 76.
(TT 29,469 hrs). Broken up Bombay.

71 (65)   Series VI-N. F-WJAK f/f 11 Mar 61. I-DABA
           Alitalia "Regolo", del 21 Mar 61. Wfs at
Rome 21 Dec 76. (TT 30594 hrs). Moved to Venetia -
Tesseria for storage 78.

72 (66)   Series VI-N   f/f 23 Mar 61. I-DABE
           Alitalia "Rigel", del 30 Mar 61. Sold to
SAETA and del 11 Dec 76. Used for spares by SAETA and
wfu at Quito. (TT 30770 hrs).

73 (73)   Series VI-N   f/f 27 Apr 61. F-OBNK
           Air Algérie del 2 May 61. Re-regd 7T-VAK
6 Nov 64. Crashed while attempting forced landing
following in-flight fire in Biskra region of S.Algeria,
26 Jul 69. 35 killed. (TT 16717 hrs).

74 (68)   Series VI-N   f/f 14 Apr 61. I-DABI
           Alitalia "Sirio", del 22 Apr 61. Lsd to SAM
during 1969. Wfs at Rome 28 Feb 77. Moved to Venetia-
Tesseria for storage 78. (TT 30534 hrs).

<u>75</u> (78)   Series III   f/f 17 Jun 61. F-OBNL Air
          Algérie, del 20 Jun 61. Re-regd 7T-VAL 9 Nov 64
and later named "Gorges du Rummel". Wfu Jun 76 and
stored at Algiers. Reportedly with Algerian AF.

<u>76</u> (69)   Series VI-N. F-WJAO f/f 22 Apr 61. OO-SRF
          Sabena, del 28 Apr 61. Lsd Air Algérie
26 Nov 74 -   Lsd Air Inter 20 Apr 76-
25 Jun 76. Lsd Catair 30 Mar 77 - 13 Apr 77 and
18 May 77 - 29 Jun 77. Sold to Europe Air Service
Feb 78, in service May 78 as OO-SRF, and re-regd
F-BXOO.

<u>77</u> (71)   Series VI-N   f/f 9 May 61. I-DABU Alitalia
          "Vega", del May 61. Wfs at Rome 26 Sept 76.
(TT 30859 hrs).

<u>78</u> (87)   Series VI-N. F-WJAL f/f 8 Aug 61. OO-SRH
          Sabena, del 11 Aug 61. Retd to SNIAS
3 Sep 73 as F-WLGA. Lsd Indian AL as VT-ECH "Suryadoot"
and del 12 Oct 73. Retd to SNIAS 21 Nov 75 and
stored at Toulouse. To TU-TXR Air Afrique Jan 76
and Wfu at Dakar Jun 76, presumably for spares.

<u>79</u> (77)   Series VI-N   f/f 2 Jun 61. I-DAXU
          Alitalia "Canopo", del 9 Jun 61. Wfs
at Rome 14 Sep 76. Moved to Venetia-Tesseria
for storage 78. (TT 30002 hrs).

<u>80</u> (90)   Series VI-N   f/f 9 Oct 61. I-DAXT
          Alitalia "Polluce", del 17 Oct 61. Wfs
at Rome 19 Feb 77 (TT 30277 hrs). Moved to Venetia-
Tesseria for storage 78 and then lsd to Aerolineas
Argentinas.

<u>81</u> (102) Series VI-N   f/f 10 Jan 62. I-DABR
          Alitalia "Bellatrix", del 19 Jan 62. Wfs
at Rome 19 Oct 75. (TT 27828 hrs).

<u>82</u> (105) Series VI-N   f/f 31 Jan 62. I-DABZ
          Alitalia "Spica", del 9 Feb 62. Damaged at
Torino 30 Dec 70 on take-off and repaired. Lsd
SAM. Wfs 30 Aug 75. Sold to SAETA as HC-BAI and
del 14 Nov 75. Never entered service and wfu at
Quito.

<u>83</u> (81)   Series III   f/f 29 Jun 61. F-BJTC
          Air France "Bourbonnais", del 10 Jul 61
but lsd immediately to Air Liban as OD-ADY and
re-del 10 Aug 61. Retd to Air France as F-BKGZ
13 Dec 61 "Comte de Foix". Lsd to Royal Air Lao
as XW-PNH 26 Sep 74 until wfu mid - 75. and
stored at Bangkok. Re-regd F-BSGZ to Air France
30 Dec 75 and re-entered service from Bangkok
1 Jan 76. Dbr when hijacker blew himself up on a/c
at Ho Chi Minh City 28 Aug 76.

<u>84</u> (83)   Series III   f/f 22 Jul 61. F-BJTD
          Air France but not del as such and del as
TS-IKM to Tunis Air 31 Aug 61. Wfu at Tunis Nov 75
and stored.

<u>85</u> (113) Series VI-N   f/f 24 Mar 62. I-DABT
          Alitalia, "Denebola", del 2 Apr 62.
To SAM 1 Jun 68 and operated for ATI Jan 72. Wfs
Rome 8 Feb 77. (TT 30277 hrs). Moved to Venetia-
Tesseria for storage 78.

<u>86</u> (74)   Series VI-R F.WJ   f/f 19 May 61. N1001U
          United "Ville de Toulouse", del 10 Jun 61.
Sold 27 Jan 71 to Aero Service Corp (a division of
Litton Industries) for use as an aerial survey
aircraft. Lsd 31 May 71 - 19 May 73 LASA-
Engenharia e Prospeccoes SA (a subsidiary of Cruzeiro).
as PT-DUW. Retd to N1001U and fitted with GEMS radar
by Goodyear being re-regd to Western Geophysical Co
(another subsidiary of Litton).

<u>87</u> (76)   Series VI-R   f/f 8 Jun 61. N1002U
          United "Ville de Cahors", del 20 Jun 61.
Sold to Transavia 1 May 70 as PH-TRY "Provincie
Friesland" and del 4 May 70. Used for Frank Sinatra
tour May 75. Sold 31 Jul 76 to Atlanta Skylarks
and re-regd N777VV to their associate Independent
Air Inc. Del 4 Aug 76 but did not enter service.
Lsd to Dr. Alex Gazaui for conversion into luxury
interior for charters to entertainers and del
to Long Beach, CA 3 Jun 77. Moved to Burbank
14 Aug 77. Sold to Ron Clark Enterprises Inc.
28 Oct 77.

<u>88</u> (79)   Series VI-R   f/f 26 Jun 61. N1003U
          United "Ville de Marseille", del 2 Jul 61.
Sold to Sterling as OY-SAH 26 Nov 71 and del 27 Nov 71
"City of Copenhagen". Lsd Egyptair 1 Jun 75 - 1 Dec 75.
Sold to Airborne Freight Corp for operation by Midwest
Air Charter   78, Regd N902MW and del 20 Dec 78.

<u>89</u> (80)   Series VI-R   f/f 28 Jun 61. N1004U
          United "Ville de Paris", del 6 Jul 61.
Sold to Sterling OY-SAL "City of Stockholm" 16 Dec 71
and del next day. Lsd Egyptair 1 May 75 - 30 Nov 75.
Proposed sale to Starline (77) fell through. Lsd to
Alisarda 17 Jun 77 - 27 Jul 77 and 25 Aug 77-19 Sep 77.
Sold to Midwest Air Charter 22 May 78 and r/o
Copenhagen 10 Jul 78 in their colours, and regd
N903MW. Del ex Copenhagen 31 Jul 78 as N903MW.

<u>90</u> (84)   Series VI-R   f/f 19 Jul 61. N1005U
          United "Ville de Grenoble", del 31 Jul 61.
Sold to Sterling as OY-SAP and del 30 Jan 72. Not
cvtd for Sterling service but lsd to Filipinas Orient
as PI-C970 and del 17 Oct 72. Re-regd RP-C970   75.
Painted in Transasian colours at Hong Kong   75.
but never operated. Stored Hong Kong until flown to
Copenhagen 11 May 76 and then scrapped Jan 77
(TT 17328 hrs).

<u>91</u> (85)   Series VI-R   f/f 5 Aug 61. N1006U
          United "Ville de Saintes", del 11 Aug 61.
Sold to Sterling as OY-SBV 3 Mar 72 but not taken up
and lsd to Transavia as PH-TVZ, del on 8 Mar 72
"Provincie Zeeland". To Sterling OY-SBV 2 Dec 72,
"City of Amsterdam". Sold to UTA 7 May 74 for
Mauritanian Government as 5T-CJW.. Re-regd early 76
as 5T-MAL "Zemmour" then to 5T-RIM 25 Apr 76.

<u>92</u> (86)   Series VI-R   f/f 22 Aug 61. N1007U
          United "Ville de Coutances", del 30 Aug 61.
Lsd to Transavia as PH-TRX "Provincie Limburg" and
del 26 May 70, purchased by them 26 Oct 70. Lsd
Itavia 1 May 75 - 25 May 75. Wfs 31 Oct 75 and to
Schiphol Fire Service (TT 26546 hrs). Finally
destroyed after fire practise Dec 78.

<u>93</u> (88)   Series VI-R   f/f 11 Sep 61. N1008U
          United "Ville de Rochefort", del 20 Sep 61.
Sold to Sterling as OY-SBW "City of Paris" and del
22 Mar 72. Lsd to Transavia as PH-TVT 26 Jun 73 -
10 Sep 73. Lsd to Transavia (as OY-SBW) 28 Jun 74 -
31 Jul 74 and again 8 Aug 74 - 17 Aug 74. Lsd to
Starline 19 Sep 77 to 9 Mar 78. Sold to Airborne
Freight Corp for operation by Midwest Air Charter,
17 Nov 78, regd N904MW and del 1 Dec 78.

<u>94</u> (89)   Series VI-R   f/f 29 Sep 61. N1009U
          United "Ville de Rouen", del 7 Oct 61.
Sold to Sterling as OY-SBY "City of Rome" and del
22 Mar 72. Sold to Minerva as F-BUZC and del 18 Oct 75.

<u>95</u> (91)   Series VI-R   f/f 10 Oct 61. N1010U
          United "Ville de Strasbourg", del 20 Oct 61.
Sold to Sterling as OY-SAM "City of Brussels" and
del 7 Jan 72. Sold to Airborne Freight Corp for
operation by Midwest Air Charter 22 May 78 and
regd N905MW and del 15 Oct 78.

<u>96</u> (92)   Series VI-R   f/f 23 Oct 61. N1011U
          United "Ville de Dijon", del 26 Oct 61.
Lsd to Transavia as PH-TRH "Provincie Noord-Holland"
and del 15 May 71. Purchased by them 15 Oct 71.
Wfs 22 Jun 74 after damaging wing at Schiphol.
Sold for scrap Dec 75 and broken up Jan 76.
(TT 22,494 hrs).

<u>97</u> (93)   Series VI-R   f/f 4 Nov 61. N1012U
          United "Ville de Lille", del 11 Nov 61.
Sold to BIAS as OO-CVA 10 Mar 71 "Stad Antwerpen/
Ville d'Anvers". To SOVATOUR (op by Air-Touring)
as F-BTON 29 Jun 72, "Ile de Beauté". Also operated
by Air Fret during 1972. Lsd to Catair Dec 73
until wfs at Le Bourget   75. Flown to Dinard
for storage May 76. Sold to CRB Logemat Oct 77.
To Toulouse for use as spare   -77 by Minerve.

<u>98</u> (94)   Series VI-R   f/f 15 Nov 61. N1013U
          United "Ville d'Arles", del 23 Nov 61.
Sold to Sterling 7 Jan 72 as OY-SAN "City of London",
del 11 Jan 72. Wfs at Stockholm-Arlanda 14 Jul 73
after wing struck light stanchion. Scrapped Oct 74
(TT 17407 hrs)

99 (95)   Series VI-R   f/f 21 Nov 61. N1014U
United "Ville de Nice", del 30 Nov 61. Sold
to Sterling as OY-SAK "City of Athens", del
9 Dec 71. Lsd Transavia 31 Jul 74 - 6 Aug 74. Sold
8 Feb 78 to Minerve as F-GAPA.

100 (96) Series VI-R   f/f 28 Nov 61. N1015U
United "Ville de St. Nazaire", del 8 Dec 61.
Sold to Transavia 16 Nov 70 as PH-TRS "Provincie
Groningen", del 3 Dec 70. Lsd Air Afrique 18 May 74 -
3 Sep 74. Sold Apr 75 and returned from overhaul
2 May 75 (from MEA, Beirut) without colour scheme.
Used by Transavia until wfs 14 Feb 76 and sold to
UTA at Le Bourget 24 Feb 76 for Tchad Government.
Painted as TT-AAD early 76 and re-regd TT-AAM
Apr 76 "El Baraka du 13 Avril 1975".

101 (97) Series VI-R   f/f 6 Dec 61. N1016U
United "Ville de Nantes", del 14 Dec 61.
Sold to Sterling 28 Jan 72 as OY-SAO "City of
Helsinki". Sold to Catair Dec 72 and del as
F-BUFF 3 Mar 73. Operated by ALBATROS on behalf
of Catair, and then wfu at Orly   78. Sold
to EAS for spares Dec 78 and broken up.

102 (98) Series VI-R   f/f 12 Dec 61.
N1017U United "Ville de Cannes",
del 29 Dec 61. Sold to Transavia 30 Nov 70
and del 1 Dec 70 PH-TRU "Provincie Utrecht".
Lsd SAM Apr 72 - Jul 72. Sold 31 Aug 76 to
Atlanta Skylarks but already del 8 Aug 76 as
N555SL, regd to Independent Air (an associate
company). Sold 1 Jun 78 to Kearney & Trecker
Corp, del 1 Jul 78 as N555SL. Re-regd N2296N
16 Aug 78, but believed ntu. Re-regd N98KT
5 Oct 78.

103 (99) Series VI-R   f/f 20 Dec 61 N1018U
United "Ville de Bordeaux", del 4 Jan 62.
Sold to Sterling as OY-SAR, del 11 Feb 72. Did
not enter Sterling service but lsd to Filipinas
Orient as PI-C969 from 15 Sep 72. Wfu   74,
and later stored at Hong Kong. Returned to
Manila and scrapped 23 Oct 76.

104(100) Series VI-R   f/f 28 Dec 61. N1019U
United "Ville de Lyon", del 6 Jan 62.
Sold to Sterling as OY-SAJ "City of Oslo", del
11 Feb 72. Last flight 17 Jan 77. (TT 21,645 hrs).
Wfs 21 Jan 77 and proposed for sale to Starline
but this did not materialise. Scrapped at
Kastrup Aug 77, the cockpit section being
preserved for public relations duties.

105(122) Series III   f/f 9 May 62. F-BJTI
Air France "Navarre", del 23 May 62.
Lsd to Kingdom of Libya AL 7 Oct 67 - 18 Nov 67,
Lsd to Air Charter Intl March 71 - Oct 71. Lsd
Royal Air Maroc Feb 73. Sold to Sud Aviation
4 Sep 73 for (Etat: SGAC, SFA, Melun-Villaroche.

106(114) Series VI-N   f/f 1 Apr 62. I-DABS
Alitalia "Dubhe", del 6 Apr 62. Wfs
at Rome 1 Oct 76.(TT 28386 hrs). Moved to
Venetia-Tesseria for storage 78.

107(104) Series VI-R F-WJAL f/f 25 Jan 62. EC-ARI
Iberia "Albeniz", del 5 Feb 62. Sold to Aviaco
4 Jun 73. Wfs Sep 75 and then sold to TAC
Colombia for spares and scrapped Madrid-Barajas
Jul 77. (TT 25,243 hrs).

108(109) Series VI-R F-WJAM f/f 19 Feb 62. EC-ARJ
Iberia "Chapi", del 28 Feb 62. Sold to
Templewood Aviation 7 Aug 73 and then to FEAT
via Commander Aircraft Sales, Eire (who reserved
marks EI-AVY-NTU) as B-2501 21 Aug 73. Wfu   77.

109(110) Series VI-R   f/f 12 Mar 62. EC-ARK
Iberia "Granados", del 30 Mar 62. Slight
damage in landing at Birmingham 22 Sep 67. Sold
to Aviaco 4 Jun 73 and wfs 1976. Sold to TAC
Colombia and del 20 Nov 76 as HK-1812X, re-regd
HK-1812.

110(115) Series VI-R   f/f 6 Apr 62. EC-ARL
Iberia "Manuel de Falla", del 18 Apr 62.
Sold to Templewood Aviation 7 Aug 73 and then to

FEAT via Commander Aircraft Sales, Eire (who reserved
EI-ATR NTU) as B-2503 21 Aug 73. Lsd to Air Vietnam
May 74 - Jul 74.

111(101) Series III   f/f 28 Dec 61. F-BJTE Air
France "Grenoble", del 12 Jan 62. Later modified
to Series III/48T. Lsd Air Charter Intl 1 May 71 - 1 Apr 73.

112(106) Series III   f/f 8 Feb 62. SE-DAF SAS
"Sven Viking", del 17 Feb 62. Lsd to Thai
International during 1963. Wfs 23 Sep 74 and stored at
Stockholm-Arlanda for Swedish Air Museum. (TT 27321 hrs).
Nose reported to be at school in Dragor, nr Copenhagen.

113(107) Series III   f/f 22 Feb 62. F-BJTF Air
France "Orléanais", del 7 Mar 62. Lsd Royal Air
Maroc. Later modified to Series III/48T. Lsd Air Charter
Intl 1 Jul 71 - 1 May 73.

114(103) Series VI-R   f/f 16 Feb 62. N1020U
United "Ville de Calais", del 24 Feb 62. Sold
to Sterling as OY-SBZ "City of Reykjavik", del 3 Mar 72.
Sold to UTA for Gabon Government as TR-LWD named "Ivindo"
21 Jan 76.and del 15 Jun 76. Transferred to Air Gabon
1 Jun 77.

115(118) Series III   f/f 19 Apr 62. F-BJTG Air
France "Roussillon", del 4 May 62. Later modified
to Series III/48T. To Air Charter Intl 1 Sep 70.

116(108) Series III   f/f 19 Feb 62. OH-LED Finnair
"Sinipiika", del 23 Feb 62. Retd to Sud Aviation
19 Oct 64 and del to CEV: Bretigny as F-ZACE 1 Dec 64.

117(124) Series VI-R F-WJAN f/f 9 Jul 62. CS-TCA TAP
"Goa", del 13 Jul 62. Sold to SAN as HC-BAJ
"Santa Maria" 23 Nov 75.

118(117) Series VI-R F-WJAO f/f 15 Jun 62. PP-PDU Panair
"Antao Leme da Silva", del 19 Jul 62. Written off
on approach to Recife 6 Sep 63 due to overstressing avoiding
mid-air collision with light aircraft. Flown to Rio and
then grounded. Dismantled and auctioned 28 Apr 69 for
scrap. (TT 2422 hrs).

119(123) Series III   f/f 5 Jun 62. F-BJTJ Air
France "Bourbonnais", del 18 Jun 62. Lsd
Swissair 1 Apr 64 - 29 Mar 66 as HB-ICR. Later modified
to Series III/48T. Retd to Air France, to Air Charter Intl
1 Sep 70. Lsd briefly to Radio Europe 1 Mar 77.

120(120) Series VI-R F-WJAK f/f 25 Jun 62. PP-PDV
"Domingos Rodrigues de Carvalho", del 19 Jul 62.
Lost door in flight 31 Oct 63 en route Brasilia-Rio
and grounded for inspection. Retd to Sud Aviation - 64,
and when Panair ceased operations in Feb 65, aircraft
became property of Brazilian Govt. Retd to Brazil for lse
to Cruzeiro   1.67. Skidded off wet runway at Manaus
23 Dec 73 and burnt out - no fatalities. (TT 20740 hrs).

121(111) Series III   f/f 10 Mar 62. HB-ICS Swissair
(ordered by SAS) "Uri", del 17 Mar 62. Sold to
China A/L as B-1850 and del 10 Apr 71. Wfu at Taipei.

122(112) Series III   f/f 20 Mar 62. HB-ICT Swissair
(ordered by SAS) "Schwyz", del 29 Mar 62.
Damaged at Zürich 25 Apr 62. Sold to China A/L as B-1852
and del 12 Jan 71. Crashed into Formosa Strait nr
Penghu Island 20 Nov 71 after inflight explosion en route
Osaka-Taipei. 25 killed - believed sabotage (TT 26099 hrs).

123(116) Series III   f/f 12 Apr 62. HB-ICU Swissair
(ordered by SAS) "Aargau" del 19 Apr 62. Sold
to N.V. Huygen & Co. 20 Oct 70 (owner of TEN-BEL Touring)
for operations between Tenerife and Belgium. Due to
difficulties in obtaining landing rights in Spain aircraft
parked at Zürich 10 Nov 70 - May 71, as HB-ICU named
"Islas Canarias". Sobelair took over operations for
TEN BEL May 71 and re-regd OO-SBQ Jul 71. Sold to
Aérotechnique Internationale for Catair, regd F-BUFH, del
1 Jun 74. Lsd Aerotour from May 76 and later sold to them.

124(119) Series III   f/f 26 Apr 62. F-BJTH Air
France "Franche-Comté", del 10 May 62. Later
modified to Series III/48T. To Air Charter Intl 1 Sep 70.

125(126) Series VI-R   f/f 20 Jul 62. CS-TCB TAP
"Damao", del 24 Jul 62. Sold to SAN as HC-BAT
"La Pinta" 27 Nov 75.

126 (127) Series VI-R F-WJAN f/f 18 Jul 62 PP-PDJA Panair
"Francisco Camargo", del 17 Aug 62. Wfs on demise
of Panair   Feb 65, then lsd Cruzeiro 25 Mar 66 from
Brazilian Govt. Crashed on landing at Sao Luiz.
Maranhao 1 Jun 73, 23 killed. (TT 26382 hrs).

127 (121) Series VI-N   f/f 29 May 62. LV-PVT
Aerolineas Argentinas "Sirius", del 7 Jun 62.
Re-regd LV-HGY after del. Crashed at Cordoba-Pajas
Blanca 3 Jul 63, on landing - 4 on ground killed, no
fatalities on board. (TT 1909 hrs).

128 (158) Series 10A laid down for TWA, completed as
Series VI-N, F-WLKJ f/f 10 Dec 63. VT-DPO
Indian AL "Pavandoot", del 21 Dec 63. Written-off
at Bombay 3 Jul 73 when nosewheel collapsed on
landing and caught fire. No casualties. Subsequently
scrapped. (TT 25952 hrs).

129 (132) Series VI-R   f/f 26 Nov 62. PP-CJA
Cruzeiro, del 10 Dec 62, arr. Rio
23 Dec 62. Wfu 30 Jun 75 and stored Porto Alegre.

130 (160) Series 10A laid down for TWA, completed as
Series VI-N f/f 6 Jan 64 VT-DPP "Akashdoot"
del 17 Jan 64. Crashed on landing at Delhi-Palam
15 Feb 66. 2 Killed. (TT 5388 hrs).

131 (128) Series VI-R F-WJAO f/f 2 Aug 62. PP-PDZ
Panair "Francisco Dias de Avila".
del 16 Sep 62. Wfs on demise of Panair Feb 65, and
lsd to Cruzeiro from Govt. 25 Mar 66. Wfu 30 Jul 75
and stored at Porto Alegre, until sold to Aerotal,
del Nov 78.as HK-2212X, later HK-2212 "Agualongo".

132 (169) Series 10A laid down for TWA, completed as
Series VI-N f/f 10 Apr 64 I-DABL
Alitalia "Fomalhaut" del 15 Apr 64. To SAM 1 Apr 69
and back to Alitalia   74. Wfs at Rome 2 Dec 75.

133 (134) Series VI-R   f/f 7 Dec 62. PP-CJB
Cruzeiro del 13 Dec 62, arr. Rio 23 Dec 62.
Wfu 16 Jun 75 and stored at Porto Alegre. Sold to
Aerotal Mar 78. Del May 78 as HK-1709X for spares
use only at Bogota.

134 (179) Series 10A laid down for TWA, completed
as Series VI-N F-WLGA f/f 13 Nov 64.
VT-DSB Indian AL "Vayudoot", del 28 Nov 64.
Crashed at Bombay 4 Sep 66, 4 crew killed.
(TT 4901 hrs).

135 (137) Series 10A laid down for TWA, completed
as Series VI-N F-WJAK f/f 24 Jul 63.
YU-AHB JAT "Bled", del 27 Feb 63. Wfu   76,
and stored at Belgrade.

136 (159) Series 10A laid down for TWA, completed
as Series VI-R F-WLKI f/f 20 Jan 64. To
Sud Aviation for Sud-Lear autoland trials. Re-regd
F-BLKI 23 Jan 64. Lsd to Austrian A/L OE-LCU
"Steiermark", del 13 May 66. Damaged 21 Feb 70 by
in-flight explosion en route to Frankfurt. Wfs
14 Jan 72 and del to Euralair 15 Jan 72. Re-regd
F-BTDL 7 Mar 72. To"Société Civile de la Caravelle"
1977.

137 (129) Series VI-R   f/f 19 Nov 62. CS-TCC
TAP "Diu", del 27 Nov 62. Sold to SAN
18 Nov 75 and del as CS-TCC "La Nina". Not put into
service stored at Quito. Re-regd HC-BFN Mar 79.

138 (154) Series VI-R   f/f 25 Jul 63. N210G
Garrett Corp, del 31 Jul 63 but not flown
to USA and stored at Toulouse. Sold to Iberia
EC-AXU "Alfonso X el Sabio", del 7 Jan 65. Lsd
Aviaco May 73, and sold to them 4 Jun 73. Sold
to TAC and del as HK-1811-X 14 Feb 76. Re-regd
HK-1811,and stored at Bogota without entering
service.

139 (135) Series VI-N   f/f 5 Jan 63. YU-AHA
JAT "Dubrovnik", del 11 Jan 63. Wfu
76 and stored at Belgrade. Sold to Aerotour
78 as F-BYAI, and regd to them 8 Feb 78.

140 (164) Series 10A laid down for TWA, completed
as Series VI-R F-WJAQ f/f 29 Feb 64.
CC-CCO LAN Chile F/N 501 del 6 Mar 64. Sold to
Aerotal as HK-1778 "El Motilon".

141 (136) Series III   f/f 1 Feb 63. F-BJTK Air
France "Principaute de Monaco", Not
delivered despite CofA allocation 6 Feb 63. To GLAM/
Armée de l'Air Jan 60 with c/s F-RAFG, del 13 Sep 63.

142 (139) Series III   f/f 22 Feb 63. F-BJTL Air
France "Aunis et Saintonge", del 15 Mar 63.
Later modified to Series III/48T. To Air Charter Intl
72 -   Returned to Air France

143 (131) Series VI-N F-WJSO f/f 20 Dec 62. Sud
Aviation for Smith autoland trials. Re-regd
F-BJSO 29 Dec 62. Sold to Alitalia I-DABM
"Procione", del 27 May 64. Chartered frequently by SAM.
Wfs at Rome 18 Jan 77. (TT 24427 hrs). Moved to
Venetia-Tesseria for storage 78.

144 (143) Series III   f/f 13 Mar 63. F-BJTM Air
France "Maine", del 9 Apr 63. Later modifed
to Series III/48T. Sold to Air Burundi 20 May 75 as
9U-BTA "Mutongati".

145 (144) Series III   f/f 5 Apr 63. F-BJTN Air
France "Marche", del 18 Apr 63. Later
named "Comminges". To Royal Air Cambodge 16 Jan 69
XU-JTA "Angkorwat". Company title changed to Air
Cambodge 1970. Destroyed by Communist attack on
Phnom-Penh 22 Jan 71.

146 (140) Series VI-N   f/f 5 Mar 63. I-DABV
Alitalia "Acruz", del 14 Mar 63. To SAM
1 Nov 68. Sold to SAETA and del 25 Jan 77 as I-DABV.
Stored at Quito, and then regd HC-BDS   77.

147 (130) Series III   f/f 2 Oct 62. HB-ICV
Swissair (ordered by SAS) "Schaffhausen",
del 19 Oct 62. Crashed at Dürrenäsch 4 Sep 63
following in-flight fire en route Zürich-Geneve-
Rome, 80 killed (TT 2283 hrs)

148 (146) Series III   f/f 20 May 63. F-BJTO
Air France "Nivernais", del 30 May 63.
Renamed "Pays Basque". Later modified to Series III/48T.
To Air Charter Intl 1 Sep 70.

149 (133) Series VI-N   f/f 25 Oct 62. LV-PVU
Aerolineas Argentinas "Rigel", del 31 Oct
62. Re-regd LV-HGZ after del. To Fuerza Aerea
Argentina Jul 73 as T-92. Sold 12 May 75 via Fokker
to International Air (a broker) as N46SB and del to
Lawton, OK Jul 75 for long term storage. Transferred
to Nevada Air Tours Inc. 28 Feb 78. Sold to EAS as
F-GBMJ and del 19 Dec 78 "Valentinois".

150 (141) Series VI-N   f/f 23 Mar 63. I-DABW
Alitalia "Betelgeuse" del 1 Apr 63. To
SAM 1 Apr 69 - Nov 72. Wfs 28 Sep 76. Sold to SAETA
and del 22 Nov 76 as I-DABW. Stored at Quito.

151 (149) Series VI-N   f/f 16 May 63. YU-AHD
JAT"Opatija", del 31 May 63. Crashed at
Moganik 11 Sep 73 in violent storm on approach to
Titograd, 44 killed. (TT 23373 hrs).

152 (148) Series III   f/f 6 May 63. F-BJTP Air
France "Comtat Venaissin", del 16 May 63.
Later modified to Series III/48T.

153 (145) Series VI-N F-WJAL f/f 19 Apr 63. OD-AEE
MEA del 26 Apr 63. Destroyed in Israeli
commando attack on Beirut 28 Dec 68. (TT 14657 hrs).

154 (151) Series III   f/f 24 May 63. CN-CCY
Royal Air Maroc del 1 Jun 63. Lsd MEA
23 Jan 69 - Jul 69. Later named "Ilfrane". Wfu
at Casablanca 77.

155 (157) Series VI-N   f/f 15 Nov 63.
VT-DPN Indian AL "Gagandoot", del 26 Nov 63.
Damaged in collision with airport coach at Bombay
23 Jan 73. Wfs 29 Oct 76. (TT 33794 hrs).

156 (150) Series VI-R F-WJAN f/f 20 May 63. OE-LCE
AUA "Tyrol", del 14 Jun 63. Wfs 24 Apr 72.
and sold to Luxair LX-LGG "Princesse Marie-Astrid",
5 May 72, del ex Vienna 2 May 72. Wfu 1 May 78,
sold to SAN and del ex Luxembourg 16 Jun 78 to TAP
in Lisbon for modification.

<u>157</u> (147) Series VI-N   f/f 20 Apr 63. OD-AEF MEA
          del 6 May 63. Destroyed in Israeli commando
attack on Beirut 28 Dec 68. (TT 14286 hrs).

<u>158</u> (155) Series VI-R built for VASP. Order cancelled and
          completed for Sud Aviation, possibly as F-WLHY
f/f 27 Sep 63. Re-regd 7 Oct 63 as F-BLHY. Lsd to GLAM
for tour of South America, Aug - Oct 64, with c/s F-RAFA.
Retd to F-BLHY. Sold to Kingdom of Libya AL as 5A-DAA,
del 24 Jul 65. Company name changed to Libyan Arab A/L
1 Sep 69. Wfu   76 and stored at Tripoli.

<u>159</u> (153) Series VI-R   f/f 30 Jan 64. EC-AVZ
          Iberia "Sarasate", del 7 Feb 64. Wfs 5 Aug 72
and scrapped at Madrid-Barajas Dec 73. (TT 17,765 hrs).

<u>160</u> (173) Series VI-R   f/f 25 Jun 64. CC-CCQ LAN-
          Chile f/n 503 del 2 Jul 64. Sold to Aerotal
as HK-1780   75. Wfu   77

<u>161</u> (138) Series VI-R F-WJAL f/f 6 Feb 63. OE-LCA AUA
          "Wien", del 18 Feb 63. Wfs 28 Aug 72 and
sold to Catair 16 Jan 73, as F-BUFC, del 22 Jan 73.
Operated by ALBATROS for Catair in 78 until sold to
Euralair 78.

<u>162</u> (156) Series VI-R ordered by VASP. Then intended
          for Cruzeiro as PP-CJC(1) but NTU f/f
23 Apr 64 and lsd to Finnair as OH-LER 12 May 64 -
2 Sep 64. Regd as F-BJTD(2) to Sud Aviation 18 Sep 64.
Sold to Kingdom of Libya A/L as 5A-DAB, del 28 Jul 65.
Company name changed to Libyan Arab A/L 1 Sep 69.
Wfu   .76 and stored at Tripoli.

<u>163</u> (163) Series VI-R   f/f 25 Jun 63. EC-ATV
          Iberia "Tomas Luiz de Victoria", then
"Maestro Victoria", del 9 Jul 63. Crashed in
Atalayasa mntns,nr San Jose, Ibiza 7 Jan 72 on
route Madrid-Ibiza, 104 killed. (TT 18,427 hrs).

<u>164</u> (170) Series VI-R   f/f 18 Apr 64. CC-CCP
          LAN-Chile F/N 502, del 4 May 64. Sold
to Aerotal as HK-1779   .75.

<u>165</u> (142) Series VI-R   f/f 16 Apr 63. EC-ATX
          Iberia "Turina", del 25 Apr 63. Lsd Aviaco
May 73, then sold to them 4 Jun 73. Sold to TAC
HK-1810 Feb 76.

<u>166</u> (166) Series VI-R   f/f 20 Mar 64. OE-LCI
          AUA "Salzburg", del 28 Mar 64. Wfs 31 Dec 71
and lsd to Luxair as LX-LGF 4 Mar 72. Purchased by
them 2 Feb 73. Wfu 1 May 78, sold to SAN and del
ex Luxembourg 17 Jun 78 to TAP in Lisbon for
modifications.

<u>167</u> (188) Series VI-R F-WJAQ f/f 13 Apr 65. OE-LCO AUA
          "Karnten", del 22 Apr 65. Wfs 27 Nov 71,
sold same day to Euralair and del 28 Nov 71. Re-regd
F-BSEL 19 Jan 72.

<u>168</u> (152) Series VI-R   f/f 28 Jun 63 PP-CJD
          Cruzeiro del 8 Jul 63, arr. Rio 14 Jul 63.
Wfu 30 Oct 75 and stored at Porto Alegre.

<u>169</u> (162) Series 10B3 prototype F-WLKJ f/f 3 Mar 64
          in Finnair colours. Re-regd F-BLKJ to
Sud 3 Jul 64. Sold to Finnair OH-LSG "Jyvaskyla",
del 31 May 66, as 10B3/56T.

<u>170</u> (165) Series III   f/f 26 Feb 64. OY-KRF
          SAS "Torkil Viking", del 3 Mar 64. Lsd
to Thai (as OY-KRF) 2 Mar 68 - 30 Apr 68, 10 Mar 69 -
16 Apr 69 and 17 Dec 69 - 24 Feb 70. Wfs 18 Jan 74
and del to Catair F-BUOE 5 Feb 74. Not put in
service and sold to China A/L B-1856.   74.
Wfu   77.

<u>171</u> (196) Series VI-R F-WLKR f/f 28 Oct 65. EC-BBR
          Iberia "Padilla", del 6 Nov 65. Damaged
by fire in hangar at Madrid-Barajas 29 Sep 73, and
scrapped there Dec 73. (TT 16,293 hrs).

<u>172</u> (182) Series III   f/f 12 Dec 64. SE-DAG
          SAS "Dag Viking", del 17 Dec 64. Sold to
Flygvapen 30 Nov 70, then via Forsvaret Materielverk
1 Mar 71 and to Flygvapen 4 May 71. Extensive
modifications at Malmo and r/o with serial FV 85172
6 Sep 72. Used by Forsokscentralen with code '17',
to F13 as '81'   74.

<u>173</u> (161) Series VI-R   f/f 13 Jan 64. EC-AVY
          Iberia "Amadeo Vives", del 24 Jan 64. Wfs
2 Oct 72 and scrapped at Madrid-Barajas Dec 73.
(TT 18,367 hrs).

<u>174</u> (178) Series VI-N   f/f 4 Nov 64. OD-AEO MEA
          del 10 Nov 64. Retd to SNIAS Mar 75 F-WJAN.
Derelict at Toulouse by Mar 76.

<u>175</u> (181) Series VI-N F-WJAL f/f 22 Dec 64. OO-SRI
          Sabena del 29 Dec 64. To Sobelair 1 Oct 73.
Lsd Air Inter 23 Jun 75 - 30 Sep 75. Sold to Catair
F-BYCB 26 Apr 76. Operated by ALBATROS on behalf of
Catair in 78 until transferred to Europe Air Service
78. Retd to Sobelair as OO-SRI   78. Sold to
Minerve as F-GATZ Dec 78.

<u>176</u> (183) Series 10B1R prototype F-WLKS f/f 18 Jan 65.
          Re-regd F-BLKS 22 Apr 65. Lsd to Iberia as
EC-BDC "Hilarion Eslava", del 1 Apr 66. On completion
of lease 1 Apr 72 sold to Aviaco and re-regd EC-CAE
13 Apr 72.

<u>177</u> (168) Series III   f/f 26 Mar 64. F-BJTQ Air
          France "Champagne", del 3 Apr 64. Later
modified to Series III/48T. Renamed "Martinique" for
Caribbean services   74 - 76, and this name still
carried in European use.

<u>178</u> (167) Series III   f/f 12 Mar 64. TS-TAR
          Tunis Air, del 18 Mar 64. Wfu at Tunis   77.

<u>179</u> (187) Series VI-N   f/f 3 Feb 65. I-DABF
          Alitalia "Mizar", del 19 Feb 65. Overran
runway on landing at Marseilles-Marignane 2 Aug 69
and went into l'Etang de Berre, 55m from shore.
No casualties, aircraft recovered 8 Aug 69 and sold
as scrap. (TT 11885 hrs).

<u>180</u> (177) Series VI-N   f/f 3 Sep 64. LV-PBJ
          Aerolineas Argentinas "Antares", del
18 Sep 64. Re-regd LV-III after del. To Fuerza Aerea
Argentina Jul 73 as T-93. Sold 12 May 75 by Fokker
to International Air (a broker) as N49SB and del to
Lawton, OK Jul 75 for long term storage. Transferred to
Nevada Air Tours Inc 28 Feb 78. Sold to EAS as F-GBMK
and del 4 Feb 79.

<u>181</u> (171) Series 10B3/52T   f/f 11 Jul 64. OH-LSA
          Finnair "Helsinki", del 22 Jul 64. Carried
Kar-Air sticker briefly in   70. Modified to
10B3/54T 1978.

<u>182</u> (172) Series 10B3/52T   f/f 30 Jul 64. OH-LSB
          Finnair "Tampere", del 4 Aug 64. Modified
to 10B3/53T in 1978.

<u>183</u> (180) Series 10B3/52T   f/f 30 Dec 64. OY-STA
          Sterling del 30 Mar 65. Sold to Syrian Arab
A/L 6 Jun 71 as YK-AFC and del 19 Jun 71. Company
name changed to Syrianair   77.

<u>184</u> (194) Series 10B3   f/f 12 Oct 65. YK-AFA
          Syrian Arab A/L "8 Mars", del 22 Oct 65.
Company name changed to Syrianair   77.

<u>185</u> (174) Series 10B3/52T   f/f 21 Aug 64. OH-LSC
          Finnair "Turku", del 27 Aug 64. Modified
to 10B3/53T in 1978.

<u>186</u> (206) Series 10B3/52T   f/f 5 Apr 66. OY-STB.
          Sterling del 8 Apr 66. Sold to Syrian Arab
A/L YK-AFD and del 20 Nov 71. Company name changed to
Syrianair 77.

<u>187</u> (175) Series 10B3/52T   f/f 5 Sep 64. OH-LSD
          Finnair "Oulu", del 10 Sep 64. Modified to
10B3/53T in 1978.

<u>188</u> (203) Series 10B3/52T   f/f 7 Mar 66. OH-LSF
          Finnair "Pori", del 12 Mar 66. Modified to
10B3/53T in 1978.

<u>189</u> (176) Series 10B3/52T   f/f 14 Sep 64. OH-LSE
          Finnair "Lahti", del 23 Sep 64. Modified
to 10B3/53T in 1978.

<u>190</u> (209) Series 10B3   f/f 23 Jun 66. YK-AFB
          Syrian Arab A/L "17 Avril", del 30 Jun 66.
Company name changed to Syrianair 77.

Caravelle VI-N YU-AHA (c/n 139) of JAT (Sud-Aviation via J Wegg)

Top   : Caravelle III 7T-VAL (c/n 75) of Air Algerie
         at Brussels (NGJ Roozen)

Bottom : Caravelle III 9U-BTA (c/n 144) of Air Burundi
         at Nairobi in 1976 (PR Keating)

Top   : Caravelle VI-R RP-C970 (c/n 90) of Transasian
         at Copenhagen in August 1976 (G Boccheni)

Bottom : Caravelle VI-R HK-1811 (c/n 138) of TAC
         at Bogota in October 1976 (N Oertel)

Top   :   Caravelle III No.141 of the Armee de l'Air
            at Shannon in October 1976 (GM Nason)

Bottom :   Caravelle III 81 (c/n 172) of the Swedish Air Force
            at Arlanda in February 1977 (T Lakmaker)

<u>191</u> (185) Series III   f/f 8 Feb 65. OY-KRG SAS
            (leased from Sud) "Alf Viking", del 16 Feb 65.
Retd 17 Feb 69 and lsd to Transavia as PH-TRN 4 Apr 69 -
15 Oct 70. Lsd to JAT as YU-AJG May 72 until damaged on
landing accident at Belgrade Feb 73. Wfu and used as a
cabin crew trainer. (TT 16717 hrs).

<u>192</u> (191) Series VI-N   f/f 5 Apr 65. I-DABP
            Alitalia "Castore". del 13 Apr 65. To SAM
    71 -   72. Wfs 14 Jan 77. Sold to
Aérotour as F-BYAU and del 15 Apr 77. Written-off
after undercarriage collapse at Oujda, Morocco 9 Dec 77.
(TT 22,546 hrs).

<u>193</u> (186) Series III   f/f 3 Mar 65. SE-DAH SAS
            (leased from Sud) "Torgny Viking, del
12 Mar 65. Retd 12 Mar 69. Regd F-BRIM 30 Apr 69
for lease to Transunion. Sub-leased to Panair
13 May 69 - Jul 69. Retd to SNIAS and del to SOGERMA
28 Dec 71, for conversion as flying test-bed for
SNECMA M53. Received c/s F-ZACF after CofA expiry
May 72 and f/f with M53 18 Jul 73. To SNECMA 20 Sep 73
- 23 Jul 76. Returned to SOGERMA for conversion as
test-bed for CFM56 engine and f/f as such 17 Mar 77.

<u>194</u> (193) Series VI-N   f/f 5 Jul 65. YU-AHE
            JAT "Budva", del 9 Jul 65. Wfu
76 and stored at Belgrade, later scrapped.

<u>195</u> (190) Series III   f/f 16 Jun 65. CN-CCZ
            Royal Air Maroc del 25 Jun 65. Later named
"Zagora". Wfu at Casablanca 77.

<u>196</u> (192) Series VI-N F-WJAL f/f 23 Apr 65. OO-SRK
            Sabena del 29 Apr 65. Sold to Aérotechnique
Internationale for Catair, 5 Mar 75; re-regd F-BVPU
12 Mar 75. Lsd and then sold to Aerotour May 75.

<u>197</u> (184) Series VI-R   f/f 19 Feb 65. EC-AYD
            Iberia "Juan Cristosomo Arriaga", del
26 Feb 65. Lsd to Aviaco 1973. Sold to FEAT as
B-2505 14 Jun 74.

<u>198</u> (189) Series VI-R   f/f 25 Mar 65. EC-AYE
            Iberia "Jose Maria Usandizaga", del
2 Apr 65. Wfs 30 Dec 73. and scrapped at Madrid-Barajas.
(TT 18,052 hrs).

<u>199</u> (195) Series 10BIR   f/f 22 Jul 65. JY-ACS
            Alia "Amman", del 28 Jul 65. Sold via UTA
to Air Afrique as TU-TCN "Nouakchott", del 13 Dec 73.
Sold to Trans Europa 78 as EC-DCN.

<u>200</u> (197) Series 10B1R F-BNFE f/f 6 Dec 65. JY-ACT
            Alia "Jerusalem", del 25 Feb 66. Sold to
SATA 21 Mar 73 as HB-ICK. Lsd to CICR (Comité
International de la Croix Rouge - i.e. International
Red Cross) 23 Jul 76 - 4 Aug 76. Retd to SATA and
crashed on approach to Madeira-Funchal 18 Dec 77,
36 killed.

<u>201</u> (201) Series 10B1R   f/f 21 Jan 66. F-BNRA
            UTA del 26 Jan 66. Lsd Air Afrique as
TU-TXQ 19 Sep 75 - Apr 76. Restored as F-BNRA but
CofA suspended 28 Jul 76. Lsd Air Afrique as TU-TXQ
again 7 Oct 76 - 18 Jan 77 then wfs by UTA.
(TT 9605 hrs). Sold to l'Armée de l'Air for GLAM
with c/s F-RAFH and del Dec 77.

<u>202</u> (204) Series 10B1R   f/f 20 Apr 66. EC-BDD
            Iberia "Jesus Gurudi", del 29 Apr 66.
Crashed on Black Down, Fernhurst, Sussex 4 Nov 67
during descent to London-Heathrow, 37 killed.
(TT 3,638 hrs).

<u>203</u> (199) Series VI-N F-WJAL f/f 8 Nov 65. VT-DUH
            Indian A/L "Rashiradoot", del 28 Dec 65.
Name deleted 75. To Pushpaka 78.

<u>204</u> (200) Series VI-N F-WJAQ f/f 18 Jan 66. VT-DUI
            Indian A/L "Bharatdoot", del 26 Jan 66.
Name deleted 75. To Pushpaka 78.

<u>205</u> (208) Series VI-N   f/f 17 May 66. I-DABG
            Alitalia "Arturo", del 27 May 66. Operated
for SAM frequently May 71 - 74. Wfs 28 Feb 77.
Sold to Aerotour as F-BYAT and del 18 Mar 77.

<u>206</u> (210) Series III   f/f 22 Jun 66. F-BNKA
            Air France order but lsd immediately to
Air Inter, del 28 Jun 66, and purchased by them

23 Jan 75.

<u>207</u> (205) Series III   f/f 10 Mar 66. TS-MAC Tunis
            Air del 17 Mar 66. Wfu 77.

<u>208</u> (207) Series III   f/f 8 Apr 66. F-BNKB Air
            France order but lsd immediately to Air Inter,
del 15 Apr 66, and purchased by them 23 Jan 75.

<u>209</u> (198) Series III   f/f 6 Dec 65. LN-KLN SAS
            "Roald Viking" (leased from Sud), later
"Trygve Viking", del 17 Dec 65. Retd to SNIAS 16 Dec 69.
Lsd Transunion as F-BRUJ 15 Mar 70. Sub-leased Air
Mali Jan 71 - Mar 71, Tunis Air Apr 71 - Jul 71. Retd
Sep 71 to SNIAS and lsd Inex-Adria as YU-AJE May 72 -
Nov 72. Re-regd F-BUFM to SNIAS 21 Mar 73 and lsd to
Tunis Air 1 Apr 73 -   73. Sold to Govt
of Rwanda Mar 74 as 9XR-CH.

<u>210</u> (202) Series III   f/f 1 Feb 66. SE-DAI SAS
            "Alrik Viking", del 10 Feb 66. Sold to
Flygvapen 30 Nov 70, then to Förvarets Materielverk
20 Sep 71 for del to Flygvapen 27 Sep 71 as FV 85210
coded '21', To F13 code '82' Feb 74.

<u>211</u> (212) Series 10B3/52T   f/f 13 Jan 67. OH-LSH
            Finnair "Kuopio", del 20 Jan 67. Modified to
10B3/53T in 1978.

<u>212</u> (211) Series 10B3/52T F-WJ f/f 25 Aug 66. OY-STC
            Sterling. Not delivered and regd 29 Aug 66
to Sud as F-BOEE for lease to UTA 31 Aug 66 - 12 Mar 67.
Del to Sterling OY-STC 30 Mar 67. Lsd Finnair as OH-LSK
8 Jan 75 (delivered 17 Dec 74) - 12 Mar 76. Leased to
TAE and del from Finnair lease 13 Mar 76 and re-regd
EC-CUM, Scheduled for return 31.3.79.

<u>213</u> (213) Series VI-N   f/f 19 Oct 66. VT-DVI
            Indian A/L "Meghdoot", del 23 Oct 66. Name
deleted 75.

<u>214</u> (214) Series III   f/f 28 Nov 66. D-ABAM LTU
            (lsd from Sud Aviation) del 6 Dec 66. Retd
29 Dec 67 for sale to Air Inter as F-BNKI, del 5 Jun 68.
Destroyed by fire on ground at Orly 4 Jan 71 - cause of
fire undiscovered. (TT 9350 hrs).

<u>215</u> (220) Series 11R F-WJAL f/f 21 Apr 67. TU-TCO
            Air Afrique "Ouagadougou", del 17 Jul 67.

<u>216</u> (215) Series VI-N   f/f 21 Dec 66. VT-DVJ
            Indian A/L "Hansadoot", del 26 Dec 66. Name
deleted   75. Written off at Bombay 17 Jun 75 when
overran runway, no casualties. (TT 23542 hrs).

<u>217</u> (216) Series III   f/f 15 Feb 67. F-BNKC Air
            Inter del 15 Mar 67.

<u>218</u> (223) Series VI-N   f f/f 23 May 67. YU-AHF JAT
            "Split", del 29 May 67. Wfu   76
and stored at Belgrade. Sold to Aérocentre Feb 78
(for Aérotour) as F-BVPZ. Del 31 Aug 78.

<u>219</u> (226) Series 11R F-WJAK f/f 1 Jun 67. Re-regd
            F-BJAK for Le Bourget Salon Jun 67. TU-TCY
Air Afrique "Yaounde", del 17 Jul 67.

<u>220</u> (217) Series III   f/f 18 Feb 67. F-BNKD
            Air Inter del 25 Feb 67.

<u>221</u> (221) Series VI-R   f/f 22 Apr 67. 5A-DAE
            Kingdom of Libya A/L, del 29 Apr 67. Company
name changed to Libyan Arab A/L 1 Sep 69. Wfu
76. and stored at Benghazi, then at Tripoli.

<u>222</u> (218) Series 10B1R   f/f 1 Mar 67. F-BNRB UTA
            del 7 Mar 67. Sold to SATA Nov 72 and del
5 Dec 72. Re-regd HB-ICQ 15 Jan 73.

<u>223</u> (222) Series 10B1R F-WJAQ f/f 30 May 67. EC-BIB
            Iberia "Teobaldo Power", del 7 Jun 67.
Sold to Aviaco 17 May 72. Sold to TransEuropa.

<u>224</u> (219) Series III   f/f 25 Mar 67. F-BNKE Air
            Inter del 31 Mar 67.

<u>225</u> (224) Series 10B1R   f/f 19 Jun 67. EC-BIC
            Iberia "Emilio Arrieta", del 27 Jun 67.
Lsd to Aviaco 19 May 72. Crashed on approach to
Coruna-Alvedro 13 Aug 73, 85 killed plus 5 on ground.
(TT 10,796 hrs).

226 (228)   Series VI-R   f/f 24 Jul 67. EC-BIA
            Iberia "Padre Antonio Soler", del 28 Jul 67.
Caught fire 5 Nov 73 on ground at Madrid-Barajas and
written-off. Scrapped Dec 73. (TT 11,874 hrs).

227 (232)   Series III   f/f 16 Nov 67. F-BNKF
            Air Inter del 25 Nov 67.

228 (225)   Series 10B1R   f/f 29 Jun 67. EC-BID
            Iberia "Tomas Breton", del 6 Jul 67. Sold
to Aviaco 19 May 72. Crashed into sea on approach
to Madeira-Funchal 5 Mar 73. 3 crew killed.
(TT 9,875 hrs).

229 (233)   Series III   f/f 30 Nov 67. F-BNKG
            Air Inter del 2 Jan 68.

230 (227)   Series 10B1R   f/f 12 Jul 67. EC-BIE
            Iberia "Jeronimo Jimenz", del 20 Jul 67.
Sold to Aviaco 22 May 72.

231 (231)   Series VI-N F-WLGB f/f 20 Oct 67. VT-DWN
            Indian A/L "Shantidoot", del 2 Nov 67.
Name deleted 75. Crashed at Bombay 12 Oct 76
attempting landing after engine fire immediately
after take-off. 94 killed plus 1 on ground.
(TT 24374 hrs).

232 (229)   Series 10B1R   f/f 21 Sep 67. EC-BIF
            Iberia "Francisco Tarrega", del 29 Sep 67.
Sold to Aviaco 29 May 72.

233 (234)   Series VI-N   f/f 26 Dec 67. YU-AHG
            JAT "Ohrid", del 11 Jan 68. Wfu 76
and stored at Belgrade. Sold to Aérocentre Feb 78
(for Aérotour) as F-BYCY.

234 (247)   Series VI-R F-WJ   f/f 13 Jun 68. Ordered
            by Kingdom of Libya A/L but not del.
Stored Toulouse then lsd Luxair LX-LGE "Princesse
Marie-Astrid" 7 Mar 70 - 8 Mar 72. Retd to SNIAS
and lsd to SATA as HB-ICP 24 Mar 72 - 24 Jan 73, then
retd to SNIAS and re-regd F-BRGX. Lsd to SATA,
again as HB-ICP Aug 73 - 25 Jun 75. Restored
as F-BRGX and lsd to Catair Jul 75 -
Lsd to Air Inter Jan 76 - Sep 76. Retd to SNIAS
and cancelled from French register Mar 77.

235 (235)   Series 10B1R F-WJAM f/f 13 Dec 67. D-ABAP
            LTU del 18 Dec 67. Sold to Special Air
Transport for expected del Oct 79.

236 (244)   Series 10B1R   f/f 17 Jun 68. JY-ADG
            Alia "Bethlehem", del 21 Jun 68. Later
renamed "Aqaba". Sold to Trans Europa Apr 75 as
EC-CPI.

237 (252)   Series VI-N   f/f 25 May 69. YU-AHK
            (leased from Sud) del 7 Jun 69. Retd to
SNIAS May 73 and lsd to Indian A/L as VT-ECI 12 Dec 73
- 4 Jun 75 "Vyomdoot". Retd to SNIAS and re-regd
F-BRGU 26 Aug 75. Stored Toulouse until lsd Minerve
Jun 76.

238 (236)   Series 10B3/52T, later converted to
            10B3/54T. F-WLGB f/f 24 Jan 68. OY-STD
Sterling, del 23 Feb 68. Leased to TAE 3 Mar 75 as
EC-CMS until 31 Mar 79.

239 (245)   Series 10B1R F-WLGC f/f 12 Jul 68. D-ABAW
            LTU del 19 Jul 68. Sold to Special Air
Transport 28 Aug 78.

240 (230)   Series 11R   f/f 19 Oct 67. 9Q-CLC
            Air Congo "Impala", del 27 Oct 67. Later
renamed "Lubumbashi", Company name changed to Air
Zaire Oct 71. Sold via UTA French Govt. Jun 76 for
del to Armée de l'Air 9 Jul 76. Del to ETOM.82 at
Papeete, French Polynesia, 5 Aug 76 with c/s
F-RBPR, coded "PR", named "Torea".

241 (254)   Series III laid down for Air Inter but
            completed as Series VI-N f/f 21 Jan 69,
7601 Jugoslav Air Force (JRV) del 8 Feb 69. Re-serialled
74101 Sep 70. Earmarked for preservation at the new
Air Museum under completion in Belgrade.

242 (240)   Series III   f/f 18 Mar 68. F-BOHA
            Air France "Comte de Nice", del 27 Mar 68.
Renamed "Guyane" for Caribbean services then reverted
to original name. Later modified to Series III/48T.

243 (238)   Series 10B1R F-WJAM f/f 20 Feb 68. D-ABAV LTU
            del 28 Feb 68. Sold to Special Air Transport
May 79.

244 (242)   Series III   f/f 10 Apr 68. F-BOHB Air
            France "Bearn", del 19 Apr 68. Crashed
into sea off Cap d'Antibes 11 Sep 68 following
unexplained in-flight fire. 95 killed. (TT 985 hrs).

245 (243)   Series III   f/f 3 May 68. F-BOHC
            Air France "Aquitaine", del 17 May 68. Later
modified to Series III/48T. Renamed "Guyane" then
"Guadeloupe" for Caribbean services   74 -
76. Reverted to original name.

246 (237)   Series III   f/f 5 Feb 68. TS-ITU
            Tunis Air, del 16 Feb 68. Wfu   77.

247 (248)   Series 10B1R ordered by Aviaco and due for
            completion Mar 68 but not del. Stored
Toulouse. F-WJAM f/f 29 Dec 69. D-ANYL LTU del 31 Dec 69.
Retd to SNIAS 21 Jan 74 but LTU remained owner until
16 Feb 74 when aircraft sold to Trans Europa as EC-CIZ,
del Apr 74.

248 (239)   Series III   f/f 7 Mar 68, F-BNKH Air
            Inter (lsd from Sud). del 15 Mar 68.

249 (241)   Series 10B3/52T, later converted to 10B3/54T.
            F-WJAK f/f 2 Apr 68. OY-STE Sterling   del
9 Apr 68. Wfs 13 Jan 75 and sold to Central African
Republic as TL-ABB, del 28 Jan 75. In Apr 77 named
"Empereur Bokassa 1er".

250 (251)   Series 10B1R ordered by Aviaco and due for
            completion Apr 68 but not del. Stored
Toulouse until disposed of to Iberia (with part ownership
by SNIAS) and with pool operation of Aviaco and Trans
Europa., f/f 11 Feb 70.
EC-BRJ del 3 Mar 70 to Trans Europa "Renacuajo III".
Operated in Iberia colours   72.

251 (250)   Series 11R   f/f 9 Oct 68. 9Q-CLD
            Air Congo "Bukavu", del 18 Oct 68. Later
renamed "Bujumbura", but again named "Bukavu" by 1976.
Company name changed to Air Zaire Oct 71. Sold via
UTA Jun 76 to French Govt. for del to Armée de l'Air
Aug 76. Del 9 Aug 76 to ETOM.82 at Papeete, French
Polynesia, with c/s F-RBPS, coded "PS", named
"Maire Nui".

252 (249)   Series III   f/f 25 Sep 68. F-BNKJ
            Air Inter (lsd from Sud). del 12 Oct 68.

253 (253)   Series 10B1R ordered by Aviaco and due for
            completion May 69 but not del. Sold to SATA
            f/f 2 Mar 70 HB-ICN del 6 Mar 70.

254 (246)   Series III   f/f 5 Aug 68. CN-CCT Royal
            Air Maroc, del 9 Aug 68. Later named "Tafraout".
Wfu   77 at Casablanca.

255 (257)   Series 10B1R ordered by Aviaco and due for
            completion May 69 but not del. F-WJ.. f/f
5 Dec 69. OY-SAY Sterling (lsd from SNIAS), del
12 Dec 69. Retd to SNIAS 15 Jan 71 and lsd to SATA
HB-ICO, del 12 Feb 71.

256 (256)   Series III   f/f 21 Nov 69. F-BNKK
            Air Inter (lsd from SNIAS), del 2 Dec 69.
Damaged at Orly 30 Sep 77 by grenade exploded by
hijacker.

257 (258)   Series 10B3/54T   f/f 3 Feb 69. OY-STF
            Sterling, del 12 Feb 69. Proposed for lease
to Sterling Philippines 25 Mar 75 and painted with
their titles but not del to them until 4 Oct 75.
Re-regd RP-C123 Nov 75 "Lapo-Lapo".

258 (263)   Series III   f/f 3 June 70. F-OCPJ
            SNIAS. Lsd Tunis Air 8 Jun 70-   70.
Re-regd to F-BJGY NTU and regd F-BSRY for lease to
Air Inter 24 Dec 70. Forward section damaged by bomb
explosion when parked at Bastia, Corsica 22 Mar 74
and declared written-off.

259 (261)   Series 10B3/54T   f/f 7 Mar 69. OY-STG
            Sterling, del 14 Mar 69. Used cockpit section
from c/n 63. Lsd Finnair OH-LSI 1 Apr 74 - 1 Jul 76.

260 (259) Series III   f/f 19 Jan 70. F-BNKL
           Air Inter (lsd from SNIAS), del 24 Jan 70.

261 (255) Series 11R   f/f 21 Jul 69. EC-BRX
           Iberia (lsd from SNIAS) "Renacuajo I",
del 30 Jul 69. Operated in pool with Aviaco and
Trans Europa. Repainted in Trans Europa colours
        73.

262 (264) Series 10B3/54T   f/f 28 Mar 69. OY-STH
           Sterling, del 4 Apr 69.

263 (260) Series 10B1R ordered by Aviaco for pool
           operation with Iberia and due for
completion by Jun 69. F-WJAN f/f 12 Jan 70 OY-SAZ.
Sterling (lsd from SNIAS) del 16 Jan 70. Retd to
SNIAS 27 Oct 71 and reverted to F-WJAN for lease to
French Govt. for postal services Nov 71 - Dec 71. Lsd
to LTU as D-ABAF(2) 13 Jan 72 - 19 Jan 76. Re-regd
F-OCKH Jan 76, then F-WJAK 18 Mar 76. Lsd to Trans
Europa as EC-CYI Dec 76.

264 (262) Series 11R   f/f 11 Sep 69. EC-BRY
           Iberia (lsd from SNIAS) "Renacuajo II",
del to Trans Europa 19 Sep 69 for pool operation with
Aviaco. In Iberia service   72, then reverted
to Trans Europa colours. Sold via UTA Oct 76 to French
Govt. for del to Armee de l'Air. Del to ETOM.82 at
Papeete, French Polynesia, with c/s F-RBPT, coded "PT",
named "Teva".

265 (265) Series 10B3/54T   f/f 6 May 69. OY-STI
           Sterling del 14 May 69.

266 (266) Series 10B3/56T   f/f 18 Apr 70. OY-STK
           Sterling del 6 May 70. Caught fire on take-
off at Teheran 15 Mar, and written-off, 15 killed
(TT 13773 hrs).

267 (267) Series 10B3/56T   f/f 10 May 70. OY-STL
           Sterling del 19 May 70. Crashed nr
Al Fujayrah 14 Mar 72 on approach to Dubai en route
Colombo-Copenhagen. All 112 on board killed (TT 6657 hrs).

268 (268) Series 10B3/56T   f/f 12 Jun 70. OY-STM
           Sterling del 16 Jun 70.

269 (269) Series 12/56T F-WJAN (but also quoted as F-WJAK)
           f/f 29 Oct 70. OY-SAC Sterling del 18 May 71.

270 (270) Series 12/56T   f/f 1 Feb 71. OY-SAA
           Sterling del 12 Mar 71. Proposed lease to
Transunion in summer 74 fell through but chartered
by them from 19 Apr 74 until Sep 74.

271 (271) Series 12/56T   f/f 22 Mar 71. OY-SAB
           Sterling del 26 Mar 71. Lsd to Catair as
F-BVPY 14 Jan 75 then leased to Air Inter 9 Apr 76 -
30 Nov 76. Retd to Sterling as OY-SAB.

272 (272) Series 12/56T   f/f 14 May 71. OY-SAD
           Sterling del 7 Jun 71.

273 (273) Series 12/58T   f/f 22 Feb 72. OY-SAE
           Sterling del 28 Feb 72.

274 (274) Series 12/58T F-WTOA f/f 4 May 72. F-BTOA
           Air Inter (lsd from SEFIPROM) del 20 Oct 72.

275 (275   Series 12/58T F-WJAL f/f 15 May 72. OY-SAF
           Sterling del 19 May 72.

276 (276) Series 12/58T   f/f 10 Jun 72. OY-SAG
           Sterling del 19 Jun 72.

277 (277) Series 12/58T   f/f 1 Dec 72. F-BTOB
           Air Inter (lsd through SEFIPROM) 20 Dec 72.

278 (278) Series 12/58T   f/f 10 Jan 73. F-BTOC
           Air Inter (lsd through SEFIPROM) del 17 Apr
73.

279 (279) Series 12/58T   f/f 12 Feb 73. F-BTOD
           Air Inter (lsd through SEFIPROM) del
20 Feb 73.

280 (280) Series 12/58T   f/f 8 Mar 73. F-BTOE
           Air Inter (lsd through SEFIPROM) del
16 Mar 73.

Late Additions/News

18   Reportedly with Algerian AF.

31   Struck fence taxying at Frankfurt 12 Mar 79 and
     declared a w/o.   Broken up between 23 Apr 79 and
18 May 79.

62   Sold to Midwest as N901MW and del 27 Jun 79.

88   Was painted in Wright Airlines colours in
     anticipation of operating from Cleveland to Detroit
(Wright and Midwest are apparently closely connected).

91   Crashed into Atlantic Ocean off Senegal 27 May 79
     and w/o.

129 Sold to Midwest as N907MW and due for del Aug 79.

140 Reported w/o in Colombia 20 Jul 79.

155 Broken up Bombay.

160 In service as "Isleno".

168 Sold to TAC as HK-2287X and del 19 Apr 79.

195 CoA exp May 77.   WFU and canc late 1978.

240 F-RBPR reportedly w/o Tahiti Jun 79.

241 Sold May 79 to R Collier/Aerot'ex of Dinard as F-BVSF,
     del Jul 79.

257 To TAE as EC-DFP Feb 79.

270 Reg F-BVTB applied for Transunion service.   After
     initial test flight, removed in favour of OY-SAA.

NORD 260/262

Max Holste was a little-known French designer until success came in 1952 with the MH-1521 Broussard, a rugged single-engined utility transport similar in many ways to the DHC Beaver. Nearly 400 were built, mostly for military use but also in small numbers for export. By 1957 the Reims-based design team was working on a larger twin-engined transport, the MH-250 Super-Broussard, which was flown in 1959 with Pratt & Whitney R-1340 power plants. But with the development of light turboprops a new version was soon drawn up as the MH-260, with two Turboméca Bastan IIIA engines. The prototype, c/n 001 (F-WJDV), made its first flight in July 1960 and after initial tests was fitted with more powerful Bastan IV turboprops, becoming the MH-260-01 in the process. This was a period when the DC-3 replacement myth was being developed, and the new Max Holste aircraft was judged to be a good prospect in the expected market for feederliners (the French word 'avion d'apport' being coined on the occasion).

Of the ten MH-260s (later Nord 260) built with government subsidy (therefore implying participation of the nationalised industry in the programme), only nine were actually flown. Three were temporarily operated by the Norwegian operator Widerøe's Flyveselskap A/S, and other aircraft were evaluated by the French domestic airline Air Inter in 1963. An order for six by Ansett of Australia was not confirmed, and most aircraft were finally delivered to the CEV for use as hacks or for various experiments, occasionally resulting in modifications.

Other versions were considered - the MH-270 with Armstrong Siddeley P.182 power-plants, and the MH-280 with Lycoming T.53s - but they were not proceeded with. Also related to the breed was the Continental GIO-470A-powered MH-350 Broussard Major, a twin-engine intermediate between the original MH-1521 and the MH-250.

But as work progressed on improved variants of the basic Super Broussard design, Max Holste had been getting into financial difficulties. An initial co-operation agreement in October 1959 with Nord Aviation for production of the MH-260 soon led to a complete take-over of the whole programme by Nord, with the resulting change of designation from the MH- to the N-prefix, the MH-260 being often referred to as the Nord 260. (The Max Holste organisation at Reims also gave way to a new concern, Reims-Aviation, which was to become the busy European licensee for USA Cessna products.)

Serious attempts to market the Super Broussard (or Nord 260 from late 1960) failed, but fortunately design of the MH-262 pressurised cabin development was under way. The type number 261 was a project for an unpressurised Bastan V powered military transport which was not developed.

The Nord/MH-262, soon to become known as the Nord 262, was in fact a new aircraft with its circular-section fuselage, and the prototype made its maiden flight late in 1962. In May 1963, with the first production aircraft flying, the Bastan IVB engines were changed to Bastan VIs. By June 1964 four N-262s were flying as the Federal Aviation Agency was on the verge of selecting the famous DC-3 replacement designs. At that time it was known that no single type could replace the famous Douglas workhorse, but the five top-scoring designs finally announced were (in order) a Fairchild 27-seater, the Nord 262, the Potez 840 and two projects by Lear and Republic (the last being a 14-seater only). The choice of the two French aircraft came as good news, and hopes of large orders soared. In fact, the Potez 840 failed to attract orders, and the anticipated sales for some 200 Nord 262s were never to be achieved. By early 1965, however, a go-ahead was given for an initial production batch of 40, increased to 60 in 1966.

The first order was placed by Air Inter in June 1964 for four aircraft, which entered service on French domestic routes. The first Air Inter Nord 262 (c/n 4) was also the first aircraft from the production line at Bourges (the previous aircraft had been built at Chatillon-sous-Bagneux and the Nord 260 batch in Toulouse). Air Inter was not entirely satisfied with the aircraft, and late in 1968 they were transferred to Rousseau Aviation, which was later taken over by TAT, the fast-growing French third-level operator. Of course, the most publicised contract was the breakthrough on the American market, with eight (later 12) aircraft ordered by Lake Central Airlines, taken over by Allegheny in July 1968. The American 262s were delivered in 1965-66.

Third customer was Japan Domestic Airlines which, pending delivery of its own three aircraft, leased Nord 262 c/n 2. The Japanese Nord 262s were later operated in the Philippines. But after what may be considered as a fair success initially, sales were slow, being generally for single examples: Alisarda, Air Madagascar, Air Ceylon, Linjeflyg, Cimber Air, Commute-Air and Air Comores. Some options were not taken up, and only French military orders kept the production line busy. The initial batch of the Nord 262A version for the Armée de l'Air was transferred to Aéronavale in 1971, and the Armée de l'Air obtained Frégates instead, the Bastan VII-powered version. Both services placed follow-on orders, nearly half the total production going to the military services, with additional aircraft for French Government agencies.

Other aircraft were sold to African governments (such as Upper Volta, Gabon and the Congo) or as calibration aircraft. But with a few exceptions no more orders for brand-new aircraft were obtained after 1971, and the last six unsold aircraft on the production line were purchased by the French Air Force on 3 September 1975, the line being closed down with Fregate c/n 110 delivered in March 1977.

Many explanations have been proposed for this disappointing slump in sales following such initial promise for an aircraft which was first hailed - perhaps with too much emphasis - as the long-awaited DC-3 replacement. First, it must be said that the Nord 262 had a smaller capacity than the DC-3, and second the Nord 262 was launched when third level transport and the market for a short-haul airliner was still comparatively limited. Prospective customers were small local operators, and sales to each of them in large numbers could not be expected. It has also often been said that the Nord 262 was introduced too early on the market but, in retrospect, this argument now appears less convincing.   As the production list shows, however, the second-hand market for the Nord 262 has been very active, reflecting in the process the constantly changing aspect of short-haul transport.

The first version of the Nord 262 was the Nord 262A powered by the Bastan VI, the Bastan IVB of the early aircraft having been discarded.   The Nord 262B designation was assigned to the four aircraft specially modified at Air Inter's request, also fitted with Bastan VI turboprops.   The name Frégate (Frigate, apparently in connection with other ship names adopted for the second generation French airliners such as the Caravelle and the Galion airbus project) referred to the Nord 262C version with 1,145 hp Bastan VII power plants, the Nord 262D being the Armée de l'Air variant of the Frégate.   The single Nord 262E was c/n 2 fitted with the Bastan VI and experimental modifications.   Thirtythree aircraft out of the total production were Frégates, either C or D variants.

But by 1975 a Nord 262 American conversion was announced as the Mohawk 298.   The Bastan VI engines were replaced by United Aircraft PT6A-45 engines at the request of Ransome Airlines, which had seven aircraft originally acquired by Allegheny through its merger with Lake Central.   A new life was then given to more 262s, with four ex-Linjeflyg aircraft bought by Allegheny and three ex-Filipinas Orient aircraft purchased by American Jet Industries.   Growing interest from American operators (such as Swift Aire and Altair) led to hopes of resuming production, and licence production by Grumman American was even mentioned.   The last three unsold aircraft (c/n 99, 101 and 102) were purchased by Allegheny and Ransome, and the rejuvenated Mohawk 298 aircraft entered service with Allegheny early in 1977 in the busy commuting area of New York, with main base at Harrisburg.

With growing interest in maritime reconnaissance aircraft, Aérospatiale made proposals in 1977 for a maritime and coastguard version referred to as the Nord 262 A-II (with 1,140 hp Turboméca Bastan VI C-4s) which could be delivered by 1980 if production of the Nord 262 was to be resumed.   This, it was hoped, would help to sell more basically commercial aircraft, including the American conversions. A converted aircraft was demonstrated to the French Navy, but it was finally decided not to press the proposal, and Nord 262 production finally ended with aircraft c/n 110 delivered to the Armée de l'Air for the CIET training centre.

Specifications

High-wing twin turboprop airliner with 26-28 seats.

1   Nord 262A

Span: 21.90m
Length: 19.28m
Height: 6.21m
Wing area: 55 sq m
Operating weight empty: 6,860 kg
Maximum take-off weight: 10,300 kg
Power plants: 2 x 1,065ehp Turboméca Bastan VI C
Cruising speed: 365 km/hr at 1,500m
Take-off distance: 1,160m
Cruise altitude: 5,850m
Range: 1,126 km with payload of 2,086 kg or 322 km with 2,900 kg

(Statistics released by SNIAS revealed that average flight duration of all Nord 262As was 41 minutes as compared with 62 minutes for the Frégate.)

2   Nord 262C Frégate

Span: 22.60m
Length: 19.28m
Height: 6.21m
Wing area: 55.8 sq m
Operating weight empty: 7,225 kg
Maximum take-off weight: 10,800 kg
Power plants: 2 x 1,145ehp Turboméca Bastan VII C
Cruising speed: 408 km/hr at
Take-off distance: 1,070m
Service ceiling: 8,690m
Range: 1,825 km (with maximum load but FAA reserves)
       2,400 km (absolute with no reserves)
Maximum speed at height: 418 km/hr at sea level
Passenger configurations: 26 plus 2-3 crew plus baggage
                          Maximum 29 passengers or 29 troops
Variants (sub-types): Nord 262C (commercial, eg SFA, Gabon Govt, East African Community) with military counterpart - 18 for Armée de l'Air = Frégate Srs D)

Production List

C/n 001   MH-250 F-WJDA.   F/f 20 May 59.

C/n 01   MH-260 F-WJDV.   F/f 27 Jul 60.   To CEV/EPNER, coded "GG".

C/n 1   MH-260 F-WJSN.   F/f 29 Jan 62.

C/n 2   MH-260 F-WKRB?   F/f 02 May 62.   To F-BKRB Nord Aviation (CofA 10 Aug 62).   CofA expired Feb 65.

C/n 3   MH-260 F-BKRH.   To Nord Aviation (CofA 13 Sep 62).   To Turboméca 1969.

C/n 4   MH-260 F-BKRS.   To Nord Aviation (CofA 27 Nov 62).   CofA expired Nov 64.

C/n 5   MH-260 F-BLEA.   To LN-LMB Widerøe del 15 Dec 62. Retd to Nord Aviation.

C/n 6   MH-260 F-WLGP.   To F-BLGP Nord Aviation (CofA 23 Apr 63).   To LN-LMG Widerøe 1963.   Retd to Nord Aviation   .   To CEV coded "MA".

C/n 7   MH-260 F-BLHN.   To Nord Aviation (CofA 14 May 63). CofA expired Dec 67.   To CEV coded "MB"? (Astazou engines).

C/n 8   MH-260 F-BLHO?   To LN-LME Widerøe   . Retd to Nord Aviation as F-BLHO (CofA 03 Jun 64). CofA expired Nov 64.   To CEV coded "MC".

C/n 9   MH-260 F-WLHP.   To CEV coded "ME".

C/n 10   MH-260.   Never flown.   Completed without engines and equipment and used for spares at Melun-Villaroche.

C/n 01   N-262 F-WKVR.   F/f 24 Dec 62.   To CEV as F-ZADM coded "DM".

C/n 1   N-262 F-WLKA.   F/f 06 May 63.   To F-BLKA Nord Aviation (CofA 25 Jun 65).   CofA expired May 66. Cvtd to Nord 262CS as F-BLKE del 15 Oct 70 to SNIAS.   To F-WLKE.   Retd to F-BLKE SNIAS (CofA 10 Dec 71) for communication duties.

C/n 2   N-262 F-WLHQ.   F/f 16 Aug 63.   To F-BLHQ Nord Aviation (CofA 18 Aug 64).   Lsd to Japan Domestic 1965.   To 5R-MCC Air Madagascar (CofA 17 Sep 66).   Cvtd to Nord 262E.   To F-BNGB Nord Aviation/SNIAS (CofA 09 Jan 69).   To Rousseau Aviation del 27 Feb 70.   Written off 31 Dec 70 between Algiers and the Balearic Islands.

C/n 3   N-262 F-WLHR.   F/f 26 Mar 64.   Cvtd to Nord
        262A prototype.   To F-BLHR Nord Aviation (CofA
30 Nov 67).   Retd to F-WLHR Dec 67.   To CEV as F-ZVOH
coded "OH" del 03 Apr 73.   To F-BLHE CEV/EPNER 1974.
Retd to F-ZVOH coded "OH".

C/n 4   N-262 F-WLHS.   F/f 08 Jun 64.   To F-BLHS Air
        Inter del 21 Jul 64.   Cvtd to Nord 262B11.   To
Rousseau Aviation Nov 68.   To TAT (Locaero).

C/n 5   N-262 F-BLHT.   F/f 06 Aug 64.   To Air Inter
        del 12 Aug 64.   Cvtd to Nord 262B11.   To
Rousseau Aviation Nov 68.   Written off 12 Nov 73 at
Craon, Mayenne.

C/n 6   N-262 F-BLHU.   F/f 14 Sep 64.   To Air Inter del
        18 Sep 64.   Cvtd to Nord 262B11.   To Rousseau
Aviation Nov 68.   To TAT (Pretabail).   To ZS-IZX Avna
del 08 Mar 74.   To National A/W "Vryheid"(?)

C/n 7   N-262 F-BLHV.   F/f 08 Jan 65.   To Air Inter del
        21 Jan 65.   Cvtd to Nord 262B11.   To Rousseau
Aviation Nov 68.   To TAT (Locaero).

C/n 8   N-262A14 F-BNDE.   F/f 22 Jul 65.   To Nord
        Aviation.   To JA8646 Nitto/Japan Domestic del 09
Nov 65.   To PI-C967 Fairways Orient A/L Dec 74.   Reregd
RP-C967   .   To N87TC American Jet Industries Dec
75.   To N420SA Swift Aire   77.

C/n 9   N-262A12 F-WLHX.   F/f 20 May 65.   To N26201
        Lake Central A/L del 17 Aug 65.   To Allegheny
A/L "Yvonne" 01 Jul 68.   To CF-BCU BC A/L del 28 Mar
70.   To Pacific Western Sep 70.   Retd to N26201
Ransome A/L Feb 72.

C/n 10   N-262A12 N26202(1).   F/f 15 Jul 65.   To Lake
         Central A/L del 05 Sep 65.   To Allegheny A/L
"Claudette" 01 Jul 68.   Cvtd to Nord 262A44.   To
7T-VSR STA/Air Algérie, bought 14 Oct 70.

C/n 11   N-262A12 N26203.   F/f 25 Aug 65.   To Lake
         Central A/L del 17 Sep 65.   To Allegheny A/L
"Nicole" 01 Jul 68.   To Ransome A/L.

C/n 12   N-262A12 N26207.   F/f 02 Oct 65.   To Lake
         Central A/L del 14 Oct 65.   To Allegheny A/L
"Yvette" 01 Jul 68.   Cvtd to Nord 262A44.   To 7T-VSS
STA/Air Algérie, bought 14 Oct 70.

C/n 13   N-262A12 N26208.   F/f 30 Oct 65.   To Lake
         Central A/L del 20 Nov 65.   To Allegheny A/L
"Celeste" 01 Jul 68.   To Ransome A/L.

C/n 14   N-262A12 N26209.   F/f 09 Dec 65.   To Lake
         Central A/L del 17 Dec 65.   To Allegheny A/L
"Michele" 01 Jul 68.   Cvtd to Nord 262A44.   To 7T-VSQ
STA/Air Algérie, bought 23 Mar 71.

C/n 15   N-262A14 JA8652.   F/f 15 Jan 66.   To Nitto/
         Japan Domestic del 08 Feb 66.   To PI-C968
Fairways Orient A/L Dec 74.   Reregd RP-C968   .
To N88TC American Jet Industries Oct 75.   To Calco
Leasing Corp   .   To N417SA Swift Aire   76.

C/n 16   N-262A12 N26210.   F/f 23 Dec 65.   To Lake
         Central A/L del 13 Jan 66.   To Allegheny A/L
"   " 01 Jul 68.   To CF-BCR BC A/L del 15 Apr 69.
To Pacific Western Sep 70.   Retd to N26210 Ransome A/L
Feb 72.

C/n 17   N-262A12 N26211.   F/f 08 Jan 66.   To Lake
         Central A/L 28 Jan 66.   To Allegheny A/L
"Brigitte" 01 Jul 68.   Cvtd to Mohawk 298 prototype
and reregd N29812, f/f 07 Jan 75.

C/n 18   N-262A12 N26212.   F/f 15 Feb 66.   To Lake
         Central A/L del 11 Mar 66.   To Allegheny A/L
"Colette" 01 Jul 68.   Cvtd to Nord 262A44.   To 7T-VST
STA/Air Algérie, bought 14 Oct 70.

C/n 19   N-262A12 N26213.   F/f 08 Mar 66.   To Lake
         Central A/L del 02 Apr 66.   To Allegheny A/L
"Monique" 01 Jul 68.   Cvtd to Nord 262A44.   To 7T-VSU
STA/Air Algérie, bought 14 Oct 70.   Written off 24 Jan
79.

C/n 20   N-262A12 F-BLHX.   F/f 23 Mar 66.   To Nord
         Aviation.   Lsd to Alisarda   .   Retd to
Nord Aviation/SNIAS   .   Cvtd to Nord 262A37.   To
SFA del 16 Apr 69.

C/n 21   N-262A15 F-WNLI.   F/f 17 Apr 66.   To F-BNLI
         Nord Aviation (CofA 26 Apr 66).   Canc Nov 67.
Cvtd to Nord 262A21.   To F-BOFQ Nord Aviation/SNIAS
(CofA 06 May 68).   To Europe Aéro Service del 26 Nov
69.   To OY-BDD Cimber Air del 16 Sep 71.   Lsd to Saudia
03 Dec 74-01 May 75.   Lsd to TAT 1975/76.   To N7885A
Altair A/L del 10 Jan 77.

C/n 22   N-262A14 JA8663.   F/f 04 May 66.   To Japan
         Domestic del 13 Aug 66.   To PI-C966 Fairways
Orient A/L Dec 74.   Reregd RP-C966   .   To
N89TC American Jet Industries   75.   To N419SA Swift
Aire   77.

C/n 23   N-262A12 N26215.   F/f 17 May 66.   To Lake
         Central A/L del 03 Jun 66.   To Allegheny A/L
"   " 01 Jul 68.   To CF-BCS BC A/L del 17 Mar 69.   To
Pacific Western Sep 70.   Retd to N26215 Ransome A/L Feb 72.

C/n 24   N-262A12 N26217.   F/f 01 Jul 66.   To Lake
         Central A/L del 15 Jul 66.   To Allegheny A/L
"   " 01 Jul 68.   To CF-BCT BC A/L del 22 Apr 69.
To Pacific Western Sep 70.   Retd to N26217 Ransome A/L
Feb 72.

C/n 25   N-262A21 I-SARL.   F/f 27 Jun 66.   To Alisarda
         del 04 Jul 66.   To OY-BDL Cimber Air   70.   To
D-CIMB Cimber Air Apr 71.   Retd to OY-BDL Cimber Air,
reregd 01 Nov 74.   To N481A Altair A/L Oct 75.

C/n 26   N-262A24 F-WNTT.   F/f 10 Oct 66.   To F-BNTT
         Nord Aviation/SNIAS (CofA 18 Nov 66).   To
Rousseau Aviation del 10 Nov 66.   Written off 29 Dec 73
at Dôle, Jura.

C/n 27   N-262A20 F-BNMO.   F/f 25 Aug 66.   To Nord
         Aviation/SNIAS.   Cvtd to Nord 262A25.   To
Rousseau Aviation del 31 Mar 67.   Written off 05 Dec 71
at Lannion (Côtes du Nord).

C/n 28   N-262A28 F-WNMP.   F/f 04 Nov 66.   To French Navy
         del 11 Jul 67.   Cvtd to Nord 262A29.   Full c/s
F-YCKY painted in civil regn style for short period on
wings and fuselage before being deleted.   Currently "28"
(French Navy Service Central de l'Aéronautique serial).

C/n 29   N-262A22 F-BNKX.   F/f 15 Sep 66.   To Nord
         Aviation.   To 4R-ACL Air Ceylon del 24 Mar 67.
To F-WNDD   .   Cvtd to Nord 262AG43.   To G-AYFR Dan-
Air del Jul 70.   To F-BTDQ SNIAS   72.   To Rousseau
Aviation del 22 Feb 72.   To N26227 Ransome A/L Apr 75.
Cvtd to Mohawk 298 as N29808 after del 30 Dec 76 to
Allegheny A/L.

C/n 30   N-262A32 F-WNDA.   F/f 01 Dec 66.   To F-BPNS
         Nord Aviation/SNIAS.   To SFA del 16 May 67.

C/n 31   N-262A26 F-WNDB.   F/f 30 Dec 66.   To SE-CCR
         Linjeflyg del 06 May 67.   WFU 15 Feb 74.   To
N26224(1) Ransome A/L del 20 Nov 74.   Cvtd to Mohawk
298 as N29802 after del 16 Dec 76 to Allegheny A/L.

C/n 32   N-262A26 SE-CCS.   F/f 31 Jan 67.   To Linjeflyg
         del 19 May 67.   WFU 16 Feb 74.   To N26225
Ransome A/L del 04 Dec 74.

C/n 33   N-262A27 F-WNDD.   F/f 30 Mar 67.   To OY-BCO
         Cimber Air del 02 Aug 67.   Sold to Fokker-VFW 27
Jun 74 but lsd back to Cimber Air.   Lsd to Delta Air

Transport-NLM 07 May 76-16 Aug 76.   To N274A Altair A/L del 26 Jul 77.

C/n 34   N-262A21 I-SARP.   F/f 14 Apr 67.   To Alisarda del 21 Apr 67.   To OY-BDM Cimber Air   70. To (D-CAMY) then D-CIMA Cimber Air (CofA 14 Apr 71). Retd to OY-BDM Cimber Air (CofA 30 Dec 74).   To N488A Altair A/L del 30 Apr 75.

C/n 35   N-262A32 F-WNDB.   F/f 16 Jun 67.   To F-BPNT Nord Aviation/SNIAS.   To SFA del 19 Jun 67.

C/n 36   N-262C50P F-WPXA.   F/f 09 Jul 68.   To F-BPXA SNIAS (CofA 01 Feb 72).   To "XT-MAJ" Upper Volta AF del 05 Apr 74.

C/n 37   N-262A30 F-WNDC.   F/f 03 Jul 67.   To D-CADY W Krauss/IFG del 10 Aug 67.   To Cimber Air Mar 72. WFU 02 Jul 74.   To OY-BLV Cimber Air   76.   Lsd to Delta Air Transport-NLM   76.   Lsd to TAT   78.   To Altair A/L Jun 78 as N275A.

C/n 38   N-262A20 F-WOFA.   F/f 25 Jul 67.   To F-BPNU Nord Aviation/SNIAS.   To SFA del 28 Jul 67. Cvtd to Nord 262A32.

C/n 39   N-262A32 F-BPNV.   F/f 10 Oct 67.   To Nord Aviation/SNIAS.   To SFA del 01 Mar 68.   Written off 14 Aug 75 at St Yan French National Civil Aviation Training School (Dept de Saône-et-Loire).

C/n 40   N-262A32 F-BPNX.   F/f 05 Nov 67.   To Nord Aviation/SNIAS.   To SFA del 28 Feb 68.

C/n 41   N-262A33 F-WOFC.   F/f 30 Nov 67.   To 5R-MCU Air Madagascar del 02 Sep 68.   To F-BVPP Air Alsace del 15 Feb 75.   To N418SA Calco Leasing Corp for Swift Aire del 06 Jan 77.   Written off 10 Mar 79.

C/n 42   N-262A32 F-WOFD.   F/f 31 Dec 67.   To F-BPNY Nord Aviation/SNIAS (CofA 25 Jun 68).   Cvtd to Nord 262A36.   To Air Comores del 24 Mar 69.   To Rousseau Aviation del 17 Aug 71.   To N26228 Ransome A/L Aug 75.   Cvtd to Mohawk 298 as N29811 for Allegheny A/L.

C/n 43   N-262A28 F-WOFE.   F/f 31 Aug 68.   To French Navy del 23 Dec 68.   Originally c/s F-YCKB, wears "43" S C Aéro number.   While wearing F-YCKB caught in violent storm (cunimb) Aug 69 near Meaux (Dept de Seine et Marne) and bellylanded following failure of both turboprops.   Eighteen passengers and crew escaped unhurt.   Repaired on the spot and took off from nearby field only eight days later.

C/n 44   N-262A34 F-ZKJL.   F/f 13 Mar 68.   To French AF as F-RBOA coded "OA" del 08 Jan 69.   Written off 21 Jan 71 near Privas, Ardeche.

C/n 45   N-262A34.   F/f 05 Apr 68.   To French AF as F-RBOB coded "OB" del 24 Mar 69.   To French Navy del 14 Sep 71.   S C Aéro "45" F-Y..   Current.

C/n 46   N-262A34.   F/f 07 May 68.   To French AF as F-RBOC coded "OC" del 28 Nov 68.   To French Navy del 02 Sep 71.   S C Aéro "46" F-Y..   Current.

C/n 47   N-262A27 OY-BKR.   F/f 25 Jul 68.   To Cimber Air del 15 Nov 68.   Sold to Fokker-VFW 27 Jun 74. Lsd back to Cimber Air.   To N7886A Altair A/L del 25 Feb 77.   Written off 08 Apr 77 near Reading, PA, and proposed reregn N276A NTU.

C/n 48   N-262A36 F-WOFZ.   F/f 22 Jul 68.   To F-OCOG Nord Aviation/SNIAS (CofA 03 Oct 69).   To Air Comores del 07 Oct 69.   To F-BSTN SNIAS (CofA 19 Apr 71). To Rousseau Aviation del 17 Aug 71.   To TN-ACS Lina Congo Mar 75.   Cvtd to Mohawk 298 as N29824 after del 26 Jul 77 to Allegheny A/L.   Written off 12 Feb 79.

C/n 49   N-262A35 F-WOFX.   F/f 27 Nov 68.   To F-BOHH SFA del 06 Jan 69.   Cvtd to Nord 262A32.

C/n 50   N-262A38 F-WOFC.   F/f 30 Jan 69.   To F-OCNQ Nord Aviation (CofA 15 Apr 69).   Lsd to Tunis Air del 17 Apr 69.   Reregd TS-LIP and bought 13 Apr 70. Cvtd to Mohawk 298 as N29817 after del to Allegheny A/L Feb 76.

C/n 51   N-262A34.   F/f 18 Feb 69.   To French AF as F-RBOD coded "OD" del 20 Feb 69.   To French Navy del 07 Oct 71.   S C Aéro "51" F-Y...   Ten thousandth French Naval Personnel passenger of "Lialog" (Aéronavale Nord 262 air transport service) flown on 15 May 74, aircraft then being F-YDCA.

C/n 52   N-262A34.   F/f 21 Feb 69.   To French AF as F-RBOE coded "OE" del 03 Apr 69.   To French Navy del 28 Sep 71.   S C Aéro "52" F-Y...

C/n 53   N-262A34.   F/f 21 Mar 69.   To French AF as F-RBOF coded "OF" del 22 Apr 69.   To French Navy del 04 Oct 71.   S C Aéro "53" F-Y...

C/n 54   N-262A30 F-WNDA.   F/f 12 Mar 69.   To D-CIFG IFG del 29 Mar 69.   To Cimber Air Mar 72.   WFU 02 Jul 74.   To OY-TOV Cimber Air (CofA 08 Nov 74).   To N487A Altair A/L del 12 Mar 75.

C/n 55   N-262A40 F-ZVMH coded "MH".   F/f 15 Apr 69 ENSA del 16 Apr 70.

C/n 56   N-262A26 F-WNDA?   F/f 02 May 69.   To SE-CCT Linjeflyg del 30 Oct 69.   To N26226 Ransome A/L Cvtd to Mohawk 298 as N29813 for Allegheny A/L.

C/n 57   N-262A42 OY-IVA.   F/f 13 May Danish CAA del 30 Sep 70.   (Calibration aircraft.)

C/n 58   N-262A41 F-ZVMJ coded "MJ".   F/f 03 Jun 69.   To CEV del 24 Mar 71.

C/n 59   N-262A29.   F/f 25 Jun 69.   To French Navy del 15 Dec 69.   S C Aéro "59" F-Y...

C/n 60   N-262A29.   F/f 16 Jul 69.   To French Navy del 16 Feb 70.   S C Aéro "60", acceptance 16 Feb 70 and flown from Le Bourget to St Raphael 24 Feb 70 for checking and setting of weather radar by French Navy test centre, entering service with the Section de Liaison de Dugny (Dugny support flight).

C/n 61   N-262A29.   F/f 01 Sep 69.   To French Navy del 02 Apr 70.   S C Aéro "61" F-Y...   By Jun 75 carried insignia of 3S and 56S (for a short period).

C/n 62   N-262A29.   F/f 15 Sep 69.   To French Navy del 07 Apr 70.   S C Aéro "62" F-Y...

C/n 63   N-262A29.   F/f 23 Oct 69.   To French Navy del 04 May 70.   S C Aéro "63" F-Y...

C/n 64   N-262D51 F-ZJYV.   F/f 15 Apr 70.   To French AF del 22 Jul 70 with CEAM coded 118-IT (F-SDIT). By late 1977 as F-RBAA coded "AA" with French AF.

C/n 65   N-262A29.   F/f 22 Jan 70.   To French Navy del 04 Jun 70.   S C Aéro "65" F-Y...

C/n 66   N-262D51.   F/f 10 Jul 70.   To French AF del 30 Mar 72, mostly as F-RBAB coded "AB".

C/n 67   N-262A41 F-ZVMI coded "MI".   F/f 28 Apr 70.   To CEV del 03 Jun 70.

C/n 68   N-262D51.   F/f 02 Sep 70.   To French AF del 16 Jun 71, mostly as F-RBAC coded "AC".   By Jun 74 was operated on behalf of ELA 44 (Escadrille de Liaisons Aériennes 44).   Was "CR" (F-SCCR) by Aug 77.

C/n 69   N-262A26 SE-FUA.   F/f 24 Mar 70.   To Linjeflyg del 21 Sep 70.   WFU 16 Feb 74.   To N26222 Ransome A/L del 30 Oct 74.

C/n 70   N-262A29.   F/f 04 May 70.   To French Navy del
         15 Oct 70.   S C Aéro "70" F-Y...

C/n 71   N-262A29.   F/f 03 Jun 70.   To French Navy del
         29 Oct 70.   S C Aéro "71" F-Y...

C/n 72   N-262A29.   F/f 23 Jun 70.   To French Navy del
         22 Dec 70.   S C Aéro "72" F-Y...

C/n 73   N-262A29.   F/f 21 Jul 70.   To French Navy del
         25 Feb 71.   S C Aéro "73" F-Y...

C/n 74   N-262C61 F-BSUF.   F/f 15 Dec 70.   To SFA (for
         calibration) del 28 Jul 71.

C/n 75   N-262A29.   F/f 09 Sep 70.   To French Navy del
         22 Mar 71.   S C Aéro "75" F-Y...

C/n 76   N-262D51.   F/f 08 Oct 70.   To French AF del 09
         Jul 71, mostly as F-RBAD coded "AD".

C/n 77   N-262D51.   F/f 26 Nov 70.   To French AF del 19
         Jul 71, mostly as F-RBAE coded "AE".   Following
storage was redelivered to ELA 44 (Escadrille de
Liaisons Aériennes).   By Jun 77 was recoded "CS"
(F-SCCS).

C/n 78   N-262D51.   F/f 12 Jan 71.   To French AF del 16
         Jul 71, mostly as F-RBAF coded "AF".

C/n 79   N-262A29.   F/f 25 Jan 71.   To French Navy del
         10 Feb 72.   S C Aéro "79" F-Y..

C/n 80   N-262D51.   F/f 09 Feb 71.   To French AF del 19
         Aug 71, mostly as F-RBAG coded "AG".   Following
storage, to ELA 44 (see c/n 77).   Was coded "CT" (F-SCCT)
by Oct 78.

C/n 81   N-262D51.   F/f 19 Feb 71.   To French AF del 24
         Aug 71, mostly as F-RBAH coded "AH".   Was 070-MA
(F-SDMA) with ferry unit by May 77.

C/n 82   N-262C62 F-WNDC.   F/f 15 Jun 71.   To "TR-KJA"
         Gabon AF del 30 Jul 71.

C/n 83   N-262D51.   F/f 18 May 71.   To French AF del 30
         Sep 71, mostly as F-RBAI coded "AI".   Later at
disposal of ELA 00/43 for communications work from HQ
Bordeaux.

C/n 84   N-262A27 OY-BDR.   F/f 02 Apr 71.   To Cimber Air
         del 30 Apr 71.   To N486A Altair A/L del 03 Feb
75.

C/n 85   N-262A29.   F/f 14 Apr 71.   To French Navy del
         10 Nov 71.   S C Aéro "85" F-Y..

C/n 86   N-262D51.   F/f 02 Jun 71.   To French AF del 21
         Oct 71, mostly as F-RBAJ coded "AJ".

C/n 87   N-262D51.   F/f 17 Jun 71.   To French AF del 30
         Sep 71, mostly as F-RBAK coded "AK".   Recoded
"CU" (F-SCCU) - see c/n 68 and 77.

C/n 88   N-262D51.   F/f 28 Sep 71.   To French AF del 30
         Jan 72, mostly as F-RBAL coded "AL", at disposal
of ELA 00/43 (see c/n 83).

C/n 89   N-262D51.   F/f 05 Oct 71.   To French AF del 27
         Jan 72, mostly as F-RBAM coded "AM".

C/n 90   N-262C62 "TR-KJB".   F/f 24 Sep 71.   To Gabon AF
         del 11 Nov 71.

C/n 91   N-262D51.   F/f 18 Nov 71.   To French AF del 10
         Feb 72, mostly as F-RBAN coded "AN".

C/n 92   N-262D51.   F/f 15 Dec 71.   To French AF del 15
         Mar 72, mostly as F-RBAO coded "AO".

C/n 93   N-262D51.   F/f 11 Jan 72.   To French AF del 10
         Apr 72, mostly as F-RBAP coded "AP".

C/n 94   N-262D51.   F/f 28 Mar 72.   To French AF del 19
         Jun 72, mostly as F-RBAQ coded "AQ".   To 070-MB
(F-SDMB) (see c/n 81).

C/n 95   N-262D51.   F/f 18 May 72.   To French AF del 26
         Jul 72, mostly as F-RBAR coded "AR".

C/n 96   N-262C63 F-WNDA.   F/f 28 Apr 72.   To 5Y-DCA East
         African DCA del 13 Jun 72.   To Kenyan DCA.

C/n 97   N-262C64 "TR-KJC".   F/f 20 Jun 72.   To Gabon AF
         del 12 Jan 73.

C/n 98   N-262C65 "XT-MAK".   F/f 22 Aug 72.   To Upper
         Volta AF del 13 Nov 74.

C/n 99   N-262A .   F/f 10 Oct 72 and stored.   To
         Allegheny A/L del 15 Jan 76 for conversion to
Mohawk 298 as N29814.

C/n 100 N-262A45 F-WNDA.   F/f 29 Sep 73.   To F-BVRV Air
         Alsace del 31 May 74.   To N26224(2) Ransome A/L
del 27 Apr 78.   To Altair A/L.

C/n 101 N-262A .   F/f 09 Oct 74 and stored.   To
         Allegheny A/L del 15 Jan 76 for conversion to
Mohawk 298 as N29816.

C/n 102 N-262A20.   Stored then N26202(2).   F/f 29 Nov
         76.   To Ransome A/L del 14 Dec 76.

C/n 103 N-262C66 F-WNDB.   F/f 16 Jan 74 and stored.   To
         "TN-216" Congo AF del 17 Jul 75.   To "TN-230".

C/n 104 N-262C67.   F/f 26 Feb 74 and stored.   To F-ODBT
         SNIAS (CofA 27 Nov 75).   Lsd to Bouraq   .
To F-BYCT SNIAS (CofA 02 Dec 76).

C/n 105 N-262D51.   F/f 24 Jun 74 and stored.   To French
         AF del 20 Oct 76, mostly as F-RBHL coded "HL"
with CIET 340.

C/n 106 N-262D51.   F/f 22 Nov 74 and stored.   To French
         AF del 09 Jun 76, mostly as F-RBHM coded "HM"
with CIET 340.

C/n 107 N-262D51.   F/f 20 May 75 and stored.   To French
         AF del 23 Sep 76, mostly as F-RBHN coded "HN"
with CIET 340.

C/n 108 N-262D51.   F/f 26 Mar 76.   To French AF del 23
         Jun 76, mostly as F-RBHO coded "HO" with CIET 340.

C/n 109 N-262D51.   F/f 15 Sep 76.   To French AF del 23
         Feb 77, mostly as F-RBHP coded "HP" with CIET 340.

C/n 110 N-262D51.   F/f 29 Nov 76.   To French AF del 11
         Mar 77, mostly as F-RBHQ coded "HQ" with CIET 340.

Note on Military Call-Signs

1   French Air Force

Armée de l'Air aircraft were delivered to GAEL (Groupe
Aérien d'Entraînement et de Liaison) based at Villacoublay
near Paris, with call-signs in the F-RBOA block.   They
were transferred to the Aéronavale (with the exception of
the ill-fated F-RBOA c/n 44).   By the time the Nord 262Ds
were delivered, a reorganisation of the unit had taken
place and the aircraft were coded in the F-RBAA block, the
first aircraft initially flying with the CEAM.   The last
two letters of the call-sign were painted on the nose.
Gradually aircraft have been put at the disposal of
regional HQ, keeping their original identities, but
recently new codes have been allocated within Escadrilles
de Liaison Aériennes (eg ELA 44 with F-SCCA block).

2   French Navy

Aéronavale aircraft are used both as communications
and transport aircraft (the so-called "Lialog" or Liaisons
Logistiques scheme having been set up to provide the Navy

with its own air transport service) and as trainers as well. The following blocks of call-signs have been or are still used:

F-YCAA + F-YCBA - Escadrille 2S (now F-YDAA + F-YDBA)
F-YCCA + F-YCEA - Escadrille 3S (now F-YDCA + F-YDDA)

F-YCKA - Escadrille 11S/Section de Liaison de Dugny, which is now Section de Soutien de Dugny with F-YEFA block
F-YFFA + F-YGFA - Escadrille 55S (twin-engine conversion unit), now F-YDOA

MERCURE

The Dassault company (now Dassault-Bréguet), a major manufacturer of fast-selling military jets, had always tried to break into the civil market - and the Falcon jet sales bear testimony to their initial success. At a higher level, however, ventures into commercial transport have up to now proved disappointing. The reason probably lies in the fact that a military aircraft manufacturer will have to try especially hard to be successful in the air transport field and that, in the face of world-wide competition, it is difficult to start building large airliners from scratch.

By 1964 Marcel Dassault was evincing interest in designing a short-haul airliner (see the Mystere 30 project) with a capacity of about 40 passengers. This project was not developed, but a few years later - in 1967 - work began on another short-haul transport, this time a wide-bodied airliner for 150 passengers. The twin-engined Dassault airliner was designed to fill a gap which had been found to exist on the market by 1969, with only the Boeing 727-200 as a competitor.

The go-ahead decision was taken with considerable support from the French Government, which was providing some 55 per cent of the programme cost, including tooling for production. The name Mercure (which had already been selected a few years earlier for an undeveloped airliner project of a smaller capacity) was chosen, Marcel Dassault himself stating jokingly that, when looking for the name of an ancient god, he had found only one to have wings on his helmet and ailerons on his ankles! This was Mercury of course.

Some international co-operation was set up for the production of the Mercure, parts and assemblies of the aircraft being manufactured also in Turin (FIAT-Aeritalia, 10 per cent of the programme), Seville (CASA with 10 per cent, being mostly the forward fuselage section), Haren-Gosselies (the Belgian concern SABCA having a 6 per cent share) and a minor Swiss contribution (2 per cent) from FFA-Emmen; Canadair also joined the programme at a later date.

Two prototypes and two static test airframes were ordered, and Mercure 01 made its maiden flight from Mérignac in May 1971. A modification programme took place in autumn of the same year, and flight tests were resumed until more modifications were incorporated late in December. With this final aerodynamic touch-up, the aircraft flew again on 12 April 1972.

With all 01 aerodynamic refinements and changes incorporated, the first flight of the second prototype (02) took place five months later, this time at Istres.

A brand-new assembly hall had been erected specifically for Mercure production next to the BA.125 "Charles Monier", which is also the main site in France for tests and trials, the station housing the CEV southern branch - with Mediterranean weather permitting all year round operations. The major sub-assemblies for the manufacture of the Mercure were to be flown in by Aeromaritime Super Guppy.

The first prototype, which had been demonstrated in June 1972 at the Turin Air Show, landed later in the same month at Orly for airport operational feasibility evaluation. This prototype was also engaged in autoland trials in view of future Air Inter operation with CAT. III certification. ATT (atterrissage tout temps or blind-landing autoland operations) by Air Inter are very necessary on French domestic flights such as the Paris-Lyon route if the airline is to remain competitive with fast trains in winter and poor weather. This was the reason why Air Inter had initiated regular ATT services with its Caravelle fleet, and with the airline's order for ten Mercures ATT operation remained equally vital.

French certification of the Mercure was originally intended to be obtained by October 1973 but was considerably delayed while possible restrictions - temporary at least - on autoland operation were announced. This, coupled with production delays, caused serious concern to Air Inter to the extent that the airline even considered bringing legal action against the manufacturers. (By the time the aircraft finally entered commercial service in June 1974, CAT.III certification still had to be obtained.)

Following air transport certification and CAT. II certification on 12 February 1974, an urgent need was felt to demonstrate the aircraft's ability, as clouds were gathering over the future of the Mercure. The French Government also made it clear that no further financial aid would be granted. The second prototype therefore made a demonstration flight with a full load on the Paris-Casablanca route (with heavy test instrumentation and some 130 passengers on board).

At last the long-expected delivery of Mercure No 1, the first production aircraft, took place in May 1974, opening revenue service on the Paris-Lyon route on 4 June. By July 1974 the first four production Mercures were flying. Meanwhile, No 5 was carrying out CAT.III operation trials, and CAT.IIIA certification was at long last obtained on 30 September 1974, aircraft No 5 and No 6 being the first fully operational ATT Mercures to enter Air Inter service.

The first four aircraft were returned to Dassault for a short period, being in turn brought up to date with CAT.III operational provisions. (Note: ATT decision height is 15m -

5O ft - with runway visual range of 150m - 500 ft.   Equipment includes Bendix autopilot and Thomson-CSF HUD.)   No 1 was redelivered to Air Inter only much later, in 1975, as the aircraft was non-standard and had to undergo lengthy modifications.

The tenth and last aircraft was delivered to the airline in December 1975, as sales prospects for the Mercure were looking very uncertain. When it was realised that further orders could not be expected, the Mercure affair appeared a serious failure.   The ten production aircraft were reportedly sold at a loss (which is quite understandable with so small a production for such a large airliner), while the taxpayer was now forced unwillingly to provide Air Inter with an additional subsidy;   indeed, with a substantial increase in the expected cost of spares for the ten aircraft in service, French government support was obtained over a plan of several years.   The Mercure failure once more gave fuel to advocates of a nationalisation of Dassault-Bréguet, an often-debated political issue in French affairs.

Since early 1973, however, there had been a new Dassault proposal called the Mercure 200 (as opposed to the Mercure 100 initial production batch).   Proposed as a slightly stretched version, initially with JT8D-17s and later GE/SNECMA CFM-56 jets, the Mercure 200 had a gross weight increased to 65,000 kg with accommodation for 147-176 passengers - later 160-180 - depending on airline needs.   But the new project was soon to cause another political storm, as Dassault turned to American manufacturers (McDonnell-Douglas and Lockheed) to develop the aircraft in co-operation.   The 1975 preliminary talks led to serious discussions of the project in 1976 with MDD only, with the French Government imposing a 40 per cent share of the national aerospace industry on production of the new aircraft. In August 1976 an agreement was reached between MDD, Dassault-Breguet and SNIAS, with plans being made to set up the assembly line at Toulouse under SNIAS control.

In October 1976, while the second Mercure prototype was flying across the Atlantic to be demonstrated at an MDD-held symposium for airline executives, the project name was changed to the ASMR Transport (Advanced Short to Medium Range).   No final decisions had been taken at the time of going to press.

On the other hand, the earlier Mercure 100 is flying with Air Inter without any problems and can be said to be a very reliable aircraft. It is well known that if you do not build many aircraft, they will be more expensive, and of course when they are expensive, they do not sell!   A difficult point indeed, but at a time when the replacement of Air France Caravelles was becoming a serious problem, the Mercure 100 should have been more seriously considered.

Specifications

AMD Mercure 01 Prototype

Span: 30.55m
Length: 34m
Height: 11.36m
Wing area: 116 sq m
Empty weight: 27,900 kg
Maximum take-off weight: 52,000 kg
Disposable load: 134 passengers and additional freight for 4,580 kg
Power plants: 2 x 6,803 kgp Pratt & Whitney JT8D-11

Cruising speed: 946 km/hr at 5,800m
Service ceiling: 10,700m
Maximum range: 1,770 km (134 passengers with fuel reserve)

Typical stage was given as 800 km with 134 passengers and 4,100 kg fuel reserve.   FAR 25 take-off distance 1,660m.

Mercure Production Aircraft (Mercure 100)

Span: 30.55m
Length: 34.84m
Height: 11.36m
Operating empty weight: 31,600 kg (150 passengers)
Maximum take-off weight: 56,500 kg
Maximum payload: 14,900 kg
Power plants: 2 x 7,030 kg thrust Pratt & Whitney JT8D-15
Maximum cruising speed: 932 km/hr
Range with 150 passengers and 8,500 lb (3,840 kg) fuel reserve: 1,120 nm (2,070 km)

Typical short haul 500 nm (925 km) stage with 150 passengers and 7,830 lb (3,550 kg) fuel reserve is flown in 68 minutes.   FAR 25 take-off distance 1,920m.

Interior layout easily interchangeable with up to 3,000 kg of freight in addition to the baggage of the 150 passengers, or 6,000 kg while using all the cargo hold capacity.   Fuel capacity 18,400 l in integral wing tanks.

Mercure 200 Project (1973 proposal)

Basic 146 passenger version
Span: 31.94m
Length: 35.86m
Height: 11.36m
Wing area: 118 sq m
Maximum take-off weight: 59,000 kg
Maximum payload: 16,000 kg
Power plants: 2 x Pratt & Whitney JT8D-17 each rated at 7,250 kgp (as compared with the 7,030 kgp JT8D-15 on the earlier Mercure 100)
Estimated performance: High speed cruise Mach 0.82
                       Long range cruise Mach 0.76
                       Range with maximum payload 2,410 km at 719 km/hr and 9,144m
                       Take-off distance 2,800m

Mercure CFM-56 Project (1976 proposal)

It is planned that the second Mercure prototype (F-BTMD) should test the power plant in flight and become the evaluation prototype for a possible production batch.

Span and height: as for Mercure 100
Length: approx 36.8m
Maximum take-off weight: 64,000 kg
Power plants: 2 x 9,966 kgp General Electric/SNECMA CFM-56 or possible uprated variants
Range with maximum payload: 3,700 km (basic 147 passenger version)

Production List

C/n 01   F-WTCC.   F/f 28 May 71.

C/n 02   F-WTMD.   F/f 07 Sep 72.   To F-BTMD Avions Marcel Dassault (CofA 08 Mar 73).

C/n 1   F-WTTA.   F/f 19 Jul 73.   To F-BTTA Air Inter del 16 May 74.

C/n 2   F-BTTB.   To Air Inter del 29 May 74.

C/n 3   F-BTTC.   F/f 19 Jun 74.   To Air Inter del 08 Jul 74.

C/n 4   F-BTTD.   F/f 22 Jul 74.   To Air Inter del (CofA 10 Sep 74).

C/n 5   F-BTTE.   To Air Inter, del   (CofA 15
        Oct 74).

C/n 6   F-BTTF.   F/f 19 Nov 74.   To Air Inter, del
        (CofA 21 Nov 74).

C/n 7   F-BTTG.   To Air Inter, del   (CofA 13
        Jan 75).

C/n 8   F-BTTH.   To Air Inter, del   (CofA 21
        Feb 75).

C/n 9   F-BTTI.   To Air Inter, del   (CofA 28
        Mar 75).

C/n 10 F-BTTJ.   To Air Inter, del 19 Dec 75.

GERMAN AIRCRAFT

Following the collapse of France in the dark days of June 1940 and in spite of Churchill's proposal for a Franco-British union, the German terms for an armistice were accepted and the cease-fire took place on 25 June 1940.   Only northern France and the Atlantic coast to the Spanish border were included in the Occupation Zone by German forces.

In the following years, the relations of the French Vichy Government with Germany were to become somewhat ambiguous - even though very little choice was left to the defeated country. On 23 July 1941 the so-called Franco-German Wiesbaden Agreement made provisions for the manufacture of German aircraft, sub-contract work and overhaul activities on behalf of the Luftwaffe by the French aeronautical industry. Plans called for the manufacture of 2,000 German aircraft in the occupied zone, in exchange for 1,000 French military types to be built in still unoccupied territory for delivery to the ill-fated Vichy armed forces - with use in the French colonies in mind.

In a quick reaction to the "Torch" landings in French North Africa, Southern France was overrun in November 1942, and the whole aeronautical industry was compelled to work for the Germans.   With the gradually changing spirits of the French and the increase of underground Resistance, production by French workers and engineers can only be described as painfully slow, often turning into a cunning play. Continual acts of sabotage soon took place, often disguised as accidents on industrial mishaps.   It must also be admitted however that despite such unwilling French contribution to the German war effort, French-manufactured aircraft were indeed delivered - to the Germans - so ruthless was the military occupation.

But after the French Liberation, beginning in the summer of 1944, and in spite of major industrial difficulties, the new Air Ministry found it practical to carry on with the production of German aircraft for which tooling and equipment or spares were available.   This was a sound idea and should not be underestimated in the history of the post-war rebirth of the French aeronautical industry; many of the successful post-war French aircraft were of German origins or developments of originally German designs - the Morane-Saulnier MS-500/502 (Fieseler 156), for instance, the N1000/1002 (Bf 108 Taifun) and N1101 (Me-208) communications aircraft and the SIPA S.10 trainer and its derivatives stemming from the Arado 396.   Some types, on the other hand, were soon abandoned: the AAC 5 and 6 and SNCAC NC-900 versions of the Fw-190, some Ju-88s built in small numbers and used by the French forces, not to mention prototypes test-flown but never ordered into production, such as the He-274 bomber (AAS 01) or the Blohm und Voss Bv-144 transport.

In the transport role, however, large numbers of Ju-52 trimotors were built as the AAC 1, making a serious contribution to the development of French air transport in the difficult months following the end of the war.

The French-built trimotors did not see extensive service in commercial use, but large numbers were kept in the Armee de l'Air and Aeronavale inventories for many years, phasing out of the type taking place only as late as 1960.

The NC-701/702 series, SNCAC derivatives of the Siebel 204, were also built in quantity. The twin-engined communications aircraft was also kept in French military service for a long time, the last aircraft not being withdrawn from use until 1964.   Apart from domestic operators, a small number were exported (to Sweden and Poland) for use as aerial mapping aircraft.

Both the AAC 1 and NC701/702 were a familiar sight in France, and they should not be omitted from a history of French air transport since 1945.

Another German twin-engined transport had been planned for production in France, namely the high-wing tricycle-undercarriage Messerschmitt Me-164.   The design team of Caudron at Issy-les-Moulineaux would have been responsible for production in France, but progress from 1942 was so slow that the aircraft was finally abandoned.

Last but not least, another large German type was built during the Occupation, the Dornier Do-24T flying-boat, of which 48 were manufactured by the former CAMS works at Sartrouville between 1942 and August 1944.   Production was resumed under French control after the Liberation, and SNCAN delivered some 40 more flying-boats of the type to the French Navy to be used in the transport and training rôle until 1953. Contrary to reports printed elsewhere, no French Do-24Ts were transferred to Spain, even though the Spanish Air Force did show interest in acquiring a few Aéronavale survivors in 1953. Being a military type only, the Do-24T is beyond the scope of this monograph but was worth mentioning at a time when there are still talks of a possible revival of a Do-24 derivative 40 years after the first flight of the original prototype.

In conclusion, it can be said that at least in the aeronautical field the dark years of 1940-1944 were not entirely lost.   French designers had been out of touch with modern development, most factories were ruined by 1944, but it was soon possible to start production lines of German aircraft.   In some cases, this was to prove very useful in the immediate post-war years.   On the other hand, it must also be said that the introduction of these former

German types into French service was more often plagued by technical difficulties due to the careful sabotage of spares and components manufactured before the German withdrawal.

But all in all, this unexpected German contribution to French aviation was welcome.

AAC 1

Very little has been published on the French-built Ju-52 trimotors, and yet this extremely reliable, simple and rugged transport aircraft formed the backbone of immediate post-war military and civil air transport of France. Ironically, with production nearing the 450 mark, it was also the transport built in the largest numbers in France (excluding the military production of SNCAC Martinet and Dassault Flamant, the Caravelle was second only with 282 machines built).

During the occupation of France local production of the Ju-52 had been launched under German control in the former Amiot factory at Colombes, in the suburbs to the northwest of Paris. Starting late in 1942 with a monthly output of ten aircraft, production was soon raised, and 321 aircraft were produced in 1943. Immediately after the Liberation of France, in the summer of 1944, it was decided to carry on with production of the original Ju52/3mg 10E basic type. The French designation became AAC 1, the first type produced by the Ateliers Aéronautiques de Colombes (Colombes Aeronautical Works), production of another German type, the Focke-Wulf 190 being also planned as the AAC 5 and 6 but not finalised. The French name chosen for the Ju-52 was Toucan, after the Brazilian bird. In French service however the aircraft was referred to almost exclusively as the Junkers 52.

The first eight AAC 1s were put into Air France service as early as December 1944-January 1945, and deliveries to the airline were increased regularly throughout 1945, while many more were allocated to the French Air Force and Navy. In two years more than 400 aircraft were built, the last AAC 1 (c/n 415) being taken on strength by the Armée de l'Air in 1948. Some 150 appeared with civil operators, at least temporarily. To this total must be added another batch of Armée de l'Air trimotors (with c/n 1001-1038), and the few aircraft of German origin used by the military with the original 'Werknummer'.

Air France received no fewer than 83 aircraft, with a few more allocated but not taken up. The AAC 1 was promptly used at home and overseas, including French West Africa and Madagascar in particular; the lack of adequate communications in that large island required an appropriate airline network, and the rugged construction and ease of maintenance of the trimotor proved very helpful.

Back home, night-mail services had also been resumed by the Département Postal d'Air France with three specially modified aircraft (F-BAJS, F-BAKK and F-BALN) on the long run between Paris and southwest France (Bordeaux-Toulouse-Pau). Between 1,200 and 1,500 kg of mail was carried, usually in rudimentary and therefore dangerous conditions. Hair-raising operations and a few incidents and taxying accidents culminated in the crash of F-BAJS during the night of 27-28 June 1946.

A small number of aircraft were transferred or loaned to foreign operators, appearing on the registers of the Lebanon (one later ending up in the Yemen), Yugoslavia, Czechoslovakia and Portugal. The aircraft transferred to Yugoslavia joined original Ju-52s salvaged by the Yugoslavs and apparently used both for military and commercial operation, with the establishment and development of a national State air company, JAT. Recent examination of the Ju-52 preserved at the National Air Museum near Belgrade has shown that this aircraft is in fact a French-built AAC 1 as shown by SNCASO and AIA labelled components. Unfortunately, the identity of the aircraft, which was in military marks and coded "7208" when withdrawn from use, has not yet been ascertained; it is believed to be c/n 222 but not confirmed.

At home however the large Air France AAC 1 fleet was not to last very long. From 1945-46 transfers were made, and during the following years more and more were returned to the French Government or taken up by the Armee de l'Air (approximately eight in 1947, 26 in 1948 and five in 1949), not to mention the many aircraft which crashed. In fact, only about half a dozen machines were kept in Madagascar and Africa.

Other civil operators received the AAC 1 from 1946: TAI, SANA, Air Atlas and a handful of smaller and short-lived companies. The AAC 1 also saw service with various government agencies, but again many were taken up by the French Air Force to the extent that the familiar corrugated metal trimotor soon disappeared from the civil register.

No great effort regarding possible export sales was made, however, following the OFEMA demonstration tour of French aircraft to Latin America in the summer of 1946 (the AAC 1 was not included), but an order for BMW engines was placed by Brazil where original Junkers 52s were in service. Finalisation of the order was uncertain, especially since the South American Ju-52s were re-engined with American power plants.

The AAC 1 was an old aircraft, and it is said that an old soldier never dies; the type was kept in French military service (both Armée de l'Air and Aéronavale) well into the 1950s, proof of an exceptional longevity of life. Reliability and rugged construction were appreciated overseas - in French Indochina, over Africa and the Sahara - and the last military aircraft were withdrawn from use only at the end of 1960. Military use of the AAC 1 had been much longer than commercial use, and one AAC 1 (c/n 216) has miraculously survived to be included in the growing Musée de l'Air Collection at Le Bourget, and c/n 363, transferred to the Germans in 1958 for a nominal sum, went to Munich for museum display. An interesting postscript is the transfer of 13 Armée de l'Air machines to the Portuguese Air Force in December 1960 for further operation of the elderly type.

There is a strong suspicion that the Ju-52 preserved by the Imperial War Museum at Duxford might be a French AAC 1 or at least incorporate some components of French manufacture.

Definitely not what may be called a French transport aircraft, this indigenous version of the immortal Junkers trimotor nevertheless made a significant contribution to the re-birth of French air transport in the immediate post-war years.

Specification

Low-wing trimotor, 20 passengers with crew of three. Fixed conventional landing gear.   Slots and landing flaps.

Span: 29.25m
Length: 18.90m
Height: 6m
Empty weight: 6,400 kg
Loaded weight: 11,000 kg
Power plants: *3 x BMW 132Z, 725 hp at 2,050 rpm on take-
off
435 hp at 1,750 rpm cruising speed
Maximum speed: 290 km/hr
Cruising speed: 220 km/hr at 915m
246 km/hr (at sea level?)
Range: 1,200 km (fully loaded with 2,940 litres fuel capacity)
Endurance: 6hr 30min
Service ceiling: 4,500m
Payload: 2,460 kg
* Some AAC 1s were fitted with engines other than the BMW 132, but no details have been found.

In late military service the specification was slightly altered; crew of four, empty weight 6,600 kg.

Maximum speed: 210 km/hr
Cruising speed: 190 km/hr
Payload: 1,350 kg over 1,000 km, or 13 fully-equipped soldiers or paratroops

Production List

Introductory Remarks

a   Construction numbers of early aircraft were usually painted on aircraft, with an "0" prefix (eg 01, 003, 028, 070 for c/n 1, 3, 28 and 70 respectively).

b   When known, significant dates in military service are given as follows: date of official acceptance/ date SOC (eg Jun 48/26 Jul 60).   No attempt has been made to give military identities except in a few cases.

c   Some French AF acceptance dates appear to be the month in which the aircraft were rolled out of the factory rather than the date of introduction into military service proper, as clearly shown with some aircraft transferred from civil operators.

C/n 1   F-BAJA.   To Air France del 08 Dec 44.   To AIA 21 Jan 48 for French AF Jan 48.   (Canc Apr 48.) In service until 13 Jan 61.   "01" had already been in military service when commandeered and operated by French AF transport unit III/15 circa Apr 45.

C/n 2   F-BAJB.   To Air France del 13 Dec 44.   Written off 04 Oct 47 at Pau.

C/n 3   F-BAJC.   To Air France del 15 Dec 44.   To AIA 31 Jul 47 (conversion to military marks).   Also commandeered during early 1945 with III/15 Apr 45 as 03, later 003 in Air France marks.

C/n 4   F-BAJD.   To Air France del 18 Dec 44.   Retd to

Govt 05 Aug 48 at Villacoublay for IGN but used as F-ZIFA when involved in accident 30 Sep 47 with possible temporary military use before operation by IGN. Repaired and canc late 1952.   Derelict at Creil for some time.

C/n 5   F-BAJE.   To Air France del 03 Jan 45.   Served with III/15 in 1945 (see c/n 1 and 3).   To AIA 02 Sep 47 for French AF May 48 until transferred to Portuguese AF 19 Dec 60.

C/n 6   F-BAJG.   To Air France del 03 Jan 45.   Based in Madagascar.   To AIA for French AF 23 Mar 48/08 Mar 59.

C/n 7   F-BAJH.   To Air France del 17 Jan 45.   Based in Madagascar.   To AIA 05 May 48.   Restored as F-BBOF (CofA 18 Mar 54) for Préfecture de Police.   Canc May 55 and WFU at Le Bourget after take-off accident at Gonesse 17 Mar 55.   Was derelict at Le Bourget.

C/n 8   F-BAJI.   To Air France del 22 Jan 45 "Lt A Texier".   To AIA 03 Jan 48 for French AF.   SOC date not traced.

C/n 9   To French AF Jan 48/15 Jul 59.

C/n 10   To French AF, later French Navy.

C/n 11   To French AF.

C/n 12   To French AF.

C/n 13   F-BAJJ.   To Air France del 27 Jan 45.   To AIA 22 Jan 48.   To French AF.   Acceptance date given as Nov 54 (!) /SOC 15 May 60.

C/n 14   F-BAJK.   To Air France del 27 Jan 45.   Based in Madagascar from Aug 48.   Canc Sep 55.   Presumed transferred to SGACC.   WFU in Madagascar.

C/n 15   F-BAJL.   To Air France del 01 Feb 45.   Based in Madagascar.   Canc Jan 53.   Presumed transferred to SGACC.   WFU in Madagascar.

C/n 16   F-BAJM.   To Air France "Oradour" del 06 Feb 45. To AIA 12 Mar 48 for French AF Mar 48.   Destroyed 15 Oct 57.

C/n 17   F-BAJN.   To Air France del 09 Feb 45.   To AIA 22 Apr 48 for French AF.   SOC 20 Oct 55.

C/n 18   F-BAJO.   To Air France del 26 Jan 45.   To AIA 05 Mar 48.   To French AF acceptance date Jul 55(!). SOC 31 Dec 60.

C/n 19   F-BAJP.   To Air France del 23 Jan 45.   Written off 10 Sep 45 at Le Bourget.   Reportedly commandeered and with III/15.

C/n 20   F-BAJS.   To Air France del 01 Feb 45 (CofA 17 Apr 45).   Operated on postal service by CEPM.   Flew into high-tension cables in night crash near Pau 27/28 Jun 46 and dbf (two wounded, one killed, one died later).

C/n 21   To French AF.

C/n 22   To French AF.

C/n 23   To French AF.

C/n 24   No information traced.

C/n 25   No information traced.

C/n 26   To French Navy (served with French West Africa flight Dakar as "DK-1").

C/n 27   No information traced.

C/n 28   To French Navy.

C/n 29   To French AF

C/n 30   To Frency Navy

C/n 31   To French military, presumably French AF.

C/n 32   To French Navy.

C/n 33   To French Navy.

C/n 34   No information traced.

C/n 35   No information traced.

C/n 36   F-BAJU.   To Air France del 03 Apr 45.   To AIA
         21 Jan 48 for French AF.   SOC 22 Nov 60.

C/n 37   F-BAJV.   To Air France del 03 Apr 45.   To AIA
         27 Aug 47.   To French AF acceptance date Jan
51(!)/SOC 13 Jan 61.

C/n 38   F-BAJX.   To Air France del 03 Apr 45.   To AIA
         19 Aug 47 for French AF.   Destroyed 12 Apr 57.

C/n 39   To French Navy.

C/n 40   No information traced.

C/n 41   To French Navy.

C/n 42   To CEV, later to French Navy.

C/n 43   To French AF.

C/n 44   F-BAJT.   To Air France del 05 Apr 45.   Written
         off 08 Aug 46 at Le Bourget.

C/n 45   To French AF Mar 45 to 22 Nov 60.

C/n 46   To French AF.

C/n 47   To French AF.

C/n 48   To French AF Jul 48.   Transferred to Portuguese
         AF 26 Nov 60.

C/n 49   To French AF.

C/n 50   To French AF.

C/n 51   To French AF.

C/n 52   To French AF.

C/n 53   To French AF.   Acceptance date given as early as
         Oct 44(!)   Transferred to Portuguese AF 26 Nov
60.

C/n 54   To French AF.

C/n 55   F-BAKM.   To Air France del 25 Apr 45.   Written
         off 20 Mar 47 at Freetown, Sierra Leone.

C/n 56   F-BAKN.   To Air France del 25 Apr 45.   WFU Feb
         53 due to corrosion and to DGER for spares or
ground training.

C/n 57   F-BAKO.   To Air France del 30 Apr 45.   Dbr 04
         Feb 46 at Mahon, Balearic Islands.

C/n 58   F-BAKP.   To Air France del 01 May 45.   Written
         off 05 Mar 47 at Le Bourget.

C/n 59   F-BAKQ.   To Air France del 04 May 45.   To AIA
         02 Sep 47 for French AF Sep 47/SOC 26 Aug 60.

C/n 60   F-BAKR.   To Air France del 01 May 45.   To AIA
         20 Oct 47 for French AF.

C/n 61   No information traced.

C/n 62   No information traced.

C/n 63   To French AF.   SOC 13 Jan 61.

C/n 64   To French AF.

C/n 65   No information traced.

C/n 66   To French AF May45/16 May 60.

C/n 67   To French AF.

C/n 68   To French AF.   SOC 15 Jul 59.

C/n 69   To French AF.   SOC 08 Sep 55.

C/n 70   To French AF.

C/n 71   F-BAKS.   Intended for Air France but NTU and del
         To French AF Service du Materiel.   Destrcyed 23
Jun 51.

C/n 72   F-BAKT.   To Air France del 30 May 45.   To AIA 12
         Jan 49 for French AF.   072 also given for air-
craft F-RBDA French AF, Indochina 1952, but not confirmed.

C/n 73   F-BAKU.   To Air France "Beiteddine" del 31 May
         45.   Became LR-AAC Jun 46.   To Yemen Govt 13 Apr
48.

C/n 74   F-BAKV.   To Air France del 15 Jun 45.   Written
         off 07 Jun 47 at Dakar-Yoff.

C/n 75   F-BAKX.   To Air France del 04 Jun 45.   Later
         loaned to Service des Telecommunications et de
Signalisation.   To AIA 12 Feb 49 for French AF.

C/n 76   F-BAKY.   To Air France del 05 Jun 45.   To AIA 23
         Jun 48 for French AF.   Destroyed 18 Oct 55.

C/n 77   F-BAKZ.   To Air France del 06 Jun 45.   To YU-...
         JAT del 15 Aug 45.   (Known aircraft transferred
to Yugoslavia are c/n 77, 78 and 316 - but it is not
known which one has been preserved near Belgrade after
withdrawal from JRV service.   Also it is not known
whether the trimotor destroyed in a crash at Rijeka/Fiume
was an original Ju-52 or an AAC 1.)

C/n 78   F-BALA.   To Air France del 01 Jun 45.   To YU-...
         JAT del 18 Aug 45.

C/n 79   F-BALB.   To Air France del 16 Jun 45.   To AIA 19
         Feb 48 for French AF.   SOC 1953.

C/n 80   F-BALC.   To Air France del 16 Jun 45.   To AIA 19
         Aug 48 for French AF to 11 Mar 59.

C/n 81   To French AF.   SOC 06 Apr 60.

C/n 82   To French AF.   SOC 24 Jun 55.

C/n 83   To French AF

C/n 84   To French AF.   Acceptance date given as early as
         Jan 45.   SOC Aug 59.

C/n 85   To French AF.

C/n 86   To French AF Jan 45 to 21 Mar 62, possibly the
         last AAC 1 in Armée de l'Air service.

C/n 87   To French AF Jun 45 to 18 Jan 60.

C/n 88   To French AF.

C/n 89   To French AF Jun 45 to 14 Sep 55.

C/n 90   F-BALE.   To Air France del 02 Jul 45.   Based in
         Madagascar.   Crashed 10 Apr 53 at Miandrivazo,
Madagascar.   Commandant Dedieu, one of the oldest Air

France pilots at that time, and pioneer of French civil air transport, died in the crash.

C/n 91 F-BALF.   To Air France del 02 Jul 45.   Written off 01 Jul 47 at Eseka, 100 km WSW of Yaounde, French Cameroon, after flying into hill on flight from Bangui to Douala.

C/n 92 F-BALG.   To Air France del 19 Jul 45.   To AIA 23 Aug 48 for French AF.   SOC 18 Oct 55.

C/n 93 F-BALH.   Intended for Air France but NTU and transferred to SCRAS.   To French AF.   SOC 11 Dec 56.

C/n 94 F-BALI.   To Air France del 04 Jul 45.   To AIA 22 Mar 48 for French AF.

C/n 95 F-BALJ.   To Air France del 04 Jul 45.   To AIA 05 Feb 48 for French AF.   SOC 20 Oct 55.

C/n 96 F-BALK.   To Air France del 04 Jul 45.   Written off 02 Feb 46 at Belo, Madagascar.

C/n 97 F-BALL.   To Air France del 04 Jul 45.   Transferred almost immediately to SCRAS.

C/n 98 F-BALM.   To Air France del 09 Jul 45.   Transferred almost immediately to SCRAS.

C/n 99 F-BALN.   To Air France del 05 Jul 45 (CofA 29 Aug 45).   Operated on postal service by CEPM. To AIA 21 Apr 49 for French AF.   Canc Jun 49.

C/n 100 F-BALO.   To Air France del 05 Jul 45.   Based in Madagascar.   WFU 17 Jul 51.

C/n 101 F-BAKK.   To Air France del 19 Jul 45.   To AIA 05 Nov 48 for French AF.   SOC Dec 59.

C/n 102 F-BAKL.   To Air France del 09 Jul 45.   Written off 23 Nov 45 at Toulouse.

C/n 103 F-BALD.   To Air France del 06 Aug 45.   Canc 08 Jul 48 as transferred to CATRE when retd to Govt at Villacoublay for IGN.   Restored 1949 to SGACC, still operated by IGN.   Canc late 1951.   To French AF.

C/n 104 F-BAMO.   To Air France del 20 Jul 45.   To AIA 29 Nov 48 for French AF.   SOC 15 Jul 59.

C/n 105 F-BAMP.   To Air France del 27 Jul 45.   To AIA 30 Jul 47 for French AF.   SOC 07 Mar 56.

C/n 106 F-BAMQ.   To Air France del 06 Aug 45.   Based in Madagascar.   Written off 01 Jan 52 between Andapa and Sambava.

C/n 107 F-BAMR.   To Air France del 25 Jul 45.   To AIA 10 Feb 48 for French AF.

C/n 108 F-BAMS.   Intended for Air France but severely damaged prior del.   Repaired and to French AF.

C/n 109 F-BAMT.   To Air France "Safa" del 27 Jul 45. Became L-RAMT Feb 46, then LR-AAJ (CofA 16 May 46).   To Syrian AF 09 Jul 49.

C/n 110 F-BAMU.   To Air France del 06 Aug 45.   To AIA 26 Nov 48 for French AF.   Acceptance date given as Mar 46).   SOC 12 Jun 55.

C/n 111 F-BAMV.   To Air France del 28 Jul 45 but almost immediately retd to DTA.

C/n 112 F-BAMX.   To Air France del 31 Jul 45.   Based in Madagascar and later at Brazzaville.   WFU Feb 52.

C/n 113 F-BAMY.   To Air France "El Arz" del 31 Jul 45. Became L-RAMY Feb 46, then LR-AAI (CofA 16 May 46).   To Syrian AF 09 Jul 49.

C/n 114 F-BAMZ.   To Air France del 30 Jul 45.   Retd to Govt 20 Sep 45 and canc.   To French AF Feb 47 to 29 Mar 60.

C/n 115 F-BANA.   To Air France del 30 Jul 45.   To AIA 22 Mar 48 for French AF Mar 48 to 14 Sep 60.

C/n 116 To French AF.

C/n 117 To French AF Jul 45 to 06 Apr 60.

C/n 118 To French AF.   Destroyed 21 Aug 51.

C/n 119 To French AF.

C/n 120 To French AF Jul 45 to 16 May 60.

C/n 121 To French AF Oct 47 to 06 Apr 60.

C/n 122 No information traced.

C/n 123 Written off 12 Jun 50 near Ambatondrazaka, Madagascar.

C/n 124 No information traced.

C/n 125 To French AF Jun 48 to 26 Jul 60.

C/n 126 To French military, presumably French AF or CEV (coded "H" in Aug 53).

C/n 127 To French AF, transferred to Portuguese AF 10 Dec 60.

C/n 128 To French AF.   SOC 26 Sep 55.

C/n 129 To French AF 1945 to 13 Jan 61.

C/n 130 To French AF Aug 45.   SOC Dec 59.

C/n 131 To French AF.   SOC 26 Aug 60.

C/n 132 No information traced.

C/n 133 To French AF Aug 45 to 22 Nov 60.

C/n 134 No information traced.

C/n 135 No information traced.

C/n 136 F-BANB.   To Air France del 05 Sep 45.   Written off 05 Jun 47 near Gémenos (St. Baume Mountain, Dept Bouches du Rhône) on flt from Nice to Paris.

C/n 137 F-BANC.   To Air France del 25 Sep 45.   Based in Madagascar and later at Brazzaville.   WFU Dec 51.

C/n 138 F-BAND.   To Air France del 25 Sep 45 and almost immediately canc.

C/n 139 F-BANE.   To Air France del 03 Oct 45 and almost immediately canc.

C/n 140 F-BANF.   To Air France del 07 Nov 45 (regd to SGACC?).   To AIA 09 Feb 49.

C/n 141 F-BANG.   To Air France del 05 Oct 45.   To Service des Télécommunications et de Signalisation Mar 46.   To SALS for Centre National de St-Yan Nov 49. WFU 07 Oct 52.

C/n 142 F-BANH.   To Air France del 02 Oct 45.   To AIA 12 Oct 48 for French AF Oct 48 to 09 Sep 59.

C/n 143 F-BANI.   Intended for Air France but NTU.   Kept by DTI for French AF.

C/n 144 F-BANJ.   To Air France del 22 Oct 45.   Based in Madagascar.   WFU Jun 53.

C/n 145 F-BANK.   To Air France del 10 Oct 45.   Based in

Madagascar.   Accident Antalaha, Madagascar, 05 Dec 52.   WFU Jan 53.

C/n 146 F-BANL.   NTU.

C/n 147 F-BANM.   To Air France del 22 Oct 45.   To AIA 14 Oct 48 for French AF.   SOC 13 Jan 61.

C/n 148 F-BANN.   To Air France del 26 Oct 45.   To CS-ADA Aero Portuguesa del 10 Apr 46.   To F-BDYK Air France retd Dec 47.   Based in Madagascar. Apparent SGACC ownership Jan 49-Aug 49.   Canc Jul 57. WFU in Madagascar.

C/n 149 F-BANO.   To Air France del 16 Oct 45.   Retd to Govt and written off 10 Nov 45 at Le Bourget.

C/n 150 F-BANP.   To Air France del 25 Oct 45.   Written off 13 Jan 46 at Le Bouscat, near Bordeaux, Dept de la Gironde.

C/n 151 F-BANQ.   Intended for Air France but kept by Service du Matériel.   To SOCOTRA as F-BANQ (CofA 20 Jan 47).   To Aigle Azur 24 Apr 47.   To Roger Colin/ CATI Sep 48.   Written off 21 Aug 49 in Red River at Lao Kay.

C/n 152 F-BANR.   To Air France del 20 Nov 45.   To AIA 27 Aug 47.   To French AF May 48 to 16 May 60.

C/n 153 F-BANS.   NTU.

C/n 154 To French Navy.

C/n 155 To French AF Aug 47 to 24 Jun 55.

C/n 156 No information traced.

C/n 157 To French AF.   SOC 08 Sep 55.

C/n 158 To French AF.

C/n 159 No information traced.

C/n 160 To French AF.

C/n 161 To French AF Feb 45 to 09 Sep 59.

C/n 162 To French AF Feb 45 to 09 Sep 59.

C/n 163 To French AF.

C/n 164 To French AF.

C/n 165 To French AF Oct 45 to 22 Nov 60.

C/n 166 To French AF.

C/n 167 To French AF Feb 48 to 26 Jul 60.

C/n 168 To French AF Nov 45 to 13 Jan 61.

C/n 169 To French AF.   SOC 18 Feb 55.

C/n 170 To French AF Nov 45 to 26 Jul 60.

C/n 171 To French AF.

C/n 172 To French AF Dec 45.   SOC 1955.

C/n 173 To French AF.   SOC 09 Aug 56.

C/n 174 To French AF.   Destroyed Indochina.

C/n 175 To French AF.   SOC 20 Oct 55?

C/n 176 To French AF.

C/n 177 To French AF.

C/n 178 To French AF.   SOC 03 Feb 60.

C/n 179 To French AF Jan 46 to 17 Mar 58.

C/n 180 F-BDYI.   To ONERA (CofA 22 Mar 50) (ex military built 1945).   To CEV 16 Feb 56 and canc. Restored Jan 57 for Société Transfrigorifique du Soudan. Canc May 59.

C/n 181 To French AF Jul 47 to 16 May 60.

C/n 182 To French AF.

C/n 183 No information traced.

C/n 184 To French AF.

C/n 185 To French AF.

C/n 186 To French AF Jan 46 to 09 Sep 59.

C/n 187 No information traced.

C/n 188 To French AF Aug 48 to 16 May 60.

C/n 189 To French AF.

C/n 190 To French AF.   SOC 17 Jan 55.

C/n 191 To French AF.

C/n 192 To French AF.   Destroyed 05 Jun 48.

C/n 193 To French AF.

C/n 194 F-BBZF.   To SANA (CofA 14 Jun 44).   To DTI del 18 Dec 47.   Canc Nov 49.   To French AF Mar 48.

C/n 195 F-BBYJ.   To Air France (CofA 16 Apr 46, regd to SGACC).   To AIA 02 Feb 49 for French AF.   SOC 01 Apr 60.

C/n 196 F-BBZG.   To SANA (CofA 14 Jun 46).   To French AF Mar 48 to 22 Nov 56.

C/n 197 F-BBZH.   To SANA (CofA 01 Jun 46).   Canc late 1947.   To French AF 1948.   SOC 26 Jul 60.

C/n 198 No information traced.

C/n 199 To French AF.

C/n 200 To French AF.

C/n 201 To French AF.

C/n 202 F-BBZI.   To SANA (CofA 01 Jun 46).   Canc late 1948 (officially to French AF but more probably written off Nov 46 at St Auban-sur-Erguerre).

C/n 203 F-BBZJ.   To SANA (CofA 14 Jun 46).   To DTI del 05 Apr 48 for French AF until 18 Mar 59.

C/n 204 F-BBZK.   To SANA "Kerjean" (CofA 14 Jun 46). Canc Nov 49.   To French AF Mar 48.   SOC 16 May 60.

C/n 205 To French AF Feb 47 until transferred to Portuguese AF 02 Dec 60.

C/n 206 To French AF.   SOC 08 Sep 55.

C/n 207 To French AF.   SOC 15 Jul 59.

C/n 208 No information traced.

C/n 209 To French AF.

C/n 210 To French AF.

C/n 211 No information traced.

C/n 212 To French AF.

Top   : Caravelle VI-R N1007U (c/n 92) of United Air Lines
         at New York-Idlewild in September 1961 (T O'Dell)
Bottom : Caravelle 10A 'Horizon' F-WJAO (c/n 63)
         (Sud-Aviation via J Wegg)

Nord 262B ZS-IZX of AVNA (K Smy via J-P Dubois)

Top   :   Progenitor of the Nord 260 and 262 series, Max
            Holste 260 F-WJDV in Royal Air Maroc colour
            scheme (Charles W Cain)

Centre :   Nord 262 No.45 of Aeronavale at Lille-Lesquin
            in May 1974 (M Rohart)

Bottom :   Nord 262D No.91, coded AN (F-RBAN) of the Armee
            de l'Air at Dijon in 1972 (B Regnier via S Blandin)

Dassault Mercure 100 F-WTMD in Air Inter colours at Paris-Le Bourget in May 1977 (Avions Marcel Dassault - Breguet Aviation via J-P Dubois)

Top   : AAC 1 of Aeronavale, coded 4.S.6 (E.C.P.Armees)
Bottom : AAC 1 of the Armee de l'Air at Dijon in November
          1955 (Blandin/Regnier)

Top   : NC 701 Martinet No.39 of the Armee de l'Air
           (E.C.P.Armees)

Centre : NC 702 Martinet No.303 of Aeronavale, coded
           BZ.14, at Bizerte in use with the Section de
           Liaison (via J Mutin)

Bottom : NC 701 Martinet F-BAOR in May 1947 (via J Mutin)

Top   : NC 702 Martinet F-BAOO stored at Le Bourget in
          June 1955 (JMG Gradidge)

Bottom : NC 701 Martinet SE-KAE of the RiketsAllmanha
          aero-survey company at Stockholm-Bromma in 1962
          (LE Lundin via J-P Dubois)

Top   : The Blohm und Voss Bv 144, built by Breguet
         (S.C.A. via J Delmas)
Bottom : SE-200 F-BAIY on Berre Lake, west of Marseilles
         (E.C.P.Armees)

C/n 213 To French AF Apr 46 to 13 Jan 61.

C/n 214 To French AF Nov 47 to 14 Sep 60.

C/n 215 To French AF Apr 46 to 22 Nov 60.

C/n 216 To Frency Navy. Served with the French West Africa flight at Dakar coded "DK-2". Later to Escadrille 55S as "55S.32", allocated French Navy S C Aéro number "216" in 1961. WFU and derelict at Villacoublay during mid-60s. By 1975 in storage at Dugny-Le Bourget reserve collection of national Musée de l'Air. Displayed in the open at Le Bourget since May 78 painted as c/n 334, French AF, coded "DG" as sported when flying with Groupe de Transport "Béarn" (GT 1/13) in French Indochina. (The true c/n 334 "DG" was flown by General Lissarague, current Curator and Director of Musée de l'Air.)

C/n 217 No information traced.

C/n 218 No information traced.

C/n 219 To French Navy.

C/n 220 No information traced.

C/n 221 F-BBYA. To STA (CofA 26 Aug 46). Canc Feb 50 (intended for French AF but reported as written off 1948 in Indochina).

C/n 222 F-BBYB. To STA (CofA 26 Aug 46). Canc late 1948, to military markings. Believed overhauled late 1949 by AIA de Clermont (Clermont-Ferrand, Dept du Puy de Dôme) for later transfer to Yugoslavia (with possible intermediate service with French AF). This may be the aircraft preserved in Belgrade.

C/n 223 F-BBYC. To STA (CofA 26 Aug 46). Written off 03 Mar 48 in Mediterranean between Oran and Perpignan.

C/n 224 F-BBYD. To Air Ocean (CofA 14 Jun 46). Impounded 18 Jun 47 at Le Bourget. To Govt Feb 49 and canc. To military markings.

C/n 225 F-BBYE. To LASO France (CofA 27 Jun 46). Canc late 1947. To French AF del 11 Sep 47.

C/n 226 F-BBYF. To Air France (CofA 20 Jun 46, regd to SGACC). To Autrex 24 Aug 50. Written off 26 Jul 51 at Lao Kay.

C/n 227 F-BBYG. To Air France (CofA 13 Jun 46, regd to SGACC). Crashed 30 Apr 47 at Niamey (Niger).

C/n 228 F-BBYH. To Air France (CofA 20 Jun 46, regd to SGACC). To Service des Télécommunications et de Signalisation. Written off 30 May 47 at Aoulef, Central Sahara, Algeria.

C/n 229 F-BBYI. To Air France (CofA 17 Jun 46, regd to SGACC). To SALS for Centre National de Mitry Mory. Canc late 1951.

C/n 230 F-BBZL. To LASO France (CofA 27 Jun 46). To Air Atlas . To Autrex Dec 49. Written off 11 May 53 at Hanoi.

C/n 231 F-BBYK. To TAI (CofA 25 Jun 46). Written off 06 Jan 47 on Mont Ventoux.

C/n 232 F-BBYL. To TAI (CofA 25 Jun 46). Written off 26 Oct 46 at Marignane.

C/n 233 F-BBYM. To TAI (CofA 25 Jun 46). Canc late 1947. To French AF Oct 48 to 04 Sep 58.

C/n 234 F-BBYN. To TAI (CofA 25 Jun 46). Canc late 1947. To French AF. Later transferred to Portuguese AF on 10 Dec 60.

C/n 235 F-BBYO. To Avions Bleus "Martine" (CofA 15 Oct 46). To DTIA for French AF late 1948.

C/n 236 F-BBYP. To Aérocargo (CofA 21 Oct 46). Canc late 1947. To French AF. Destroyed 14 Sep 55.

C/n 237 F-BBYQ. To CGT (CofA 05 Jul 46). To French AF(?) Oct 48.

C/n 238 F-BBYR. To CGT (CofA 05 Jul 46). Canc late 1948. To French AF.

C/n 239 F-BBYS. To CGT (CofA 19 Jul 46). To French AF Jun 48 to 01 Apr 60.

C/n 240 F-BBYT. To Air Nolis (CofA 27 Jun 46). To Air Atlas . Canc late 1951 and scrapped.

C/n 241 F-BBYU. NTU. To French AF May 46 to 26 Jul 60.

C/n 242 To French AF.

C/n 243 To French AF.

C/n 244 To French AF.

C/n 245 To French AF.

C/n 246 F-BBZA. To Air Nolis (CofA 27 Jun 46), also operated by its subsidiary LASO. To SANA "Kerguern" 1947. Canc late 1947. To French AF del 11 Sep 47. SOC 13 Jan 61.

C/n 247 F-BBZB. To SOCOTRA (CofA 10 Jul 46). Retd to DTIA Feb 48.

C/n 248 F-BBZC. To SOCOTRA (CofA 30 Jul 46). Written off 02 Jan 48 at Casablanca.

C/n 249 F-BBZD. To CGT (CofA 19 Jul 46). To French AF Oct 48.

C/n 250 F-BBZE. To Aigle Azur (CofA 30 Sep 46). Damaged at Calvi 05 Apr 47, repaired. Canc Jan 48. To French AF May 48 to 26 Jul 60.

C/n 251 To French AF.

C/n 252 To French AF May 46 to 16 May 60.

C/n 253 To French AF.

C/n 254 To French AF.

C/n 255 To French AF Jun 46. Transferred to Portuguese AF 10 Dec 60.

C/n 256 To French AF. SOC 15 Jul 59.

C/n 257 To French AF.

C/n 258 To French AF Aug 46. Transferred to Portuguese AF 02 Dec 60.

C/n 259 To French AF Oct 45 to 22 Nov 60.

C/n 260 No information traced.

C/n 261 No information traced.

C/n 262 No information traced.

C/n 263 No information traced.

C/n 264 No information traced.

C/n 265 To French Navy.

C/n 266 To French AF Oct 45 to 13 Jan 61.

C/n 267 To French AF Jun 46 to 06 Oct 58 when transferred

82

to French Navy.

C/n 268 To French AF.

C/n 269 To French AF.

C/n 270 To French AF Oct 48 to 16 May 60.

C/n 271 To French AF.

C/n 272 To French AF Jun 46 to 08 Sep 55.

C/n 273 To French AF Jun 46 to 08 Sep 55.

C/n 274 To French AF.

C/n 275 No information traced.

C/n 276 To French AF Jan 48 to 26 Jul 60.

C/n 277 To French AF.

C/n 278 To French AF.

C/n 279 To French AF Sep 46 to 08 Sep 55.

C/n 280 To French AF.

C/n 281 F-BCAA.   To Air Ocean (CofA 23 Jul 46).   Written
off 10 Oct 46 at Sefrou, Atlas Mountains.

C/n 282 F-BCAB.   To SOCOTRA (CofA 30 Jul 46).   To DTIA
Feb 48 for French AF.

C/n 283 F-BCAC.   To Compagnie de Transports Aériens
Languedoc Roussillon (CofA 26 Jul 46).   To Aigle
Azur 31 Jan 47.   Canc late 1947.   To French AF.   SOC
22 Nov 60.

C/n 284 F-BCAD.   To CTA Languedoc Roussillon (CofA 29
Jul 46).   Written off 01 Nov 46 at St Leger,
Haute Vienne.

C/n 285 Intended for STA but not traced (error for 286 or
damaged prior to del?).   To French AF 27 Jun 57.

C/n 286 F-BCAE.   To CTA Languedoc Roussillon (CofA 19
Jul 46).   To STA 1947.   To French AF Sep 48 to
26 Aug 60.

C/n 287 F-BCAF.   To Société des Transports Aériens Alpes
Provence (CofA 13 Mar 47).   Canc Feb 50.   To
French AF (acceptance date given as Jul 46).   SOC 01 Apr
60.

C/n 288 F-BEPD.   To DTIA/CATRE (CofA 16 Mar 48) (ex
military, built 1946).   Retd May 48 to military
marks at end of short-term civil programme.

C/n 289 To French AF Dec 46 to 03 Feb 60.

C/n 290 To French AF Dec 46 to 20 Oct 55.

C/n 291 To French AF Sep 46.   Transferred to Portuguese
AF 02 Dec 60.

C/n 292 To French AF Sep 46 to 09 Jan 57.

C/n 293 To French AF.

C/n 294 To French AF.   SOC 26 Aug 60.

C/n 295 To French AF Oct 45 to 20 Oct 55.

C/n 296 To French AF 1947 to 13 Jan 61.

C/n 297 To French AF.   Crashed 14 Dec 48 (believed
Nigeria).

C/n 298 To French AF.   Crashed 05 Dec 57 (believed Atar,
French Sahara).

C/n 299 To French AF.

C/n 300 To French AF Jul 48 to 14 Sep 60.

C/n 301 To French AF.

C/n 302 To French AF Oct 46 to 29 Aug 55.

C/n 303 To French AF.

C/n 304 To French AF Oct 45 to 14 Sep 60.

C/n 305 To French AF.

C/n 306 To French AF.   Destroyed 09 Jul 51.

C/n 307 To French AF.

C/n 308 F-BCHE.   To Air Ocean (CofA 07 Nov 46).
Impounded 18 Jun 47 at Le Bourget.   To Govt Feb
49 and canc.   WFU.

C/n 309 F-BCHF.   To Avions Bleus (CofA 05 Dec 46).
Canc Feb 49.   To French AF.

C/n 310 F-BCHB.   To Aérocargo (CofA 18 Sep 46).   Canc
Nov 47.   To CEV and later French AF (reported
with CEV as early as 09 Oct 47).

C/n 311 F-BCHD.   To SOCOTRA (CofA 18 Oct 46).   Canc late
1946.   Written off 23 Oct 46 at Baherya, Egypt.

C/n 312 F-BCHA.   To TAI (CofA 12 Sep 46).   Canc late
1947.   To French AF.

C/n 313 F-BCHC.   To CGT (CofA 28 Sep 46).   To French AF
Sep 48.   Destroyed 23 Apr 58.

C/n 314 F-BCHI.   To Aigle Azur (CofA 15 Nov 46).   Canc
late 1947.   To French AF.

C/n 315 F-BCHJ.   To Air Atlas "Ville d'Oujda" (CofA 06 Nov
46).   To French AF Aug 50.   Based at Brazzaville
and later in Madagascar.   WFU Sep 56.

C/n 316 F-BCHK.   To Air Atlas "Ville de Marrakech" (CofA
05 Nov 46).   To Avions Bleus Nov 48 in damaged
condition.   To YU-ACE JAT Mar 51.   Crashed near Rijeka
(Yugoslavia) on flt from Rijeka/Fiume to Zagreb 29 Jun 51.

C/n 317 F-BCHL.   To Air Atlas "Ville de Rabat" (CofA 09
Dec 46).   Canc late 1951 and scrapped.

C/n 318 F-BCHM.   To Air Atlas (CofA 09 Nov 46).   Canc
late 1951 and scrapped.

C/n 319 F-BCHN.   To Air Atlas (CofA 21 Oct 46).   Canc
Aug 52 and scrapped.

C/n 320 F-BCHO.   To Air Atlas "Ville de Safi" (CofA 06
Nov 46).   Canc Nov 49.   To French AF(?)

C/n 321 F-BCHP.   To Air Atlas "Ville de Mogador" (CofA
10 Dec 46).   To French AF Aug 50.   Based at
Brazzaville and later in Madagascar.   WFU Jul 54.

C/n 322 F-BCHQ.   To Air Atlas "Ville de Meknes" (CofA 26
Oct 46).   Written off 30 Aug 47 at El Ajeb.

C/n 323 F-BCHH.   To Aerocargo (CofA 27 Nov 46).   Written
off 14 Jul 47 at Vichy.

C/n 324 F-BCHG.   To TAI (CofA 25 Oct 46).   Canc late
1947.   To French AF.   SOC 18 Sep 56.

C/n 325 To French AF Mar 47.   Transferred to Portuguese
AF 19 Dec 60.

C/n 326 To French AF Oct 46 to 16 May 60.

C/n 327 To French AF.   Crashed 10 Oct 56 (believed Morocco).

C/n 328 To French AF.

C/n 329 No information traced.

C/n 330 To French AF.

C/n 331 To French AF.

C/n 332 To French AF.

C/n 333 To French AF.

C/n 334 To French AF. Was coded "DG" with GT1/34 or possibly 1/64 (as shown by two-letter codes reflecting change of unit designation). See c/n 216.

C/n 335 To French AF.

C/n 336 To French AF.

C/n 337 No information traced.

C/n 338 No information traced.

C/n 339 To French AF.

C/n 340 To French AF.

C/n 341 To French AF Dec 46. Crashed 02 Apr 57 believed at Djanet, French Sahara.

C/n 342 To French AF Dec 46 to 13 Jun 55.

C/n 343 To French AF.

C/n 344 To French AF.

C/n 345 To French AF.

C/n 346 No information traced.

C/n 347 No information traced.

C/n 348 To French AF Dec 48 to 13 Jan 61.

C/n 349 To French AF.

C/n 350 To French AF Dec 46 to 23 Nov 60.

C/n 351 To French AF. SOC 20 Oct 55.

C/n 352 To French AF.

C/n 353 To French AF.

C/n 354 To French AF. SOC 13 Jan 61.

C/n 355 To French AF Mar 47 to 09 Sep 59.

C/n 356 To French AF Mar 47 to 08 Dec 60.

C/n 357 To French AF Oct 45. Transferred to Portuguese AF 19 Dec 60.

C/n 358 To French AF Mar 47 to 16 May 60.

C/n 359 To French AF. SOC 20 Oct 55.

C/n 360 To French AF.

C/n 361 To French AF Jul 47 to 13 Jan 61.

C/n 362 To French AF Apr 47 to 08 Sep 55.

C/n 363 To French AF. Transferred to Munich museum in Germany 28 Mar 58 and preserved.

C/n 364 To French AF. SOC 02 Feb 56.

C/n 365 To French AF Apr 47 to 26 Jul 60.

C/n 366 To French AF May 47 to 26 Jul 60.

C/n 367 To French AF 1947. SOC date not known.

C/n 368 To French AF.

C/n 369 To French AF.

C/n 370 To French AF.

C/n 371 To French AF.

C/n 372 To French AF Oct 45 to 16 May 60.

C/n 373 To French AF.

C/n 374 To French AF.

C/n 375 To French AF Jun 47 to 16 May 60.

C/n 376 To French AF Jun 47 to 06 Oct 58 when transferred to French Navy.

C/n 377 To French AF.

C/n 378 To French AF.

C/n 379 To French AF.

C/n 380 F-BCHT. To SOCOTRA (CofA 04 Jul 47). Retd to DTIA Feb 48. Reported with French AF. To French Navy.

C/n 381 F-BCHX. To TAI (CofA 30 May 47). Canc late 1947. To French AF. SOC 20 Oct 55.

C/n 382 F-BDYA. To Aigle Azur (CofA 09 Jul 47). Canc late 1948. To French AF.

C/n 383 F-BCHZ. To Avions Bleus (CofA 17 Jun 47). Canc late 1947. To French AF.

C/n 384 F-BCHR. To STA Alpes Provence (CofA 23 Jul 47). Canc Jun 48, to military markings.

C/n 385 F-BDYB. To Avions Bleus (CofA 08 Jul 47). Canc Feb 49. To French AF.

C/n 386 F-BCHU. To SOCOTRA (CofA 14 Aug 47). Retd to DTIA Feb 48.

C/n 387 F-BCHY. To TAI (CofA 09 Jul 47). To Aigle Azur . Canc Dec 48. To French AF.

C/n 388 F-BDYJ. To Avions Bleus (CofA 26 Sep 47). Canc Jan 49. To French AF.

C/n 389 To French AF, built Jul 47. Acceptance date wrongly quoted as Oct 45. Loaned to Préfecture de Police 1951 and operated by them as F-BDYM (CofA 18 Apr 51) for about four years before reverting to military marks. Canc Mar 55.

C/n 390 F-BDYD. To LASO France (CofA 14 Aug 47). Canc Feb 50. To French AF (acceptance date as early as Aug 47). SOC 26 Jul 60.

C/n 391 F-BCHV. To SOCOTRA (CofA 17 Nov 47). Retd to DTIA Feb 48. To French AF. SOC 18 Jan 60.

C/n 392 To French AF. Transferred to Portuguese AF 19 Dec 60.

C/n 393 To French AF 1947 to 22 Nov 60.

C/n 394 No information traced.

C/n 395 No information traced.

C/n 396 No information traced.

C/n 397 To French AF. SOC 18 Jan 60.

C/n 398 F-BDYF. To GECAT (CofA 20 Aug 47). Canc Feb

50.   To French AF.   Destroyed 29 Jun 57.

C/n 399 F-BDYG.   To GECAT (CofA 16 Sep 47).   Canc Feb
50.

C/n 400 F-BDYC.   To SOCOTRA (CofA 17 Nov 47).   Canc Feb
50.

C/n 401 F-BDYH.   To SANA del Oct 47.   Written off 13
Nov 47 near St Claude, Jura.

C/n 402 No information traced.

C/n 403 To French AF Mar 48 to 13 Jan 61.

C/n 404 F-BDYE.   To SANA (CofA 06 Aug 47).   To DTI, lsd
to Air Fret Transimax late 1949.   Written off 12
May 51 at Rouigo, Algeria.

C/n 405 To French AF.

C/n 406 To French AF.

C/n 407 To French AF.

C/n 408 To French AF Mar 48 to 20 Oct 55.

C/n 409 To French AF Sep 47 until transfer to French Navy
06 Oct 58.

C/n 410 To French AF.

C/n 411 To French AF.

C/n 412 To French AF.

C/n 413 To French AF.

C/n 414 To French AF.

C/n 415 To French AF Dec 48 to 22 Nov 60

None of the aircraft from the so-called 'parallel
line' (presumed reconditioned aircraft) went into civil
operations, and they were mostly delivered to the French
AF, some being taken on strength as late as 1949.

C/n 1001 To French AF (served with CIET 1951).

C/n 1002

C/n 1003

C/n 1004 To French Navy (served in French West Africa by
1948-49).

C/n 1005

C/n 1006

C/n 1007

C/n 1008

C/n 1009

C/n 1010

C/n 1011

C/n 1012

C/n 1013

C/n 1014

C/n 1015

C/n 1016

C/n 1017

C/n 1018

C/n 1019

C/n 1020

C/n 1021

C/n 1022 To French AF (Indochina 1952).

C/n 1023

C/n 1024

C/n 1025

C/n 1026

C/n 1027

C/n 1028

C/n 1029

C/n 1030 To French Navy, acceptance flight 19 Nov 49.

C/n 1031 To French Navy.

C/n 1032 To French Navy, acceptance flight at Bourges
factory 29 May 51.

C/n 1033

C/n 1034

C/n 1035

C/n 1036

C/n 1037

C/n 1038

In addition to the French-built or French-refurbished
aircraft, some Ju-52s are known to have been operated
under their original German construction numbers.
Although not within the scope this monograph, those known
are mentioned below:

Wk-Nr 0645 To French AF.   Accident 24 Jul 47.

Wk-Nr 0703 To French AF.   In service by Oct 50.

Wk-Nr 1109 To French AF.   In service Base Aérienne 119 by
Oct 56.

Wk-Nr 3376 Ex-Luftwaffe aircraft captured in Norway.
With RAF Enemy Aircraft Servicing Unit at
Fühlsbüttel, serialled VM928.   Transferred to the French
03 May 46.

Wk-Nr 3412 As for 3376, serialled VM932 and also transferred 03 May 46.

Wk-Nr 313558 To French AF, with Groupe de Liaisons
Aériennes 49 in Oct 51.

Wk-Nr 501367 As for 3376 and 3412 above, serialled VM918.
Del to France 27 Apr 46.

NOTE ON PRESERVED AIRCRAFT

Several AAC.1s appear to have been preserved, but only
two identities have been confirmed at the time of going to
press: c/n 216 preserved in France and c/n 363 in Germany.
As regards the aircraft preserved in Yugoslavia (believed
c/n 222) and in the Imperial War Museum collection at
Duxford, only French-built components have been traced but
no aircraft c/n plate.   Whatever the amount of
cannibalising and rebuilding, both aircraft are at present
considered to be French-built Ju-52s, pending further

evidence. Furthermore, judging from assumed Portuguese Air Force serial/AAC.1 c/n tie-ups, at least two more preserved aircraft might finally be revealed as French-built; they are the ex Portuguese Air Force "6315", now with the Air Force Museum (Museu do Ar) at Alverca, and "6320" (formerly a paratroop trainer conversion) preserved at BA.2 Ota, also in Portugal.

While on the subject of unsolved problems, details are required on a French Ju-52 with call-sign F-ZCAA in Sep 54 and another which was F-ZIRG, date unknown. It is a little-known fact that F-Z... call-signs were often used by French Air Force aircraft in early post-war years, but this practice was soon discontinued.

NC-701 / 702 MARTINET

When, shortly before the Second World War, Klemm (later Siebel) engineers designed the Fh 104A "Hallore" five-seater and in 1941 its Si-204 derivative, they probably did not realise that more than 350 aircraft would be built in France from 1944 onwards under a French designation. As already described, the Germans had tried to force the aeronautical industries of occupied countries to work for them, but usually ineffectually. The types chosen for this programme were usually communications and training aircraft, an obvious choice as the countries involved were far from being enthusiastic about contributing to the German war effort. The Si-204 military trainer and light transport was one of the aircraft chosen for manufacture outside Germany, and post-war production was continued in France and Czechoslovakia (179 being built during 1945-49 in the latter country as the AERO C-3, with several sub-types).

In France, SNCAC was ordered to build 455 Siebel 204s, and the first aircraft to be assembled in the Bourges factory appeared late in 1942, 168 being accepted by the Luftwaffe before the Liberation brought production on behalf of the Germans to a halt. A few machines originally built for the Luftwaffe were taken over by the French Air Force, and it was decided to resume production immediately under the French designation NC-700, which had also been applied to about a score of salvaged "German" aircraft. With the "liberated" production line starting with c/n 1, only a single NC-700 short-lived prototype (F-BAIP) was flown, and with Renault 12S-00 power-plants (derivatives of the German AS-411), the NC-701 variant immediately followed in the summer of 1945.

The NC-701 was the glazed unstepped-nose equivalent of the German Si-204D. Later in the following year the NC-702 appeared, with a conventional stepped windscreen nose (equivalent to the Si-204A and NC-700 prototype). The French name "Martinet" (black swift) was adopted, and both versions were produced until late in 1949, when the 350th Martinet was turned out by SNCAN, which had meanwhile incorporated the SNCAC concern. To these must also be added another batch of refurbished Siebels c/n 1000 to 1020.

The bulk of the NC701 and 702 production was for French military use, mostly with the French Air Force, but also in numbers for the Navy. The twin-engined eight-seater was used by the military services as late as 1963 before being finally withdrawn from use. The Centre d'Essais en Vol also used some Martinets and, fortunately, one of these (NC-702 c/n 282) has been preserved. Another Martinet, c/n 57, has been privately preserved in a poor state at

Allennes-les-Marais in northern France.

Some NC-701s and 702s also saw limited use under civil registration. Air France was the first commercial operator of the type with about a dozen aircraft, six of them being used for a short period frcm June 1946 for postal work, but these were grounded following the accident of F-BBFA in July 1946. Several early production NC-701s intended for Air France delivery were not taken up and went instead to the Armée de l'Air.

A few more were delivered to small operators but saw only limited service, being returned to the Air Force. In 1946 the Société Africaine des Transports Tropicaux (SATT), usually referred to as Compagnie Aéro-Africaine and based in Algiers, received two aircraft which were used for about a year (mostly on the route to Zinder, in Niger) before being transferred to the CEV. In 1947 four more operators obtained Martinets for commercial use. CATA (Cie Algérienne de Transports Aériens), also based in Algeria, received four NC-702s, the three surviving aircraft being returned to the Armée de l'Air in the following year. Three aircraft went to STAR (or Société des Transports Aériens Régionaux) and six to Air Atlas, but all had been cancelled from the civil register by early 1950. The only aircraft delivered to the Société des Avions Taxis d'Indochine, c/n 140, was written off at Calcutta, presumably on its delivery flight.

It can be said, therefore, that the Martinet did not make a significant contribution to immediate post-war air transport, being used mostly as an interim measure. Being an aircraft with too small a capacity, it was soon supplanted by the operators mentioned above with heavier aircraft such as the DC-3 (or Lockheed Lodestar with Aéro-Africaine).

Other civil registered Siebels (as they were often called) were used by SNCAC, one by SNCASE, three by the French civil aviation administration, others being also allocated to SNECMA and the aeronautical research establishment ONERA.

But, curiously, a small number were exported to Poland and Sweden for photo-mapping duties and survey work. The five Polish aircraft were delivered to LOT in 1947, and two went to the Swedish Rikets Allmänna Kartverk (with appropriate SE-K.. registrations) in 1949. The Swedes also received three more ex-Armée de l'Air aircraft as late as 1963, and two have been preserved. There is little doubt that the use of the type for survey work by the French IGN as early as 1945 may have helped in both cases. The NC-701 had been the initial IGN equipment and six were delivered, being gradually replaced by more appropriate aircraft.

In conclusion, it can be said that there were few commercial prospects for an aircraft which had been originally designed by Siebel as the standard advanced crew trainer. The Si-204 had a small capacity as a transport, and when the French carried on with SNCAC production they apparently did not intend to use the NC-701 and 702 other than in military operation, the small number appearing on the civil registers (out of nearly 400 being flown in France) being merely for interim use.

Specification

Advanced twin-engined military trainer with secondary rôle as light communications transport. Originally produced in Germany as Siebel Si-204 (Si-204A-1 streamlined opaque nose, Si-204D-1 transparent nose with full dual controls with provision for five pupils as a radio trainer). Wartime production in France (Farman, Hanriot, Bloch and SNCAC factories being involved, with wings produced in Holland by Pander), and post-war by SNCAC as NC-701/702. Passenger accommodation (NC-702) eight.

Span: 21.28m
Length: 12.57m (NC-701)
       12.81m (NC-702)
Height: 3.30m
Wing area: 46 sq m
Empty weight: 3,965 kg
Loaded weight: 5,600-5,735 kg
Payload: 750 kg
Power-plants: 2 x 590 hp Renault 12S-00 (later SNECMA 12S)
Maximum speed: 350 km/hr
Cruising speed: 315 km/hr at 2,000m
             325 km/hr at 3,000m
Take-off distance: 350m
Service ceiling: 7,500m
Range: 1,700 km (with 1,140 l fuel)
     1,200 km (with 890 l fuel)

Original NC-701 f/f Oct 44 with German Argus As 411 engines.

Production aircraft: Two sub-types, both with Renault/ SNECMA 12S engines
a NC-701 trainer, f/f Apr 45. Little used sub-type marks included:

Version I (Mark I) - crew and radio trainer
Version II (Mark II) - trainer for naval use
Version V (Mark V) - navigational trainer with accommodation for four and crew of two
Version VI (Mark VI) - photo aircraft (French AF conversion in 1948)

b NC-701 communications aircraft, f/f Jul 46. Little used sub-type marks included:

Version III (Mark III) - communications with crew of two, eight passengers and additional freight capacity
Version IV (Mark IV) - communications with crew of two, seven passengers and space for radio equipment

NC-702 f/f Nov 46.

Production List

The following points should be borne in mind in the following production list:

1 Dates for military aircraft respectively apply to acceptance dates and SOC date. But a quick glance at the production list will reveal French AF acceptance dates to be totally erratic except for some early and late production machines. Those dates being from official sources have been mentioned nevertheless. Recent researches have however shown (at least in the case of aircraft c/n 1002, 1003, 1004 and 1009) that the given "acceptance date" was actually the date of first general overhaul undergone by the aircraft and first recorded in the individual aircraft files. This no doubt should apply to most Martinets in the production list.

2 In some cases, date of transfer of SOC aircraft to "Administration des Domaines" (shortened to "Domaines") is also mentioned. The administration is responsible for all government-owned property and also takes care of public auction sales of property and material made surplus to requirements or withdrawn from use. In the case of the NC Martinet, aircraft were usually disposed of for scrap.

3 As before, no attempt has been made to include detailed military histories. The Martinet has formed a major research project in the publication of the French Branch of Air-Britain, "Le Trait d'Union", since March 1974, and all relevant known information has been published in detail and is still regularly updated. Accident dates are included here, even when only slight easily-repairable damage was incurred, together with French military callsigns when known.

There are a few exceptions, however, when it was felt of interest to include remarks such as service overseas, and significant codes for French Navy and CEV aircraft.

4 As for the AAC.1, some early machines appear to have used an "0" prefix (eg c/n 4 witnessed as "04" in 1947, or c/n 26 and 46 as "026" and "046" respectively as late as 1958).

5 All aircraft mentioned as "no information traced" were either French Navy or CEV.

C/n 1 Presumed type NC-700. F-BAIP. To SNCAC (CofA 04 Oct 45). Either destroyed at Toussus in hangar collapse due to weight of snow during winter 1946-47 or crashed Feb 47.

C/n 2 Presumed type NC-701. To French AF May 45 to 20 Jan 62. Accident as F-SCFC 22 Apr 48.

From c/n 3 onwards, type NC-701

Early aircraft (c/n 1-14) were reportedly fitted with Argus AS-411 TA-1 power-plants, Renault 12.S00 retrofit taking place later.

C/n 3 To French AF. Used operationally with invasion stripes May 45 against German-held bases along the Atlantic coast. Operated by GLAM. Flew into radio mast at Ste Assise (Dept de Seine et Marne) 22 Dec 45 and destroyed.

C/n 4 F-BAJY. To Air France (CofA 16 Feb 45). To French AF 17 Dec 45. Accident 20 Aug 47, repaired. Crashed 04 Oct 54 (seven killed) and SOC.

C/n 5 To French AF 28 Oct 44. Ground collision with Nord 1101 c/n 127 at Châteauroux (Dept Indre) 12 Aug 48 and repaired. Accident 26 Mar 57. WFU 26 May 62 and finally SOC 17 Sep 62.

C/n 6 To French AF. Served operationally in Mar 45 against German-held Atlantic coast bases. Destroyed 16 Dec 45 at Massy-Palaiseau (Dept de Seine et Oise), four killed.

C/n 7 F-BAOF. Intended for Air France but NTU. To French AF, date unknown. Accident 12 Jun 47. WFU 13 Feb 61 and SOC 02 Mar 61.

C/n 8 F-BAOG. Intended for Air France but NTU. Modified Nov 49 by SNCAC with spray equipment for ONERA icing trials and experiments. To French AF, possibly crashed 30 Sep 55.

C/n 9 F-BAOH. Intended for Air France but NTU. To French AF Mar 45. Accident at Lahr (French AF base in Germany) 19 Jan 60 as F-UJGS. SOC 26 May 62.

C/n 10 F-BAOI. Intended for Air France but NTU. To French AF. SOC 24 Aug 61.

C/n 11   F-BAOJ.   Intended for Air France but NTU.   To French AF 04 Jul 45 to 21 Mar 61.

C/n 12   F-BAOK.   Intended for Air France but NTU.   To French AF 10 Oct 45.   Collided with NC701 c/n 7 12 Jun 47 and repaired.   Accident May 62.   To Domaines 09 Sep 63.

C/n 13   F-BAOL.   Intended for Air France but NTU.   To French AF, date unknown.   Accident at Avord (Dept du Cher) 09 or 10 Jul 48 on landing.   Repaired. Transferred to CEV, date unknown, but was F-ZJLS 1950-51. Fate not known.

C/n 14   F-BAOM.   Intended for Air France but NTU.   To French AF 10 Sep 47.   Accident 29 Jan 54.   SOC 20 Jan 62.

C/n 15   To French AF.   Accident 17 Jan 48 as F-SCVA. Later transferred to CEV (was F-ZJPA in Nov 59). Photographic evidence shows fin painted as "NC 700 No. 15", but very likely an NC701.

C/n 16   F-BAON.   To Air France.   To French AF 21 May 45 initially for Resident General in Tunisia. Accident at Istres (Dept des Bouches-du-Rhône) 07 Feb 49. WFU 09 Sep 63, to Domaines 23 Nov 63.

C/n 17   To French AF.   Stalled on landing at Buc 19 Dec 45, nine killed.

C/n 18   To French Navy.   Version II, del with target towing equipment.

C/n 19   To French AF 02 Jun 47.   Accident 03 Jun 54 as F-UGGY.   SOC 12 Oct 62.

C/n 20   F-BAIQ.   To SNCASE (CofA 08 Jan 46).   Canc late 1952.   Sold to military.

C/n 21   To French AF 30 May 47.   Crashed 25 Oct 56 at St Romain-sur-Meuse as F-UHKC.   SOC 26 Apr 57.

C/n 22   To French AF 31 Jul 45.   Ground collision at night with c/n 88 at Avord 10 Feb 47.   SOC 07 Nov 62.

C/n 23   To French AF.   Crashed 03 Dec 51 when landing in fog at Base Ecole 702, Avord.   Five killed.

C/n 24   To French AF.   Photographic evidence exists of an aircraft "24" fitted with experimental side-firing machine guns (two on each side of the fuselage). Only traced in two accidents, one at Avord 15 Jan 47, the other 11 Jun 49 in unknown circumstances.

C/n 25   To French AF   .   SOC 11 Dec 63.

C/n 26   To French AF 1945.   To Domaines 11 Dec 63.

C/n 27   To French AF.   Crashed 30 Jan 46 at Moulin des Ponts, near Bourg-en-Bresse (Dept de l'Ain). Among fatalities was famous French aviatrix Maryse Hilsz.

C/n 28   To French AF 27 Dec 44.   SOC 09 Sep 63.   To Domaines 21 Nov 63.

C/n 29   To French AF 23 Dec 44.   Accident on landing at Villacoublay 23 Aug 48.   To Domaines 07 Nov 62.

C/n 30   To French AF 27 Dec 45.   SOC 21 Mar 63.

C/n 31   To French AF Jan 45 to 07 Nov 62.   To Domaines 12 Oct 62.

C/n 32   To French AF 31 Aug 45.   Accident on take-off at Sarrebrück (then Saar territory) 11 Oct 47.   SOC 01 Jul 62.

C/n 33   To French AF 14 Oct 45 to 12 Oct 62.

C/n 34   To French AF 04 Nov 45 to 26 May 62.

C/n 35   To French AF 03 Nov 45.   Ground accident at Villacoublay 25 Jan 47.   SOC 26 May 62.

C/n 36   To French AF.   No details traced.

C/n 37   To French AF.   Accident 25 Jan 56 and returned to SNCAN Feb 56 to May 57 for repairs.   Back to French AF until SOC 07 Nov 62.

C/n 38   To French AF   .   SOC 07 Nov 62.

C/n 39   Presumed to French AF and later with CEV.   Was used at Colomb-Bechar, then French Sahara, 1960-61 in connection with AS-30 missile development (aircraft being used for remote control of missiles launched from a Canberra).

C/n 40   Presumed to French AF and, if so, WFU prior to 1959.

C/n 41   To French AF 27 Dec 44 to 13 Feb 61.

C/n 42   No information traced.

C/n 43   To French AF.   Accident at Nevers-Fourchambault (Dept de la Nièvre) 18 Aug 47.   WFU and SOC prior to 1954.

C/n 44   To French AF 10 Oct 45.   Accident Dec 59.   SOC 21 Nov 63.

C/n 45   Not traced until reported serving with CEAM/French AF 1962 (probably CEV).

C/n 46   To French AF 20 Nov 45 to 12 Apr 63.

C/n 47   To French AF 12 Nov 45.   Accident at Avord 24 Jul 50.   Accident 18 May 55 at Base Aérienne 115 Orange-Caritat (Dept de Vaucluse).   SOC 10 Oct 55.

C/n 48   To French AF.   Accident 19 May 48.   Fate not known but had been WFU and SOC prior to 1954.

C/n 49   To French AF.   Force-landed at Alghero (Sardinia) 19 Apr 47.   Further details not known and presumably WFU.

C/n 50   To French AF 09 Jan 46 to 22 Dec 62.

C/n 51   To French AF 07 Jan 45.   Force-landed on Island of Porquerolles (off French Riviera) 04 Feb 47 after running out of fuel;   repaired.   SOC 09 Sep 63.

C/n 52   To French AF 07 Feb 46.   Accident 14 Jan 47 at Villacoublay.   Accident 13 Oct 47 at Lahr.   Once more on 21 Jun 55.   SOC 29 Jan 62.

C/n 53   To French AF 20 Oct 47.   Used by CEAM from 18 Feb 48 (a former Version IV aircraft for HQ communications, this aircraft was converted for overseas use).   SOC 20 Jan 62.

C/n 54   To French AF 30 Jan 46.   Ground collision with French AF UC-78 43-7493 at Avord 29 Nov 48.   SOC 21 Mar 61.

C/n 55   To French AF 24 Jan 46.   Accident 03 Dec 54 at Base Aérienne 141 Oran La Sénia (Algeria) as F-TFLW.   SOC 17 Apr 62.

C/n 56   To French AF 07 Feb 46.   To instructional air-frame at Base Ecole 721, Rochefort-sur-Mer (Dept Charente-Maritime) Jan 64.

C/n 57   To French AF.   Acceptance date not recorded but first noted in French AF service Oct 50.   WFU 18 Oct 61 and SOC 19 Nov 61.   Preserved privately in poor condition at Allènnes-les-Marais, northern France.

C/n 58   To French AF 22 Dec 49, although evidenced with French AF CIET training school as early as 09 Sep 48.   First of 16 aircraft specially cvtd for radio training with CIET, with accommodation provided for two crew, one instructor and five trainees (similar conversions denoted as CIET between c/n 59 and 110). To CEV late 1950 but apparently retd to French AF and finally SOC 09 Sep 63.

C/n 59 To French AF 21 Jul 49. (CIET.) Accident 06 Apr 56 as F-UHAR. SOC 10 Jul 56.

C/n 60 To French AF. Accident on take-off at Avord 18 Apr 47. SOC 21 Mar 61.

C/n 61 To French AF 13 Feb 45 (wrongly quoted as 1943). WFU 02 Nov 62 and SOC 01 Feb 63.

C/n 62 To French AF Dec 44. (CIET.) SOC 21 Nov 63.

C/n 63 To French AF 15 Feb 46 to 21 Mar 61.

C/n 64 To French AF 24 Jan 46. CIET from 03 Oct 51. Accident with CIET 340 12 Feb 54, retd to service 30 Mar 54 with same unit. To Communications Unit GTLA 2/60 03 Mar 55 and crashed 26 Apr 55 at Verrières (then Dept de Seine-et-Oise) as F-RAHV. Two fatalities and written off.

C/n 65 To French AF Mar 48 to 21 Nov 63.

C/n 66 To French AF 1944! Remained with manufacturers longer than most Martinets when retd for overhaul and repairs, staying with SNCAN 21 Oct 54 to 10 Oct 55, retaining French AF code "1-MA". Retd to French AF 10 Oct 55. To Domaines 11 Dec 63.

C/n 67 To French AF. (CIET.) Landing accident as F-RAZG? at Toulouse-Blagnac (Dept de la Haute-Garonne) 25 Feb 49. SOC 21 Mar 61.

C/n 68-70 were photomapping aircraft, c/n 69 being further modified by inclusion of long-range tanks. See also c/n 190.

C/n 68 F-BAOP. To Air France (CofA 06 Aug 45). To IGN. Written off 13 Oct 46 at Vichy.

C/n 69 F-BAOQ. To Air France (CofA 30 Apr 46). To IGN May 46. Written off at St Etienne 25 Oct 47.

C/n 70 F-BAOR. To Air France (CofA 09 May 46). To IGN May 46. WFU Oct 57 at Creil.

C/n 71 F-BAOS NTU. Presumed French military as evidenced in Algeria 1946 with diagonal tricolore stripes across fuselage. Slight damage incurred at Villacoublay 12 Jun 47 on take-off. Crashed in bad weather at Dourdan (then Dept de Seine-et-Oise) 12 Dec 47 and written off. Four killed.

C/n 72 To French AF   . First evidenced Aug 50. SOC 03 May 62.

C/n 73 To French AF 18 Apr 46. (CIET.) Accident 08 Aug 56. SOC 20 Jan 62.

C/n 74 Presumed to French AF. Crash-landed at Bou Yagar, French North Africa, after engine failure 26 Sep 47.

C/n 75 To French AF 19 Dec 47. (CIET.) SOC 20 Jan 62.

C/n 76 To French Navy. Version II, del with target towing equipment. This aircraft and c/n 78-80, 101-105 were cvtd from Version IV aircraft, unlike c/n 18, an original Version II aircraft.

C/n 77 To French Navy. Version II, del with target towing equipment.

C/n 78 To French Navy. Version II, del with target towing equipment. Accident 04 Jun 56 as "5S.16" and further accident 29 Mar 57 with Escadrille 4S.

C/n 79 To French Navy. Version II, del with target towing equipment.

C/n 80 To French Navy. Version II, del with target towing equipment. "3S.23" F-YCCW 1952-53. Accident mid Aug 52 at Montpellier (Dept d'Hérault).

C/n 81 To French AF. Crashed 04 Aug 46 at La Rochelle (Dept de Charente-Maritime) with conflicting mid-air collision report 23 Mar 48 (see c/n 87).

C/n 82 To French AF 27 Dec 44. Accident 14 Oct 55. SOC 20 Jan 62.

C/n 83 To French AF. (CIET.) Accident at Toulouse 16 Mar 49. Withdrawn from French AF inventory prior to 1954.

C/n 84 To French AF 02 Jul 46. Accident 13 Aug 56 as F-UHQB. Crashed 30 Nov 59 near Base Aérienne 142 near Boufarik, Beni Mered (Algeria) and written off. SOC 01 Mar 60.

C/n 85 To French AF   . SOC 09 Apr 63.

C/n 86 To French AF 1944 to 11 Dec 63.

C/n 87 To French AF. To IGN as F-BBFA(2) (CofA 27 Apr 50). In-flight collision with Stampe SV-4 c/n 614 at Vert-le-Grand, just outside Brétigny airfield, 23 Mar 48 (then Dept de Seine-et-Oise). CofA suspended 09 Oct 56 at Creil and canc 1958.

C/n 88 To French AF 08 May 46. Accident 30 Nov 51 as F-TECJ. Accident 19 Dec 59 at Cazaux (Dept Gironde) as F-SDLZ. (CEAM believed during ferry flight, judging from call-sign.) SOC 06 Aug 60.

C/n 89 To French AF. (CIET.) Accident 17 Jul 56. SOC 20 Jan 62.

C/n 90 No information traced.

C/n 91 To French AF. Crashed at Base Ecole 702, Avord, 24 Jun 48 and written off.

C/n 92 To French AF Dec 44. Accident 30 May 59. SOC 07 Nov 62.

C/n 93 To French AF Dec 44. To Domaines 21 Nov 63.

C/n 94 To French AF Dec 44. To Domaines 21 Nov 63.

C/n 95 To French AF May 46 to 12 Oct 62.

C/n 96 To French AF May 46. Accident 23 Sep 58. SOC 11 Dec 63.

C/n 97 To French AF. (CIET.) SOC 21 Nov 63.

C/n 98 To French AF Dec 46 to 07 Nov 62.

C/n 99 To CEV.

C/n 100 To French AF (CIET) Jul 45 to 12 Apr 63.

C/n 101 To French Navy. Version II, del with target towing equipment.

C/n 102 To French Navy. Version II, del with target towing equipment.

C/n 103 To French Navy. Version II, del with target towing equipment.

C/n 104 To French AF initially, Jul 45. Date of transfer to Aéronavale (as for c/n 101) and conversion not known. This suggests that other aircraft in the batch c/n 101-103 and 105 were not del direct to Aéronavale. To French Navy. Version II, del with target towing equipment. To Domaines 11 Dec 63.

Nord N.2100 Norazur (S.C.A. via J Delmas)

Top   : First production Potez 841 F-WLKR of Aero-Dienst
          with Canadian Pratt & Whitney PT.6s (Potez via
          J Delmas)

Bottom : Hurel-Dubois HD-31 F-WFKU. Note the twin fins.
          (via Charles W Cain)

Top   : Hurel-Dubois HD-32 F-WHHA (via Charles W Cain)
Bottom : Hurel-Dubois HD-34 F-WHOO for the government
         aero-survey company IGN, at Le Bourget in June
         1957 (JMG Gradidge)

SE-1010 F-WEEE (Bristol Aeroplane Company via Charles W Cain)

C/n 105 To French Navy.   Version II, del with target towing equipment.   Apparently transferred to French AF or CEV.   Crashed 13 May 54 at L'Isle-Adam (then Dept de Seine-et-Oise) after in-flight collision with civilian glider.   Two killed.

C/n 106-110 all French AF CIET conversions.

C/n 106 To French AF Jun 46.   To Domaines 11 Dec 63.

C/n 107 To French AF   .   SOC 20 Jan 62.

C/n 108 To French AF Dec 46 to 11 Dec 63.

C/n 109 To French AF Oct 48.   To Domaines 11 Dec 63.

C/n 110 To French AF Mar 49.   To instructional airframe Jan 64.

C/n 111 To French AF Jun 47.   Crashed at Plancher-les-Mines (Dept de Haute-Saône) 23 Nov 55 and written off.

C/n 112 To French AF.   Oper by GLAM when crashed and destroyed 02 Sep 46 at Jouy-en-Josas (Dept de Seine-et-Oise).

C/n 113 To French AF Dec 44.   Crashed at Istres 24 Jan 55 with three fatalities and canc from inventory 10 May 55.

C/n 114 To French AF   .   Accident 08 Jul 48 as F-SCNZ.   Crashed at Toublanc 21 Mar 55 during del flight to new unit, seven killed.   Written off.

C/n 115 To French AF Oct 46 to 21 Mar 61.

C/n 116 To French AF   to 01 Jul 63.

C/n 117 To French AF Nov 46 to 01 Jul 63.

C/n 118 To CEV.   Accident 26 May 48.   Was F-ZJCP in 1959.

C/n 119 To French AF.   Take-off accident at Seno, French Indochina, 21 May 51.   SOC prior to 1954. (Last or last but one NC-701 aircraft of initial production batch.)

C/n 120 Presumed last NC-701 of first batch, rather than first NC-702.   To French AF May 46.   Crashed 29 Apr 63 and written off.

C/n 121 to 146 are NC-702 aircraft.

C/n 121 F-BAOO.   To Air France? then to SGACC.   Built early 46 (CofA 30 Nov 46).   Canc Sep 55 and WFU at Le Bourget.

C/n 122 To French AF.   Had been SOC prior to 1954.

C/n 123 F-BBFX.   To SNCAC (CofA 06 Sep 46).   Written off 11 Oct 46 at Sete.

C/n 124 F-BBFA(1).   Postal aircraft (see also c/n 125 and 128-138).   To Air France del 21 Feb 46. Crashed 22 Jul 46 at Pontarmé, near Senlis (Dept de l'Oise), although information conflicts with other reports that aircraft crashed 29 Jul 46.

C/n 125 F-BBFB.   Postal Aircraft.   To Air France del 10 Apr 46.   Retd to Govt 31 Dec 47 for French AF/ ECMA.   Remained as F-BBFB for ONERA (CofA 29 Sep 48). To CEV 01 Dec 55.   Canc Jan 56.

C/n 126 F-BBFC.   Intended for Air France but NTU.   To French AF Jun 46.   Force-landed 01 Apr 47 near Briare (Dept de Loiret) following engine failure. Landing accident 06 Aug 47 at Mont-de-Marsan (Dept de Landes).   SOC 03 May 62.

C/n 127 F-BBFD.   Intended for Air France but NTU.

C/n 128 to 138 all postal aircraft.

C/n 128 F-BBFE.   To Air France del 11 Apr 46.   Retd to Govt 31 Dec 47 for French AF/ECMA with acceptance date given as same month (Apr 46) as being entered on French civil register!   Accident 03 Apr 54.   Crashed 30 May 63 at Luxeuil (Dept de Haute Saône).   SOC Jan 64.

C/n 129 F-BBFY.   To Air France del 13 Apr 46 (regd to SGACC).   Retd to Govt 31 Dec 47 for French AF/ ECMA (in service by Mar 50).   SOC 11 Dec 63.

C/n 130 F-BBFZ.   To Air France del 09 Jul 46.   Retd to Govt 31 Dec 47 for French AF/ECMA.   Accident 06 Jun 56 as F-SCAW.   Last reported Aug 60 and fate not traced.

C/n 131 F-BBFF.   To Air France del 09 Jul 46.   Retd to Govt 31 Dec 47 for French AF/ECMA.   To Domaines 11 Dec 63.

C/n 132 F-BBFG(1).   Intended for Air France but NTU.   To CEV (reported early 1960 in Algeria as F-ZJMD).

C/n 133 F-BBFH.   To Air France (?), then to SNECMA (CofA 07 Mar 47).   Canc and transferred to French AF/ CEV late 1951.

C/n 134 F-BBFL.   Intended for Air France but NTU.   To French AF/CEV.   Accidents 02 Jan 48 and 27 Jul 48.   To CEV, crashed 29 Jun 55 as F-ZJNF.

C/n 135 F-BBFJ.   Intended for Air France but NTU.   To CEV (1949-61 at least).

C/n 136 F-BBFK(1).   Intended for Air France but NTU.   To CEV (as late as 1959).

C/n 137 F-BBFL(1).   Intended for Air France but NTU.   To CEV, crashed 16 May 51 near Brassac (Dept de Tarn) as F-ZJPH, four killed.

C/n 138 F-BBFM(1).   Intended for Air France but NTU.   To CEV.   Was F-ZJOL in Sep 59.

C/n 139 F-BBFN.   To Air France (?)   To French AF as early as Nov 46.   On loan to CATA as F-BBFN (CofA 24 Feb 47).   Canc 1949 or 1950 and retd to French AF.   Last reported in service Aug 60 and fate not traced.

C/n 140 F-BBFO.   To SATI (CofA 08 May 47).   Written off 26 Jun 47 at Calcutta (probably on del to Indo-china).

C/n 141 F-BBFP.   To French AF Aug 46.   On loan to STAR as F-BBFP (CofA 29 May 47).   Canc Feb 50 and retd to French AF.   WFU late 1963.   To Domaines 11 Dec 63.

C/n 142 F-BBFQ.   To SATT (CofA 13 Dec 46).   Canc Feb 48.   To CEV (reported 1955-56 at least).

C/n 143 F-BBFR.   To SATT (CofA 12 Nov 46).   Canc Nov 47.   To DTI Mar 48 for CEV.   Was F-Z..F coded "F" with CEV in Sep 59.

C/n 144 F-BBFS.   To SGACC (CofA 15 Mar 47).   WFU Sep 54 at Le Bourget.

C/n 145 F-BBFT.   To French AF Dec 46.   Transferred to SGACC as F-BBFT (CofA 18 Sep 47).   Operated from SGACC Algerian base at Boufarik.   Canc Jan 56 and transferred to French AF.   To Domaines 11 Dec 63.

C/n 146 F-BBFU, NTU.   To French AF Mar 47.   Ground collision with Caudron Goëland c/n 1211 13 Jul 48 at Châteauroux.   SOC 20 Jan 62.   Type was NC-702.

C/n 147 Type NC-701 or 702 uncertain.   To French AF Aug 46.   SOC 21 Nov 63.

C/n 148 Type NC-701 or 702 uncertain.   Aircraft not

traced in service, but possibly French Navy.

C/n 149 onwards.   Production reverted to NC-701 variant
(see also note for c/n 147-148).

C/n 149 To French Navy.   To Instructional airframe,
Rochefort-sur-Mer, still coded "5S.16".

C/n 150 To French Navy.

C/n 151 To French Navy.

C/n 152 Reported to French AF, but uncertain.

C/n 153 To French AF Sep 46 to 21 Nov 63.

C/n 154 To French AF Sep 47.   Take-off accident 21 Feb
49 at Clermont-Ferrand/Aulnat (Dept du Puy-de-
Dome).   SOC 01 Jul 63.

C/n 155 To French AF.   Accident 14 Mar 49.   Reported 16
Nov 59 as F-UKEG coded "92-EG" but does not
appear in French AF files!   Details and fate not clear.

C/n 156 To French AF.   To Domaines 11 Dec 63.

C/n 157 To French AF Dec 46.   To Domaines 11 Dec 63.

C/n 158 To French AF Sep 47 to 04 Nov 62.

C/n 159 To French AF Oct 46.   Quoted as sold to Sweden
20 Dec 63 but regd to Rikets Allmänna Kartverk as
SE-KAL Jun 63.   Unofficially named "Kalle".   Last flown
23 Dec 70 and preserved for the Luftfartsverket
Collection, Stockholm-Arlanda.

C/n 160 To French AF Oct 46.   To Domaines 11 Dec 63.

C/n 161 To French AF   .   SOC 20 Jan 62.

C/n 162 To French AF 1948.   WFU late 1963 as F-SDGA.
To Instructional airframe Jan 64.

C/n 163 To French AF.   Ground accident 27 Jan 47 at
Innsbrück, Austria.   SOC 19 Mar 63.

C/n 164 To French AF Jun 47.   To Domaines 11 Dec 63.

C/n 165 To French AF Feb 47.   Made international headlines when landing in error 20 Dec 47 at Kralupy,
north of Prague, Czechoslovakia, following navigational
error.   Retd to French AF.   Accidents 03 Apr 48 and 11
Oct 54.   SOC 01 Jul 62.

C/n 166 To French AF   .   SOC 20 Jan 62.

C/n 167 To French AF Dec 46.   Crash-landed 26 Sep 58 at
Hoggar, near Tamanrasset (French Sahara) as
F-SCDM.   Crew of six escaped unhurt.   SOC 27 Oct 58.

C/n 168 To French AF.   Crashed after engine failure 28
Aug 47 on take-off from Bordeaux-Mérignac (Dept
de Gironde) and written off.

C/n 169 To French AF Jul 47 to 20 Jan 62.

C/n 170 To French AF   until 21 Mar 61.

C/n 171 To French AF Feb 47.   To Domaines 11 Dec 63.

C/n 172 To French AF Dec 46.   Quoted as sold to Sweden
20 Dec 63 but regd to Rikets Allmänna Kartverk as
SE-KAM Jul 63, unofficially named "Kammen".   Canc 20 Jan
70, CofA expired 31 Mar 70, scrapped.   Now preserved in
Luftfartsverket Collection, Stockholm-Arlanda.

C/n 173 To French AF.   Accident 02 Mar 49 as F-SCFN.
Crashed 03 Dec 51 at Base Ecole 702, Avord, in
same circumstances as c/n 23 (qv), but crew escaped
unhurt.   Written off.

C/n 174 Not traced (possibly to French Navy).

C/n 175 Not traced (possibly to French Navy).

C/n 176 To French Navy F-YFED 1952-53 coded "56S.4".

C/n 177 To French Navy.

C/n 178 To CEV.   Accident 20 Oct 54, circumstances
unknown.

C/n 179 To CEV F-ZJOM 1954, F-ZJOF 1955-63.

C/n 180 To French AF Nov 47.   Ground collision 08 May 49
At Bach-Mai (French Indochina).   To instructional
airframe 16 Dec 58.

C/n 181 To French AF.   Take-off accident 14 Nov 48 at
Nha-Trang (French Indochina) and repaired.   Fate
not traced but last known serving overseas May 56.

C/n 182 To French AF.   WFU prior to 1954.

C/n 183 To French AF Apr 47.   Served with Communications
Flight French Indochina, Saigon-Tan Son Nhut, from
23 Jul 48.   Retd 1955 to Algeria.   SOC 21 Jul 59.

C/n 184 To French AF Nov 47.   Served in French Indochina,
retd to Algeria.   Crashed 17 Jul 56 as F-UIFB
coded "B".   SOC 02 Jan 57.

C/n 185 F-BDLB.   To SNCAC (CofA 24 Mar 47).   To SP-LFA
LOT del May 47.

C/n 186 To CEV.

C/n 187 to 189 were cvtd for the Police Coloniale role -
see c/n 53 for full list of conversions.
Mods included racks for 6 x 50 kg bombs, four fixed
forward-firing MAC machine guns and two more flexible
machine guns in the fuselage.

C/n 187 To French AF Mar 48.   Formerly a Version IV aircraft, this was cvtd for overseas use and served
in Madagascar in early 1950s.   SOC 31 Dec 59.

C/n 188 To French AF.   Formerly a Version IV aircraft,
this was cvtd for overseas use (with c/n 189,
191-219, 229, 234-249 and 255-265).   Served in French Indochina and damaged by Viet-Minh mortar shell at Bach-Mai 15
Nov 48, repaired.   SOC prior to 1954 (possibly in French
Indochina).

C/n 189 To French AF?   Transferred to tri-service unit
experimental centre CIEES 343 (Centre Interarmes
d'Essais d'Engins Speciaux) May 62 Sahara.   To Domaines
11 Dec 63.

C/n 190 F-BBFG(2).   Photomapping aircraft cvtd for long
range operations (see c/n 69).   To IGN (CofA Mar
47).   WFU May 57 at Creil.

C/n 191 to 219 All Police Coloniale variants.

C/n 191 To French AF Feb 47.   To CIEES 343 (see c/n 189
above).   Was F-SDKP there late 1961.   To
Domaines 11 Dec 63.

C/n 192 To French AF Mar 48 to 19 Oct 61.

C/n 193 To French AF.   Undercarriage failure 24 Nov 48
at Ambatondvazaka, Madagascar, repaired.
Accident 23 Aug 56.   Crashed 08 Oct 56 and written off.

C/n 194 To French AF.   Accident 01 May 48 as F-UHON,
presumably in Madagascar (conflicts with report
of same date accident to c/n 204 in another source).

C/n 195 To French AF Mar 48 to 25 Mar 61.

C/n 196 To French AF.   Crashed 09 May 48 as F-UHOQ at
Maudugar, Nigeria.

C/n 197 To French AF Apr 47.   Accident 23 Oct 48 as
F-UGOB.   SOC 20 Jan 62.

C/n 198 To French AF May 47.   Accident 20 Oct 54 as
F-SCFA.   SOC 20 Jan 62.

C/n 199 To French AF Dec 44.   This aircraft and c/n 205
and 236-242 modified for specific use at Avord
School with modified cockpit layout.   Accident 20 Aug 48
at Pointe Noire, Congo, French Africa.   SOC 20 Jan 62.

C/n 200 To French AF.   Accident 03 Nov 49 at Analavelona,
Madagascar.   Presumed written off and in any
case SOC prior to 1954.

C/n 201 To French AF May 48.   Crashed 25 or 27 Jun 55 as
F-SCNO.   SOC 24 Jan 56.

C/n 202 To French AF.   Landing accident 16 Apr 49 at
Anjouan Island, Archipelago of Comores.   Possibly
written off since not traced after Apr 49.

C/n 203 To French AF Mar 48.   To Domaines 11 Dec 63.

C/n 204 To French AF but not traced until transferred to
Domaines 11 Dec 63 (see also c/n 194).

C/n 205 To French AF Dec 44.   Cvtd Avord 1951 (see c/n
199).   SOC 07 Feb 63.

C/n 206 To French AF Apr 48 to 03 Dec 58.

C/n 207 To French AF Jan 49.   Served in French Indochina.
SOC May 62.

C/n 208 To French AF.   Undershot 30 May 49 at Port
Gentil, Gabon, and repaired.   SOC prior to 1954.

C/n 209 To French AF May 48.   To Domaines 11 Dec 63.

C/n 210 To French AF Apr 48.   Served in French Indochina
and retd to Algeria.   Crashed 03 or 30 Jul 59 at
Saida Nazereg and SOC 30 Oct 59.

C/n 211 To French AF Dec 48.   Served in French Indochina
and SOC locally 18 May 55 at time of French withdrawal from Viet-Nam.

C/n 212 To French AF.   Crashed 04 Apr 50 at Pointe Noire,
Congo, French Africa, and written off.

C/n 213 To French AF.   Take-off accident 11 Jul 49 at
Phan Tiet, French Indochina, and presumably
written off.   SOC prior to 1954.

C/n 214 To French AF Sep 47.   Served in French West
Africa and later French Indochina, where stored
1954 and SOC locally 07 Dec 55.

C/n 215 To French AF.   Served in French Indochina.
Crashed 02 Jan 52 and presumed written off.   SOC
prior to 1954.

C/n 216 To French AF Jul 48 to 11 Oct 62.

C/n 217 To French AF Nov 47 to 25 Mar 55.

C/n 218 To French AF.   Served in French Africa.   Crashed
on take-off at Blida while in Algeria 25 Dec 47.
WFU prior to 1954.   Possibly transferred to CEV, aircraft being reported late in 1962.

C/n 219 To French AF Apr 48.   Served in French Indochina and Algeria.   SOC 26 May 62.

C/n 220 To French Navy.   Operated by French Navy local
communications flight at Bizerta, Tunisia, coded
"BZ-10".

C/n 221 No information traced.

C/n 222 To French AF, reported ground collision 29 Nov 48
at Base Ecole 702, Avord.   Fate not known.

C/n 223 No information traced.

C/n 224 No information traced.

C/n 225 To CEV.   Accident 30 Dec 48.   Was F-ZJTJ by mid
1964.

C/n 226 To CEV.   Reported by Sep 55.

C/n 227 F-BDLE.   To SNCAC (CofA 20 May 47).   To SP-LFB
LOT del 1947, regd 28 May 47.   WFU 15 May 48 and
to Polish AF.

C/n 228 F-BDLJ.   To SNCAC (CofA 13 Jun 47).   To SP-LFC
LOT, del 1947, regd 18 Jun 47.   WFU 15 May 48
and to Polish AF.

C/n 229 To French AF.   Police Coloniale variant.   Cross-
wind landing accident 23 or 24 Dec 47 at Aoulef,
Algeria.   Repaired.   Crashed at Brusseaux, Congo, 07 Dec
49 and written off.

C/n 230 F-BDLK.   To SNCAC (CofA 30 May 47).   To SP-LFD
LOT del 1947, regd 30 Jul 47.   WFU 15 May 48 and
to Polish AF.

C/n 231 F-BDLL.   To SNCAC (CofA 25 Aug 47).   To SP-LFE
LOT del 1947, regd 02 Sep 47.   WFU 15 May 48 and
to Polish AF.

C/n 232 F-BDLM.   To SNCAC (CofA 05 Sep 47).   To SP-LFF
LOT del 1947, regd 15 Sep 47.   WFU 15 May 48 and
to Polish AF.

C/n 233 F-BDLQ.   To IGN (CofA 25 Sep 47).   Canc Oct 57.

C/n 234 to 249 All Police Coloniale variants.

C/n 234 To French AF Nov 47.   Accident 09 Jul 49 at
Sangatanga, French Africa, as F-UGOL.   Transferred
to CEV.   Was F-ZJOM 1954.   Retd to French AF.   To
Instructional airframe Aug 64.

C/n 235 To French AF Oct 47 to 07 Nov 62.

C/n 236 To French AF Nov 47 to 20 Jan 62.

C/n 237 To French AF.   Accident 30 Jul 48 with Test and
Ferry Unit.   Had been SOC prior to 1954.

C/n 238 To French AF Apr 48 to 07 Nov 62.

C/n 239 To French AF until 30 Jan 61.

C/n 240 To French AF Oct 48 to 27 Nov 58.

C/n 241 To French AF Sep 48.   Sold to Sweden 01 Dec 62 as
SE-KAN for Rikets Allmänne Kartverke, regd Jun 63,
unofficially named "Kannan".   Canc Dec 67.   CofA expired
31 Mar 68 and scrapped.

C/n 242 To French AF Jul 48 to 06 Nov 61.

C/n 243 To French AF.   Crashed 11 Nov 49 at Bangui,
French West Africa.   Not traced after 1949.

C/n 244 To French AF Nov 48 to 25 Mar 59.

C/n 245 To French AF.   Served in French Indochina.
Crashed 11 Feb 51 at Luc Nam, Viet-Nam, due to
enemy ground fire and presumed written off.

C/n 246 To French AF Nov 48 to 20 Jan 60.

C/n 247 To French AF.   Served in French Indochina, not
traced after Sep 49.

C/n 248 To French AF.   Served in French Indochina, at
Saigon-Tan Son Nhut as F-SCKF Oct 50 to Apr 55.
SOC 20 Apr 60.

C/n 249 To French AF Feb 49.   Served in French Indochina.
SOC 05 May 59.

C/n 250 To French Navy, later French AF Apr 57 to 07 Nov 62.

C/n 251 To French Navy. Crashed 17 Sep 56 coded "5S.24".

C/n 252 To French Navy F-YCEQ coded "5S.17" 1960.

C/n 253 To French Navy "4S.12". To French AF Feb 57 to 26 May 62.

C/n 254 No information traced.

C/n 255 to 265 were last batch of Police Coloniale armed variant cvtd from Version III aircraft, initially for use in French Indochina including close support.

C/n 255 To French AF. Accident 27 Jul 51 as F-UIDM. Accident 30 Sep 56. SOC 27 Jun 57.

C/n 256 To French AF Jun 49. To Domaines 11 Dec 63.

C/n 257 To French AF May 50 to 17 Apr 59.

C/n 258 To French AF. SOC 07 Nov 62.

C/n 259 No information traced.

C/n 260 To French AF Jan 50. Served in French Indochina. SOC 02 Jan 59.

C/n 261 To French AF Feb 49. Crashed 06 Jul 57 and written off, presumed in Algeria.

C/n 262 To French AF. SOC 26 May 62.

C/n 263 To French AF Apr 50 to 21 Jun 61.

C/n 264 SE-KAE. To Rikets Allmänne Kartverk, purchased 1947. Last flown 30 Dec 70 and preserved at Swedish AF Museum, Malmslätt-Linköping.

C/n 265 SE-KAG. To Rikets Allmänne Kartverk, purchased 1949. Canc Dec 67, CofA expired 31 Mar 68 and scrapped 1968.

C/n 266 First Martinet 2 transport variant, fully equipped for passenger transport (eight plus two crew) with cabin heating and soundproofing. Other similar aircraft were c/n 267-269, 272, 278-279. To CEV (1961-64 at least).

C/n 267 Martinet 2 transport variant. F-BBFK(2). To Air Atlas (CofA 18 Apr 47). To French AF with acceptance date Apr 47 (same as date of civil CofA). Canc Dec 49. Lost in Mediterranean off Dupleix, Algeria, 31 Dec 55. Five killed.

C/n 268 Martinet 2 transport variant. F-BBFL(2). To Air Atlas (CofA 03 Apr 47). To French AF Mar 48 to 17 Apr 62 (regn canc Dec 49).

C/n 269 Martinet 2 transport variant. F-BBFM(2). To Air Atlas (CofA 21 Apr 47). Canc Dec 49 but still at Melun-Villaroche (Dept de Seine-et-Marne) mid Sep 51 as F-BBFM. To French AF. SOC 20 Jan 62.

C/n 270 To French AF Jun 47. Ground collision with a Jeep 19 Mar 48 at Berlin-Tempelhof. Undershot in landing accident at Clermont-Ferrand/Aulnat 14 Apr 49. SOC 20 Jan 62.

C/n 271 To French AF. Had been WFU prior to 1954.

C/n 272 Martinet 2 transport variant. F-BDLA. To Air Atlas (CofA 07 May 47). Canc Dec 49. To French AF. Operated in Madagascar. Take-off accident at Ivato 24 May 51 and presumed SOC as not evidenced after that date.

C/n 273 To French AF Sep 47. Accident 12 Mar 48 and 18 Mar 50 at Tunis. To Domaines 11 Dec 63.

C/n 274 To French AF Feb 47 but diverted to civilian operator CATA as F-BDLF (CofA 20 May 47). Presumed canc 1948 and retd to French AF. Accident 18 Apr 56. To Domaines 11 Dec 63.

C/n 275 F-BDLG. To CATA (CofA 20 May 47). Canc 1948? Very likely the aircraft that crashed 08 Jan 48 near Palestro, Algeria, killing nine.

C/n 276 F-BDLN. To Air Atlas (CofA 11 Jun 47). Canc Dec 49. To French AF. SOC 20 Jan 60.

C/n 277 F-BDLO. To Air Atlas (CofA 11 Jun 47). Canc Dec 49. To French AF. SOC 07 Nov 62.

C/n 278 Martinet 2 transport variant. F-BDLH. This aircraft and c/n 279 incorporated minor differences including de-icing, heating of French manufacture and additional auxiliary fuel tanks. To STAR (CofA 04 Jul 47). Canc late 1948. To French AF. To Domaines 11 Dec 63.

C/n 279 Martinet 2 transport variant (see also c/n 278). F-BDLI. To STAR (CofA 07 Jul 47). To French AF (overhauled at AIA Clermont-Ferrand Feb 49). Crashed after take-off following engine failure 13 Aug 49 at Camaret (TT 345 hrs). Cancellation appeared in BV only in Feb 50.

C/n 280 F-BDLP. To CATA (CofA 18 Jul 47). Canc late 1948. To French AF. With Escadrille de Liaisons Aériennes ELA 44 from 1954 as F-SCCA, changed to F-SCCD in May 55. Was christened "Adjudant Carlotti" (a not very common practice in military use). Crashed (still as F-SCCD) 15 Sep 59 at Nîmes-Courbessac (Dept du Gard), one killed.

C/n 281 F-BDLR. To SNECMA (CofA 02 Apr 48). Crashed 28 Jul 48 at Réau-Villaroche (Dept de Seine-et-Marne).

C/n 282 To CEV. Last aircraft in service, as F-Z.BN coded "BN". SOC 13 Dec 71 at CEV Toulouse. Transferred 17 Jan 72 to Paris area for storage and preservation.

C/n 283 To French AF. Accident 30 May 55. SOC 20 Jan 62.

C/n 284 To French AF. Accident 27 May 50 at Rabat, Morocco. WFU prior to 1954.

C/n 285 To French AF Feb 51 to 20 Jan 62.

C/n 286 No information traced.

C/n 287 To French Navy. Operated 1951-52 at least by ferry unit (SRC/ERC, Section later Escadrille de Réception et Convoyage).

C/n 288 To French AF? To CEV (circa 1955). Retd to French AF (circa 1958-59). Fate not traced.

C/n 289 To French AF Jun 48. Accident 27 Sep 49 at Agadir, Morocco. SOC 20 Jan 62.

C/n 290 To French AF. Destroyed on ground at Tananarive-Ivato, Madagascar, 28 Jan 50 and SOC.

C/n 291 To French AF. Accident 21 Nov 50 at Tananarive-Ivato, Madagascar. Retd to Europe. Last operated (May 62) as station 'hack' at Base Aérienne 257 Friedrichshafen, Federal Germany, before being SOC. To Domaines 11 Dec 63.

C/n 292 No information traced.

C/n 293 To French AF Dec 47 to 07 Nov 62.

C/n 294 To CEV.

C/n 295 No information traced.

C/n 296 No information traced.

C/n 297 To French AF Feb 52. Transferred to French Navy (at least from Oct 52 to 1953) as F-YFEF "54S.6". Retd to French AF, date not known. By 1961 was operated as calibration aircraft. Fate not traced.

C/n 298 To French AF. Crashed 28 Aug 51 at Tananarive-Ivato, Madagascar, and SOC.

C/n 299 To French AF Aug 48 to 07 Nov 62.

C/n 300 F-BDLS. To ONERA (CofA 24 Nov 48). Canc Apr 58 and reportedly used for ground trials by ONERA. Also reported transferred to CEV (see also c/n 324).

C/n 301 To French AF Dec 48. To Domaines 11 Dec 63.

C/n 302 To French AF Nov 48. Accident 12 Feb 57. To Domaines 11 Dec 63.

C/n 303 Not traced initially, then with French Navy early 1950s. Operated by Station Flight, Bizerta, coded "BZ-14". Later transferred (or retd?) to French AF. To Domaines 11 Dec 63.

C/n 304 To French Navy. Operated by Ferry Unit 1951-52 and also by HQ Flight, Algiers, coded "AL-3". Last reported Dec 57.

C/n 305 To French AF Nov 48 to 26 May 62.

C/n 306 To French AF until SOC 01 Dec 63.

C/n 307 To French AF Jan 49. Accident 29 Oct 54 as F-SCEQ. Accident 21 Nov 61. SOC 04 May 62.

C/n 308 To French AF Feb 49 to 01 Dec 63.

C/n 309 To French AF Feb 49 to 06 Nov 61.

C/n 310 To French AF Jun 50. Accident 24 Nov 54 as F-TXEH. SOC 24 Jan 55.

C/n 311 To French AF Feb 49. Accident 14 Apr 54. Accident at Base Aérienne 101, Toulouse-Francazal, 09 Sep 54 as F-SCMT. Crashed 06 Feb 63 (no details) as SOC in same year.

C/n 312 To French AF Mar 49 to 26 May 62.

C/n 313 To French AF Mar 49. Crashed 11 Oct 57 and written off.

C/n 314 To French AF. Accident 06 Oct 51 as F-RAHG. To Domaines 11 Dec 63.

C/n 315 To CEV. Last reported with EPNER 1958-59.

C/n 316 To French AF. Accident 21 Jan 51 at Colomb-Bechar, Algeria. Presumably written off and WFU prior to 1954.

C/n 317 To French AF or CEV until late 1949. To CEV/EPNE Jan 50.

C/n 318 To French AF May 49. Force-landed after running out of fuel (and following navigational error) at Strodehne in Russian-occupied territory 02 May 51 as F-SCNQ. Apparently not retd.

C/n 319 To French AF Apr 49. Accident 06 Oct 50 at Guemar, Algeria. To Domaines 11 Dec 63.

C/n 320 To CEV. Was F-Z.NU coded "NU" in Sep 59.

C/n 321 To CEV. Was ZJBG Aug 56. Last reported 1962.

C/n 322 Not traced but presumed to CEV from photographic evidence.

C/n 323 To CEV. Last reported 1960.

C/n 324 To CEV. To RF-ONERA as F-BDVS (CofA 25 Feb 54). Canc late 1958 and retd to CEV. Was F-Z..N coded "N" Sep 59 (see also c/n 300).

C/n 325 F-BDLT. To SNECMA (CofA 20 May 49). Canc May 52. Reported with French AF as F-SCDH as early as 1950-51. Later transferred to CEV and still operated Aug 66 as F-Z..B "B".

C/n 326 To French AF Jun 49. Accident 27 Jun 60 at Colomb-Bechar, Algeria, as F-SCDZ. SOC 20 Jan 62.

C/n 327 To French AF Jun 49 to 12 Oct 62.

C/n 328 To French AF May 49 to 07 Nov 62.

C/n 329 To French AF May 49 to 14 Aug 64 (last Martinet in French AF service).

C/n 330 To CEV. Accident 12 Sep 55.

C/n 331 To CEV. Was F-ZJPN coded "PN" in Apr 69.

C/n 332 To CEV. Was F-ZJPQ in 1950-51. Last reported 1954.

C/n 333 No information traced.

C/n 334 To French AF. Accident 25 May 50 at Nice (Dept des Alpes-Maritimes) as F-RAHQ (138.25 TT) and repaired. Another accident 25 Jan 51 at Le Bourget, followed by a last mishap 17 Oct 51. WFU prior to 1954.

C/n 335 To French AF Oct 49. To Domaines 11 Dec 63.

C/n 336 To French AF. Damaged beyond economical repair 25 Sep 51 as F-SCER and SOC.

C/n 337 To French AF Aug 49. Accident 02 Feb 54. To Domaines 11 Dec 63.

C/n 338 To French AF Sep 49 to 08 Mar 63.

C/n 339 To French AF   49 to 07 Nov 62.

C/n 340 To French AF until SOC 07 Nov 62.

C/n 341 To French AF Nov 49 to 06 Nov 61.

C/n 342 To French AF Dec 49. Accident 30 Nov 54. SOC 20 Jan 62.

C/n 343 To French AF Feb 50. Accident 21 Jan 54. To Domaines 09 Apr 63.

C/n 344 To French AF Mar 50. Last operated by Base Aérienne 257 Friedrichshafen (May 62). To Domaines 11 Dec 63.

C/n 345 To French AF Dec 49 to 21 Jan 62.

C/n 346 To French AF. Undershot during night-landing 20 Jun 51 at Tunis-El Aouina as F-SCFC and repaired. Accident 01 Aug 56 as F-SCDU. To Instructional airframe Jan 64 (see also c/n 110 - only Martinet in use after that date was c/n 329).

C/n 347 To CEV.

C/n 348 To French AF Sep 50 to 20 Jan 62.

C/n 349 To French AF Nov 50 to 07 Nov 62.

C/n 350 Last production Martinet (NC-702). F-BDLU. To ONERA (CofA 08 Jul 49). Canc late 1958. To CEV. Airframe preserved in youth centre at Valras-Plage, nr Béziers (Dept de l'Hérault), and acquired Jun 79 by the Aero-Phénix preservation group.

Refurbished Siebel 204s on "parallel line"

C/n 1000 To French AF Nov 46 to 20 Jan 62.

C/n 1001 To French AF Apr 48.   Accident 01 Sep 51.
          Converted to NC-701 30 Nov 54.   SOC 31 Mar 62.

C/n 1002 To French AF Jun 48 (overhaul in fact).   To
          Domaines 11 Dec 63.

C/n 1003 To French AF Jun 48.   Accident at Zenina 31 Jul
          59 as F-SCDI, repaired.   Lost without trace 31
Dec 59 and SOC 04 Jan 60.

C/n 1004 To French AF May 48 to 21 Mar 61.

C/n 1005 To French AF.   Last reported operated by SASM
          Bretigny 1949-53 and WFU prior to 1954.

C/n 1006 To French AF until SOC 26 May 62.

C/n 1007 To French AF until SOC 21 Mar 61.

C/n 1008 To French AF Nov 46.   WFU 05 Mar 62 and SOC 26
          May 62.

C/n 1009 To French AF Jul 48.   SOC 09 Sep or 21 Nov 63.

C/n 1010 To French AF, reportedly Si-204D obtained via
          the RAF (RAF serial VP336).   SOC 02 Mar 61.

C/n 1011 To French AF.   Accident 11 Jul 60 at Oued
          Hamimine, Algeria, as F-UIWE.   WFU 26 May 62
and SOC 12 Oct 62.

C/n 1012 To French AF.   Last operated as F-UHXB coded
          "12-XB" from 02 Apr 63 until WFU and transferred
to Domaines 11 Dec 63.

C/n 1013 To French AF Oct 48 to 02 Mar 61.

C/n 1014 To French AF Nov 48.   SOC 09 Sep or 21 Nov 63.

C/n 1015 To French Navy, Escadrille 11S (Communications
          and Transport unit).   Crashed 23 Sep 48 at
Blida, Algeria, and written off.   Seven killed.

C/n 1016 To French Navy?   (Not traced.)

C/n 1017 To French Navy.

C/n 1018 No information traced.

C/n 1019 To French AF Oct 46.   To French Navy (with
          ferry unit early 1950).   Retd to French AF.
Cvtd for blind flying with French manufactured radio
equipment.   SOC 07 Feb 63.

C/n 1020 To French AF.   To Domaines 11 Dec 63.

UNSOLVED PROBLEMS

Aircraft with out of sequence c/ns

The following were obtained from official French AF
sources.   In most cases they appear to be wrong tran-
scriptions of identities, but possibly may be the last
digits of German Wk numbers (especially the last two).

C/n 355 Was F-SCFC 1952-53.

C/n 357 Was F-RAHS 1953.

C/n 376 Accident 25 Sep 51 as F-SCER.

C/n 386 NC-702 accident 15 Sep 48.   Also reported 16
         Sep 48.

C/n 1848 Accident 19 Jul 56 in Algeria.

C/n 2292 In storage at Châteaudun May 62.   (If not a

German Wk-Nr, see c/n 292 which has never been traced.
A coincidence?)

Morocco

At least two NC-702s were transferred to l'Armée
Royale Marocaine and were in service at Marrakech in
1960 with military codes "HT" and "HZ".   They have
defied identification so far.   One NC-701 was also seen
with Air Maroc logotype on its fuselage at Rabat in 1947.

Crashes

The following crashes could not be related to the
Production List:

04 Aug 46 French AF NC-701 (GLAM) at La Pallice (Dept de
          la Charente-Maritime).   Three killed.

01 Sep 46 French AF (GLAM) at Jouy-en-Josas after take-
          off from Villacoublay.   Seven killed.

09 Feb 48 French Navy, Escadrille 10S, at St Raphaël
          (Dept du Var).

27 Apr 48 French Navy, Escadrille 54S, St Raphaël (Dept du
          Var).

26 Nov 48 French Navy, Station Flight Karouba, Tunisia,
          crashed at Algiers.

19 Dec 48 Unidentified aircraft presumed on manufacturers'
          trials or CEV, crashed near Issoudun (Dept
d'Indre).

03 Jun 49 French Navy, Station Flight Karouba, crashed at
          Tunis-Sidi Ahmet.

06 Dec 49 French Navy, Escadrille 10S, at St Raphaël (Dept
          du Var).

11 Jan 51 CEV, NC-701 F-ZLBD, at Toussus-le-Noble.

02 Jun 52 French AF, at Tourane, Indochina.   Three killed.

04 Feb 54 French AF, ditched off Palma, Majorca, on
          Villaccublay-Perpignan-Boufarik (Algeria) flight.
Seven escaped.

25 Jan 56 French AF, NC-701 from Dijon.   Belly-landed at
          Noiron-sous-Gevrey (Dept Côte d'Or).   Six
escaped unhurt.   Possibly repaired.

06 Apr 56 French AF, crashed at Rangiers, nr Porrentruy,
          Switzerland, on Cognac-Dijon-Federal Germany
flight.   Three killed.

13 Aug 56 French AF, frum Luxeuil.   Crashed near Rocroy
          (Dept des Ardennes).   All escaped unhurt.

23 Aug 56 French AF, in-flight collision with MS-500 c/n
          795 near Mascara-Thiersville, Algeria.   Six
killed (plus two in light aircraft).

1959   French Navy, NC-701.   SOC after aborted take-
          off frcm high altitude airfield.

Some of the aircraft involved in the accidents and
crashes mentioned above may have been repaired and flown
again.

BV 144

The Blohm und Voss Bv.144 was intended as a medium-range short-field replacement for the Ju-52 following a DLH specification of 1940. At that time Germany hoped to resume civil air transport operations within a short while, but the course of history overtook such plans. However, late design work was transferred to a French team, and two prototypes were ordered in August 1942 from the Bréguet factory at Anglet, near Bayonne in southwest France. The factory was already sub-contracting on outer wings for the Focke Wulf 189. Plans, tooling and components were delivered in 1943, and both aircraft were nearing completion at the time of the German withdrawal following the D-Day landings.

The aircraft were left behind intact, but unfortunately one of them was damaged beyond repair by over-zealous Resistance fighters. The other prototype was preserved, however, and during 1945 was taken by road to Toulouse for flight trials, which began soon after the end of the war. At the controls was Jean Gonord, an experienced former test-pilot from Latécoère whose long career was later to culminate with the Leduc ram-jet prototypes. The Bv.144 prototype was found to be extremely dis-appointing, and after a few flights only development was abandoned and the aircraft was withdrawn from use in August 1946 and scrapped following failure of the undercarriage.

An interesting feature of this all-metal transport was the variable-incidence wing (apparently not tried during test-flights), allowing short-field capability and, through a maximum nine degree change of incidence, low drag in flight.

Specification

All-metal medium-range transport (prototype only) with variable-incidence high wing. Crew of two/three, 18 to 23 passengers.

Span: 26.90m
Length: 21.80m
Wing area: 88 sq m
Empty weight: 9,265 kg
Loaded weight: 13,103 kg
Payload: 2,000 kg
Power-plants: 2 x 1600 hp BMW 801MA
Estimated performance (no official figures being available) included a maximum speed of 390 km/hr, a cruising speed of 330 km/hr; ceiling 9,100m and range 1,550m.

Very little information is available on the two prototypes (Bv.144V1 and V2, presumably).

When surveying the history of aeronautical development in any country there are always a number of aircraft designed and flown only to be scrapped. This was particularly true of post-war France, when the programmes were rather confused and often too diverse.

Trying to put flying-boats into commercial operation when the supremacy of landplanes was already being established was unrealistic - as shown with the Laté 631. Such was the case with the SE-200, a pre-war design. Even more unrealistic was the SE-1000 project, with too small a payload for such a large and costly air-craft.

It appears that by 1946 French planners still favoured the operation of large trans-atlantic flying-boats, arguing that a non-stop flight would save the then dangerous stopovers at Gander and Shannon, while enjoying luxury conditions of travel with plenty of space for accommodation, galley, bar and so on. Of course the major drawback of the flying-boats was their slow speed of approximately 300 km/hour. Clearly the French failed to recognise the shifting pattern of air transport or to realise that there was little hope in winning the trans-atlantic competition with flying-boats.

On the other hand, the Hurel-Dubois high-aspect ratio aircraft were of interest, if not unique, and extensive use by the IGN of the HD-34 in a specialised rôle is proof that it was a reliable aircraft. But any opportunity to compete seriously on the market with such an aircraft must be open to question.

The Potez 840 feederliner was also not a success. A failure to achieve sales, which was certainly not anticipated when the aircraft appeared, may be explained by the infancy at the time of third-level air transport.

Another point must also be kept in mind in explaining these unwanted prototypes as well as other French types ordered into production, namely the UK supremacy which was maintained for many years both in quality and quantity, thus offering operators both new and secondhand air-craft tailored for their specific needs. Typical of this problem was Bréguet's failure to launch its Bréguet 890/895 types, as related in another chapter of this monograph devoted to unsuccessful prototypes originally intended for military use. Light transport aircraft, either French-designed or French adaptations of foreign designs, are also dealt with in separate chapters.

SE-200

The SE-200 was one of three transatlantic flying-boats designed to the specification of a Ministére de l'Air programme of 1936. Initial design work had begun with the Lioré-et-Olivier concern under the designation of Léo H-49 ("H" for "Hydravion"), but with the nationalisation of most of the French aero-nautical industry the whole project was taken over by SNCASE as the SE-200, also later referred to as the Amphitrite.

Prototypes were ordered and construction begun at the Marignane works, not far from Marseilles, on the shore of the Berre Lake. With the outbreak of war and the collapse of France in June 1940, the programme was seriously delayed, and when German forces invaded still unoccupied southern France in November 1942 only the first SE-200 prototype was nearing completion. (There are also persistent reports which state that the flying-boat was flown with its Wright power-plants as early as September 1942.) In any case, the SE-200 was flying by August 1943 and was taken over by the Germans and flown to Friedrichs-hafen. It was destroyed in 1944 by RAF bombers while moored on Lake Constance. Subsequent airframes still at Marignane in various stages of completion were also taken care of by allied bombings.

Soon after the end of the war, however, the third airframe was found to be repairable and was flown in April 1946. But it was not to enter commercial service, and after some flight testing the whole programme was abandoned. F-BAIY, the only SE-200 to have flown post-war, was then preserved in good shape (but depleted

of engines) close to the airport at Marseilles-Marignane. Until the late fifties visitors could walk around the spacious cabin, inspect the flight deck or slip into the wing tunnels designed for in-flight access to the engines, at the cost of a nominal fee. Unfortunately this giant six-engined oddity, last witness of a now long-gone era, was scrapped - a typical example of the then prevailing lack of enthusiasm towards the preservation of interesting aircraft.

This grounded flying-boat had only survived by a few years its contemporary, the Latecoere 631, the only flying-boat of the 1936 programme to have seen limited commercial service in the post-war years (qv).

This was definitely the end of the large transatlantic flying-boats, and a very similar abortive Italian design, the Reggiane Re-2008, not to mention British and American ventures in the same field, all testified that there was no longer a future in such large "passenger liners of the air".

Specification

Six-engined all-metal commercial flying-boat; proto-types only.

Span: 52.20m
Length: 40.15m
Height: 9.73m
Wing area: 340 sq m
Power-plants: The six Gnôme-Rhône P.18s originally
         intended were not fitted
         1 6 x 1,500 hp Wright 2600 "Cyclone 14"
          (prototype)

Top   : SO.94 Corse I (SNCASO via Charles W Cain)
Bottom : SO-95 Corse II F-BBIG (SNCASO via Charles W Cain)

Top   : SO.90 No.1 F-BBAA (SNCASO via J-P Dubois)

Bottom : SO.95M Corse II of Aeronavale, coded 50.S.4,
          at Lanveoc-Poulmic in 1957 (Blandin/Regnier)

Nord N.2500 Noratlas prototype F-WFKL (Bristol Aeroplane
Company via Charles W Cain)

Top   : Nord N.2501D coded GA+232 of the West German
           Air Force (via Charles W Cain)

Centre : Nord N.2508 with Palas auxiliary wing-tip jets
           F-WFRG at Le Bourget in May 1957 (JMG Gradidge)

Bottom : Nord N.2501 No.166 of ET 3/64 at Lyon-Bron in
           April 1974 (C Boisselon via S Blandin)

Top   :   The sole Aeronavale N.2504 No.01 flying over the
            French Riviera (Aeronavale via J-P Dubois)

Bottom :   Nord N.2501 No.4 of the CEV with experimental
            radar nose (CEV via J-P Dubois)

Transall C.160 TR-LWE of Air Affaires Gabon at Bordeaux-
Merignac in May 1966 (PX Henry)

Top   : Transall C.160 No.F-44 of the 61eme escadron
          of the Armee de l'Air (via J-P Dubois)

Bottom : Transall C.160 336 of the South African Air
          Force at Bourges (USIAS via J Delmas)

Top : Dassault MD-311 Flamant No.264 of the Armee de l'Air (B Regnier)

Centre : MD-312 Flamant No.294 of Aeronavale, coded 11.S.16 - note the 'solid' nose of this version compared with the 'glass' nose of the MD-311. Taken at Odiham in September 1960 (JMG Gradidge)

Bottom : MD-315 Flamant No.143 of the Armee de l'Air, coded 319-CL, at Poitiers in 1976 (C Boisselon via S Blandin)

                 2 6 x 1,600 hp Gnôme-Rhône 14R 26/27
                    fourteen cylinder radial air-cooled
                    engines (F-BAIY)
Empty weight: 32,746 kg
Maximum weight: 72,000 kg
Maximum speed: 354 km/hr at sea-level
                378 km/hr at 2,500m
Cruising speed: 305 km/hr
Maximum range: 6,060 km

Production

   (Four ordered by the French Government in 1938
meeting the 1937-modified specification for a large
transatlantic flying boat originally issued in 1936.)
Construction of a fifth boat was undertaken, but only
two were completed and flown.

C/n 01 F-BAHE, f/f 11 Dec 42. Reportedly named (or to
       be named) "Rochambeau". Confiscated by German
authorities and left Berre Lake for Friedrichshafen 17
Jan 44, where strafed and sunk by RAF Mosquitoes 17 Apr 44,
while coded "2D+UT" with Luftwaffe marks.

C/n 2   Airframe 80% completed when destroyed in
        Marignane by allied bombing.

C/n 3   F-BAIY. Also damaged by RAF bombing at
        Marignane but salvaged after the Liberation of
France. repaired and completed. F/f 02 Apr 46. Loaned
to French Navy for engine tests. Last flown 18 Oct 49
after porpoising at Berre Lake, going underwater and returning to the surface. Preserved engineless at Marignane
on public exhibition until scrapped circa 1963.

C/n 4   Airframe 70% completed. Was to have semi-
        retractable stabilising floats. Sold for scrap
by Administration des Domaines (French Govt Auction Sales)
1950.

C/n 5   Only 10% components built. Destroyed during RAF
        raid on factory.

NORD NORAZUR

   Shortly after the end of World War 2 SNCAN
(at that time also referred to as Aéronord) was
busy producing light aircraft: the Stampe SV-4,
Messerschmitt Me-108 and Me-208 derivatives
known as the N1000 and N1100 series, as well as
the successful Norécrin (N1201/1204) series.
On the heavier side, prototypes were being
designed for the French Navy (N1400, N1500),
and production of the pre-war Caudron Goëland
had also been taken up.

   When the STA issued a specification for a
Goëland replacement, Nord immediately began
design work on a new light transport and
training aircraft, the N2100 Norazur - another
example of Nord phonetic prefix tradition, eg
Norécrin, Noralpha, Noréclair and later
Noratlas. Also produced in the same programme
were the Dassault MD-303 and SO-94. Somewhat
reminiscent of wartime high-wing assault
gliders, the prototype N2100 was of unconventional layout, being a twin-pusher with
tricycle retractable landing gear. The
Norazur was designed to meet a requirement to
carry eight to ten passengers in the light
transport rôle. With famous pre-war pilot
Georges Détré at the controls, the Norazur flew
on 30 April 1947.

   Another prototype followed with different
engines, and this was designated Nord 2101.
The Nord 2102 was the final derivative, this
time with conventional tractor propellors and a
further engine change, but it was another still-
born project typical of that period.

   Similar aircraft were already available, and
no production orders for the Norazur were
placed. The prototypes had an obscure fate
and were soon forgotten.

   Note that this type was probably more
military-orientated that its inclusion in this
chapter might suggest.

Specification

   High-wing twin-pusher eight to ten seat light
transport and trainer.

Wing span: 18m
Length: 13.5m
Height: 3.85m
Wing area: 38 sq m
Empty weight: 3,280 kg
Loaded weight: 4,630 kg
Power-plants: 2 x 420 hp Potez 8D-03 driving pusher
               airscrews
Maximum speed: 340 km/hr at 1,500m
Cruising speed: 312 km/hr at 3,000m
Range: 900-1,100 km
Service ceiling: 6,500m

Production

N2100   Designed and erected at Courbevoie, nr Paris
        (technical HQ of SNCAN). Taken by road to Les
Mureaux 14 Apr 47 for f/f 30 Apr 47, powered by Potez
8D-03s and with military roundels and unmarked. Civil
test regn F-WFAB later applied and roundels removed. To
CEV for trials late 1947.

N2101   F/f   47? (powered by 390 hp Bearn 6D-07s). It
        remains unclear whether this aircraft existed.

N2102   Project only, with tractor airscrews to have been
        driven by two SNECMA 12S power-plants.

POTEZ 840

Designed by Robert Castello of glider and Fouga Magister fame (see also CM 101 in this monograph), the Potez 840 was a 16-24 seat short-haul and executive aircraft. Work had begun on the early Potez projects 80 and 82 early in 1960. The aircraft was powered by four small Astazou turboprops, a not too common combination of engines for an aircraft in this category. The Potez 840 prototype took off for the first time in April 1961 from Toulouse, followed by a second aircraft during 1962 - both airframes having been built at Argenteuil, a Potez factory near Paris.

The new aircraft, a private venture, was judged to be a promising one, and a production batch of 25 was launched at Toulouse. Two variants were offered: one, the Potez 841, was to be powered by Pratt & Whitney/UACL PT-6-A6s (with 640 hp PT-6-A2Os being envisaged for 1965 production, followed later by 750 hp PT-6-A22 turboprops), while the other production version, known as the Potez 842, had Turboméca Astazou XIIF power-plants. Both variants were to be operated at a gross weight of 8,900 kg.

The first 841 made its maiden flight during the last days of 1964. The first 842 with indigenous power-plants (and c/n 3 on the assembly line) was intended for use by the Ministry of Public Works and Transport by mid 1965 but was actually delivered to the SFA via the SGACC.

It was hoped to sell large numbers of the aircraft; as early as 1962 optimistic reports from both Turbo-Flight, the US agents, and Potez led to the announcement of some 120 Potez liners to be ordered in the USA alone for 1967-68 delivery. Later a production line was also set up at Baldonnel, with expectations to deliver the first Irish-built aircraft by the end of 1965. In fact only four production aircraft were ever built, all at Toulouse.

The first two Potez 841s were delivered to German customers through Aero Dienst GmbH and were offered for resale only many years later. The fourth (and last) aircraft, a Potez 842, was sold to Morocco for use by the Ministry of Defence.

With no more firm orders in prospect, the Potez feederliner was now in serious trouble, as also was the firm itself. When the factory was taken over by Sud Aviation in April 1967 production arrangements were still being desperately sought after. The graceful Potez aircraft was undoubtedly a fine design, and it is difficult to explain its failure to attract orders. The fact that it may have appeared too early on the market is one possibility, and another is that it had been designed, built and flown entirely as a private venture - thus having no benefit from more or less official support (and subsidies), the only exception being the single Potez 842 ordered by the SGACC.

When in 1964 the FAA had selected five aircraft from nine submitted designs in the famous DC-3 replacement competition, the Potez had come third, which was encouraging, and although the aircraft might have expected some success with such an appeal, this was not the case.

A Potez 843, announced in 1965, with PT-6s and a deeper fuselage for both cargo and passenger operation, was not built. Earlier, the Potez 880 STOL development had been considered to meet military requirements but was dropped, as was the Potez 881 civil version and a twin-engined Potez 89.

This small turboprop transport was to be one of the last Potez designs, and the famous name, which had been one of the greatest in French aviation (with thousands of aircraft built since the end of World War 1) soon disappeared from the aeronautical field. Fortunately Potez 842 c/n 3 F-BNAN is now one of the Musée de l'Air exhibits at Le Bourget.

Specifications

Potez 840 (Prototype)

For 16 or 24 passenger operation.

Span: 19.35m
Length: 15.62m
Height: 5.47m
Wing area: 35 sq m
Weight empty: 4,512 kg
              4,985 kg with equipment
Loaded weight: 7,800 kg for 16 passenger operations
              8,500 kg for 24 passenger operations
Power-plants: 4 x 530 shp Turboméca Astazou II turboprops
Maximum speed: 540 km/hr
Cruising speed: 520 km/hr
Service ceiling: 12,000m
Range: 1,050-1,850 km

Potez 841 (4 x 558 hp P&W PT6A-6) and Potez 842 (4 x 640 hp Turboméca Astazou XII

Span: 19.60m
Length: 15.90m
Height: 5.19m
Wing area: 35.0 sq m
Empty weight: 5,440 kg (841)
             5,410 kg (842)
Loaded weight: 8,900 kg (841 as 24 passenger aircraft)
             8,900 kg (842 as 24 passenger aircraft)
Maximum payload: 1,890 kg (841 as 24 passenger aircraft)
                1,920 kg (842 as 24 passenger aircraft)
Maximum speed (VMO): 500 km/hr
Cruising speed at 6,560m: 450 km/hr (841)
                     480 km/hr (842)
Take-off distance: 600m (841)
                530m (842)
Maximum operating altitude: 7,600m
Range (with maximum load and 45 minutes' reserve):
     2,400 km (841)
     3,000 km (842)
Passenger configuration/s: 24, but offered as eight to 16 passenger corporate or VIP models

The considerably improved Potez STOL transport would have been fitted with four 917 hp Turboméca Bastan IV turboprops (Potez 880 project).

Production

C/n 01 840. F-WJSH, f/f 29 Apr 61. To F-BJSH Soc Henri Potez (CofA 27 Sep 63). WFU Jul 68 at Toulouse.

C/n 02 840. F-WJSU, f/f 17 Jun 62. To F-BJSU Soc Henri Potez (CofA 22 Sep 62). To N840HP Turbo-Flight Inc, del 24 Sep 62. Retd as F-BMCY to

Soc Henri Potez (CofA 05 Nov 64).   Intended for Air
Paris Oct 73.   To Darta Aug 74.   Then to Mercure?
To Paris Flying Club Jul 78.

C/n 03 840.   No details known.

C/n 04 840.   Used for static tests, possibly became
        841 c/n 2 later.

C/n 1   841.   F-WLKR, f/f 22 Dec 64, but roll out
        quoted as 23 Dec 64.   To D-CAER Aero Dienst
(CofA 29 Sep 65).   To Kurfiss Aviation 1973.   To
N62271   Apr 74.   Intended to become F-BVPZ
in 1974 but NTU.   WFU at Toulouse in US marks.

C/n 2   841.   D-CHEF.   To Hertie (CofA 28 Mar 66).   To
        B Hoffmann 1972.   To A Ostermann 1974 (oper by
Fredair).   To N3430L   del 01 Oct 75.

C/n 3   842.   F-BNAN, built 1965.   To SFA (CofA 19 Jul
        66).   Canc Dec 76.   Preserved at Le Bourget by
the Musée de l'Air.

C/n 4   842.   CN-MBC.   To Ministere de la Defense
        Nationale (CofA 28 Oct 66).   Transferred to
military marks CN-ALL 1967.   To N9878A Euroworld
California May 78.

HUREL·DUBOIS   HD·34

Commandant Maurice Hurel (of SO-90 prototype
fame - he had fled from occupied France on the
first flight of the new type) had already been
an inventive aeronautical designer when, after
the war, he started pioneering a series of aircraft designs which all possessed a high
aspect-ratio wing as a characteristic feature.
This resulted in an aircraft type with a very
large span, where low induced drag allowed a
heavier structural wing weight.   This high-wing
formula was first tested on the HD.10 prototype (F-BFAN, how preserved in the Musée de
l'Air collection), a small experimental aircraft of strange appearance.   Work on the same
basic design was proceeded with, and a larger
transport aircraft, the HD-31, was flown in
1953, followed by two more known as the HD-32.

On 26 November 1953 Air France announced its
intention of buying 24 HD-32 aircraft for 1956
delivery.   By early 1954 28 HD-32s were
reportedly on order, with four additional aircraft for IGN use and four more also being
ordered by May 1954 by Aigle Azur.   In mid
1955 a manufacturer's agreement was concluded
with SNCASE, who were to launch no fewer than
150 production HS-32s and also take over
the responsibility for selling the aircraft.
This production batch would have included the
HD-35 and HD-36 ASW variants for French Navy
use (the two designations presumably relating
to engine changes, either Wright or Pratt &
Whitney).   At the time of going to press, it
has not been possible to find out why the whole
project was abandoned.   Aigle Azur, mostly
involved with air transport in French Indochina,
must have cancelled its order following the
termination of the battle at Dien Bien Phu and
the Geneva agreements, while it is known that
the French Navy turned to a possible use of the
Shackleton before obtaining delivery of US-
supplied Neptunes.

The first two prototypes were pressed into
French military service in time for the Suez
crisis of 1956.   The Armée de l'Air HD-32 was
even involved in the Anglo-French expeditionary
force and as such was daubed with the black and
yellow identification stripes adopted for aircraft on these operations.

During the early stages of design, a jet-
engine version of the HD-31 was planned as the
HD-45, with SNECMA ATAR power-plants, but the
project was soon abandoned.

The third prototype, purchased by the French
Government, ended a short-lived career in a
crash in South America.

The characteristic design of the HD-31/32

might well have been forgotten had it not been
ordered in small numbers as the new HD-34 model.
But the order was not from an air transport
operator;   the eight aircraft, including the
prototype, were purchased by the French Institut
Géographique National (IGN) as a specially-
designed aerial mapping aircraft, with deliveries
taking place between 1957 and 1959.

Since the end of World War 2 the IGN has
always operated a substantial fleet of aircraft
for aerial mapping and general survey work, not
to mention other similar activities, both on
French Government and private contracts.   Its
aircraft have also flown contract missions
abroad, especially in former French colonies.
The home base of the IGN at Creil, north of Paris,
has always been a place of interest to aircraft
historians, with its collection of Boeing B-17
survey aircraft and the strange looking Hurel-
Dubois fleet.   The eight HD-34s have seen
extensive service over the years, but the end of
their active if uneventful career is close.
The first aircraft withdrawn from use was
scrapped in 1967, and two more followed in 1972
and 1973, with three more in 1976 with the
introduction by the IGN of more modern types
such as extended-range Beech King Air 200s with
belly bays for cameras.

Experience has shown that the HD-34 was a
sound aircraft, at least in the air-mapping rôle.
Reputedly slow and heavy but nevertheless
reliable, it has been operated under extreme
conditions, ranging from Greenland to desert and
tropical areas.   But it certainly could not be
classed with the many DC-3 replacement
contenders when it first appeared.   Among other
things, the outsized wing span might have been a
drawback for conventional routine commercial
operation from small airfields - although the
aircraft nevertheless had some STOL capacity,
with a 750m take-off run at full load.

A further derivative of the basic design, the
HD-37, was projected as a car-ferry version, no
doubt with the Channel crossing in mind.   The
HD-45 jet version, with a loaded weight of 41
metric tons and an estimated maximum speed of
720 km/hr, retained a similar unconventional
layout with strut-mounted jet engines, but the
Caravelle was preferred.

The Société Anonyme de Construction des
Avions Hurel-Dubois was however engaged on
further development projects of similar
sturdiness, still featuring the distinctive high
aspect-ratio wing.   An early example of Anglo-
French co-operation took place as early as 1955,
with the development costs shared by Miles and
Hurel-Dubois.   Conversion of a Miles Aerovan (a

rebuild of G-AJOF) with an HD wing was undertaken; the aircraft, registered G-AHDM, first
flew on 31 March 1957 as the Hurel-Dubois/Miles
HDM-105 but was only an experimental conversion.

Maurice Hurel, who had been a prolific pre-
war designer, with a score of CAMS flying-boat
models, was now less successful; more transport
aircraft projects progressed no further than the
drawing board. The HD-130 was a 12-15 seater,
also planned as a freight version. A military
assault transport with two Astazou turboprops
was also announced as the HD-150. No official
decision on the continuation of such aircraft
designs was taken.

After 1961 the Hurel-Dubois concern, now 15
years old, turned to less glamorous aeronautical
activities, no longer designing complete aircraft apart from the man-powered Hurel Aviette
ultralight design F-WTXS, built in 1973 for the
Kremer prize but appearing too late. This
aircraft has been preserved by the Musee de
l'Air. For instance, Hurel-Dubois designed
the Espace Universel passenger seats for Air
France airliners and launched major sub-
contract work and the design of components or
assemblies in various programmes ranging from
the Mirage III jet fighter to the Concorde
supersonic transport.

Fortunately, with the age of retirement now
coming, HD-34 F-BICR was recently flown to the
Musée de l'Air to join another ex-IGN aircraft
(a B-17 Fortress). Of the last three HD-34s
retained by the IGN (F-BICQ, F-BICU and F-BICV),
the first two are now grounded, one being used
for ground engine tests while the other is
used for spares. A somewhat unique design in
the field of large transport aircraft, and
probably the only specially designed survey aircraft to have seen extensive service as such,
the HD-34 provides an interesting aspect of
French aeronautical design.

Specifications

HD-10

An experimental aircraft, 520 kg and 40 hp.

HD-31

Prototype with twin fin and rudders. Maximum
weight 13,500 kg. Two 800 hp Wright Cyclones.

HD-32

Same airframe as HD-31. Initially twin-fin, later
single fin. All-up weight raised to 17,000 kg.

Wing span: 45m
Length: 22m
Height: 6.9m
Wing area: 100 sq m
Power-plants: 2 x Pratt & Whitney R-1830-92, 1,200 hp
             each
Empty weight: 10,350 kg
Loaded weight: 17,000 kg
Payload: 5,400 kg over 1,000 km
         4,400 kg over 2,000 km
Minimum speed: 108 km/hr
Cruising speed: 275 Km/hr
Maximum speed: 310 km/hr
Service ceiling: 5,000m

HD-321

Re-engined conversion of HD-32. Two Wright 982,C4s,
1,525 hp each. Large single fin and rudder with
vertical stabilisers on tailplanes.

Height: 8.73m
Empty weight: 12,200 kg

In Armée de l'Air service a crew of five (44
passengers over 1,600 km, 18-27 casualties in MEDEVAC
role, also used on security patrols in the Sahara).

A long-range maritime reconnaissance version was
planned for the French Navy.

HD-34

High-wing twin-engine survey aircraft with short-
field capability (750m take-off run when fully loaded)
for operation by the Service des Activités Aériennes de
l'Institut Géographique National. This therefore was a
specialised production version of original transport
prototypes but not used in the transport role.
Retractable nose-wheel.

Wing span: 45m
Length: 23m
Power-plants: 2 x Wright Cyclone R.1820 C.9 HE, 1,475 hp
             each
Empty weight: 12,300 kg
Maximum weight: 18,600 kg
Payload with crew and full fuel capacity (6,400 lit):
             1,500 kg
Cruising speed: 200-280 km/hr
Service ceiling: 6,000m
Endurance: 15-20 hours
Maximum range: 3,600 km

Note on Survey Work Equipment

Crew of three (pilot, navigator and copilot/flight
engineer) plus survey operators as necessary. Two
vertical and two side-looking ports for photo-mapping and
bomber-like glazed nose. Full IFR equipment, Doppler,
nav-aids and periscopic sight.

Survey equipment includes Wild RC 8, RC 9 or RC 10
vertical cameras, Sopelem or Aero Technika cameras for
side-looking operation, radio or laser Airborne Profile
Recorder (APR), Cyclope infra-red scanner, trailing MAD-
boom and other equipment for aerial mapping, geophysical
survey or scientific research as necessary.

Production

C/n 01 HD-31.   F-WFKU, f/f 27 Jan 53 (1,600 hp Wright
            Cyclones).   To F-BFKU DTI:Hurel-Dubois,
Villacoublay (CofA 07 Jan 55).   Canc Jan 58 but transferred for short-lived career to French Navy
experimental unit (Escadrille 10S) at St Raphaël 09 Oct
55.   This exchange involved the CEV receiving an ex-
French Navy Aquilon jet fighter, HD-35/36 ASW variants
not being ordered.

C/n 01 HD-32.   F-WGVG, f/f 29 Dec 53 (2,400 hp Pratt &
            Whitneys).   Cvtd to HD-321 c/n 01 with more
powerful (3,050 hp) power-plants.   F/f as such 11 Nov 54
(possibly with twin fins) or more likely 22 Dec 55 (
probably with single fin).   Also reported as F-BGVG but
unlikely.   Transferred to French AF as "01".   Crashed
10 May 60 at Evaux-les-Bains (Dept de la Creuse).   Part
of wreckage preserved on location as water trough for
cattle.

C/n 02 HD-32.   F-WHHA, f/f 11 Feb 55.   Cvtd to HD-321
            c/n 02.   To F-BHHA as "HD-320" (CofA 08 Oct 56).
To RF/SGACC.   Crashed 30 Oct 56 at Rio de Janeiro.

C/n 01 HD-34.   F-WHOO, f/f 26 Feb 57.   Used for
            development programme.   To F-BHOO IGN (CofA 09
Jun 58) but in IGN service as early as 17 Apr 58.

C/n 2   HD-34.   F-WICP, f/f 29 May 57.   To F-BICP IGN,
            acceptance 19 Dec 57 (CofA 10 Jan 58).   WFU Feb
72 and scrapped (date of canc 17 Aug 72)

C/n 3   HD-34.   F-WICQ, f/f Jul 57.   To F-BICQ IGN,
         acceptance 06 Feb 58 (CofA 21 Feb 58).   Has now
been WFU but kept grounded at Creil for technical reasons.

C/n 4   HD-34.   F-BICR, f/f Dec 57.   To IGN, acceptance
         17 Mar 58 (CofA 16 Apr 58).   Transferred to Musée
de l'Air, Le Bourget, 22 Jun 78 for display and
preservation (thus joining fellow IGN aircraft Boeing
B-17 F-BGSO).   TT 4,898 hrs.

C/n 5   HD-34.   F-BICS.   To IGN (CofA 02 Oct 58), in
         service from 05 Aug 58 but official acceptance
given as 15 Nov 58.   WFU May 67 and scrapped (date of
canc given as 16 May 69).

C/n 6   HD-34.   F-BICT.   To IGN (CofA 16 Dec 58).   In
         service from 15 Nov 58 but acceptance as early
as 24 Mar 59.   WFU Jan 73 and scrapped (date of canc 10
May 73).

C/n 7   HD-34.   F-BICU, f/f 03 Dec 58.   To IGN (CofA 23
         Apr 59) but acceptance 24 Apr 59.   In service 24
Mar 59.   Now WFU but kept grounded at Creil for
technical purposes (summer 78).

C/n 8   HD-34.   F-BICV, f/f 13 Mar 59.   To IGN,
         acceptance 24 Apr 59 (CofA 09 Jun 59).

SE-1010

The SNCASE SE-1000 was a "stratospheric"
(then a fashionable word) transport project of
exceptionally clean design, envisaged in late
1945 for transatlantic postal operation.   It
is of interest to note that after the war
France seemed to be fascinated by the so-called
"avion postal" category, with aircraft
specifically designed for the purpose or at
least widely publicised as such - the SO-91, 93
and 94 series being another example.   One may
wonder if it would not have been simpler to
adapt existing transport aircraft for the role,
a postal aircraft being after all nothing more
than a freighter, provided that a few minor
modifications were included (such as blind-
flying equipment for night operation).

The SE-1000 design was not built, however,
but gave birth to the SE-1010 which, with a
modified nose-section, became a high-altitude
photo-survey aircraft built specifically for
the Institut Géographique National.   Flown in
1948, the prototype was destroyed during a test
flight less than a year later, and the three
other airframes under construction were not
completed.

This elegant four-engined aircraft, which
could be compared to some extent with its
beautiful contemporary, the Republic XR-12,
incorporated several technical innovations (at
least by immediate post-war French standards),
including a pressurised cabin.   Experimental
use was considered, such as operation as an
airborne TV transmitter or for scientific and
weather research work, but development was
halted after the crash of the prototype.
Projected transport variants included the
SE-1001 (with longer fuselage and a hold
capacity of 24 cu m) and the SE-1030, with
accommodation for 40 passengers and a range of
some 4,000-4,300 km.   The SE-1030 was actually
a so-called "sub-stratospheric" version of the
1010, and speed would have been about 480 km/hr.

Specification

SE-1010

Four-engined mid-wing monoplane for high-altitude
survey work.

Span: 31m
Length: 21.81m
Height: 5.2m
Wing area: 116.3 sq m
Empty weight: 15,000 kg
Loaded weight: 25,000 kg (as survey aircraft including
                          1,300 kg of photographic
                          equipment and 1,000 kg of
                          extra equipment)
               32,800 kg (as 14 passenger version with
                          freight hold - 26.5 cu m in
                          all-cargo variant)
Payload: 3,350 kg
Power-plants: 4 x 1,210 hp Gnôme-Rhône 14R 28/29, 1,590
              hp on take-off
Maximum speed at 8,000m: 635 km/hr
Cruising speed at 8,000m: 535 km/hr
Service ceiling: 8,000-10,000m
Maximum range: 6,300 km

Production

C/n 01   SE-1010.   F-WEEE, f/f 24 Nov 48 (following IGN
         B-17 F-BEEA/B/C/D and thus denoting intended use
by the Institut Géographique National).   Crashed 01 Oct
49 near Carcès (Dept du Var) on test flight from
Marignane.   Three other airframes under construction
not completed.

Inclusion of such a subject in a monograph devoted to commercial aircraft may be unexpected. It is well known that on the whole transport aircraft designed to meet military requirements are seldon feasible for commercial operation. As a rule they can be used as freighters only, and while civil operators generally have to purchase new aircraft for the carriage of passengers, they are able to turn to ageing second-hand aircraft or surplus military types for freight operations.

The major French military transport has been the Nord 2501 Noratlas, and various attempts at civil sales were made with little response; quite evidently it was cheaper to buy surplus aircraft such as the ubiquitous Dakota, and the successful Bristol 170 was already well established on the world market. Very few Noratlases purchased brand-new appeared in civil marks during the fifties, but a number of ex-German machines have been released for civil use.

The next generation of military transports was the Franco-German Transall. Again, wide-spread commercial use of the type seems unlikely (as a comparison, the successful Lockheed Hercules has indeed been sold in small numbers to various civil operators but remains nevertheless a predominantly military transport). In the Transall case at least, operation by the French "Postale" is of interest, with a remarkable serviceability rate.

Also included in this chapter is a wartime design for a light twin-engined transport (the SO-90/95), which was actually intended for commercial use (in the postal rôle at least) but which was in fact only to serve with a military operator, the French Navy, as a communications and trainer aircraft. The series of twin-engined light transport aircraft designed by Marcel Dassault, which saw little use other than with the armed forces of France, complete the chapter.

Another chapter in this monograph deals with military prototypes not ordered into production.

SO·90/95 CORSE

The Société Nationale des Constructions Aéronautiques du Sud-Ouest (SNCASO), which had been formed in 1936 under the Nationalisation Programme, had maintained some activities after the Armistice in June 1940. A design team had moved to Cannes-Mandelieu and under the name "Groupe Technique de Cannes" had developed among other projects two commercial aircraft, the SO-30 and the SO-90. The latter was derived from the Bloch 800 (later SO-80) design of early 1940 for a light twin-engined postal aircraft, and the first flight of the prototype, with engines not previously air-tested, took place on 16 August 1943 under the most daring circumstances.

Ostensibly the SO-90 was to run its engines only and test the brakes. The Italian guards at the airfield (the French Riviera then being partly occupied by Italian troops) were convinced that it was necessary to remove some barbed wire fencing, and the aircraft took off, only to land three hours and ten minutes later and after a trans-Mediterranean flight at Philippeville in Algeria. Nine people were on board, including the pilot, Commandant Maurice Hurel, who pre-war had designed so many CAMS seaplanes and who was later to pioneer high aspect ratio wing aircraft in France (see Hurel-Dubois HD-34).

The SO-90-01 was flown back to France after the Liberation, this time sporting D-Day stripes as was common practice with most French aircraft of late 1944 and early 1945.

Production of the type was launched at Rochefort, 26 being ordered in 1946. But, typical of the peculiar French post-war industrial policy, the 25 airframes nearing completion at Rochefort were cancelled because the 325 ch Béarn 6D power-plants were not available, and the majority were soon scrapped. One notable exception was a single SO-94 variant being readied for demonstration tours abroad. An experimental variant of the SO-90 nonetheless began trials at Melun-Villaroche on 18 September 1952 with a bicycle tandem landing-gear in connection with the development of the SO-4050 Vautour bomber.

In 1945 the SO-90 had been considered for the domestic postal line then being started; this would have resulted in the SO-91, still retaining the original Béarn engines. A cheaper and immediate solution was found with the use of French-built Ju-52s, and the 25 production SO-91s which would have been ordered on behalf of the PTT (the French Postal Administration) were eventually not built. Nothing is known of a possible SO-92 variant.

The next development was the SO-93, also designed as a light transport and for postal purposes. Main features incorporated were more powerful engines of German ancestry, the Renault 12S, and a tricycle nosewheel undercarriage. The SO-93 was therefore displayed on the inaugural flight of the Postale de Nuit in October 1945, but this sole prototype was lost in a fatal crash in Argentina on 27 July 1946 during a demonstration tour to Latin America.

The SO-94 was the production variant carrying ten passengers instead of eight or 2,000 kg of freight as against 1,200 kg on the SO-93 prototype. Seventy-five were ordered, later reduced to 50 and then to 25. In fact only 15 were finally produced by the SNCASO concern, and with no commercial customers in view they were delivered to the French Navy for use as trainers in Morocco, radar conversions being known as SO-94R.

The SO-95 was the last and final version, reverting to a conventional two-wheel landing-gear with tailwheel. Following the practice of giving the names of French provinces to commercial transport aircraft, the name of Corse was chosen after the Mediterranean island of Corsica. (The name Bayonne had initially been allocated during the war to the original SO-90 prototype in a short-lived attempt to rationalise the names of French aircraft.) Forty-five Corse aircraft were built under a somewhat complicated procedure involving the SNCASO plant at Nantes and the former Blériot works at Suresnes, near Paris, and temporary commercial trials were undertaken by Aigle Azur and Air Maroc, while two aircraft were used on scheduled services in India by Air Services of India from 1949 to October 1950. But in the end the SO-95s were all delivered or transferred to the French Navy, being used by them mostly in the communications role and as trainers.

During the fifties the aircraft were mostly stationed at bases in North Africa (Agadir in Morocco, Lartigue in Algeria and Karouba in Tunisia), with gradual moves from 1961 towards the Base Aéronavale of Lann-Bihoué in Brittany. The type was a regular visitor to the UK during the late fifties and early sixties. At least 30 were still in service circa 1960-61, but withdrawals began from 1962 and the SO-95 quietly disappeared, the last aircraft being reported in 1965.

At least two aircraft survived as instructional airframes for many years at Rochefort, but sad to report, these were scrapped, leaving no preserved Corse until 1971 when French members of the International Association of Aviation Historians, Air-Britain, discovered a fuselage in poor condition in a scrapyard near Mormant (Dept de Seine-et-Marne). This was from the aircraft which had been used by the CEV for the monotrace undercarriage experiments with the single-letter code "E".

The future of the wreck remained uncertain for a long time, but fortunately it now appears that this unique aircraft will be saved from scrapping. "Ailes Anciennes", a non-profit organisation devoted to aircraft preservation and restoration (and working mostly for the benefit of the Musée de l'Air) has managed to reach an agreement with the owner of the scrap-yard. The neighbouring wreck of the SO-7010 (qv), another rare item, will be recovered for the Musée de l'Air by Ailes Anciennes, while the SO-90 is to be re-assembled on the spot. Should the owner ever consider relinquishing ownership of the airframe, the Musee de l'Air would be given priority to obtain the aircraft.

The stubby little SO-90/95 transport can no doubt be included in the "also-ran" category. Production had been complicated by taking place at various factories and, as attempts at commercial sales had been a total failure, the French Navy was given almost no choice but to operate the aircraft.

In retrospect, it appears that the main trouble with the Corse programme was that too much time was wasted by imposing French-produced engines which were hardly available, not always ready for use and uneconomical.

Specification

SO-93

Tricycle landing gear, light transport and postal aircraft for two pilots, one radio operator and eight passengers.

Span: 16m
Length: 12m
Height: 4.39m
Empty weight: 3,300 kg
Loaded weight: 5,200 kg
Payload: 1,200-1,900 kg
Maximum speed: 450 km/hr at 2,400m
Cruising speed: 375 km/hr
Service ceiling: 7,500m
Range: 1,200 km
Power-plants: 2 x 440 hp Renault 12S-00 (580 hp on take-off)

Main cabin of 12 cu m capacity (1,200 kg) or accommodation for eight passengers.

SO-94

Commercial variant with tricycle landing gear.

Span: 16.18m
Length: 12.35m
Height: 4.39m
Empty weight: 3,500 kg
Loaded weight: 6,200 kg
Payload: 2,000 kg
Maximum speed: 445 km/hr at 2,400m
Cruising speed: 375 km/hr
Service ceiling: 7,500m
Power-plants: 2 x 580 hp Renault 12S-00

Cabin capacity of 13 cu m (2,000 kg) or accommodation for ten passengers.

SO-95

Span: 17.9m
Length: 12.35m
Height: 4.29m
Empty weight: 4,024 kg
Loaded weight: 5,605 kg
Payload (useful load): 882 kg (excludes pilot, fuel/oil)
Maximum speed: 354 km/hr at sea-level
Cruising speed: 335 km/hr at 2,700m
Take-off distance (to clear 15.25m): 850m
Range: up to 1,300 km
Passenger configuration: 10-13 passengers plus two crew side by side
Postal version for 1,500 kg freight
Power-plants: 2 x   hp SNECMA 12S-02

Production

C/n 01 SO-90. F/f 16 Aug 43 from Cannes to Algeria. 350 hp Béarn 6D-07 power-plants. Production of 26 started in 1946 but scrapped after canc of order. Experimental tandem landing-gear variant f/f 03 Sep 52.

C/n 1 SO-90 F-BBAA (the series F-BBAA onwards was reserved for SO-90 production until canc).

C/n 7 SO-90 F-BBAG. Was in SNCASO livery with name "Cassiopée" on nose (apparently this name was intended for the SO-90 variant at that time).

SO-91 postal aircraft.   Production of 25 not finalised.   Tricycle landing gear.

C/n 01   SO-93.   F/f 17 Aug 45 with military roundels, later F-WBAP and F-BBAP (CofA 03 May 46).   With OFEMA demo tour to Latin America Aug 46 (together with NC-702, N1101, MS-502 and Stampe SV-4C).   Crashed 23 Jul 46 at Mangininhos, Argentina, on demo flt.

SO-94.   F/f 06 Mar 47.   Initial production order of 75 gradually cut down to 15 and del to French Navy, mostly as SO-94R radar trainers, remainder as SO-94M military transports.

C/n 1   SO-94.   To French Navy.   Served with Escadrille 10S at disposal of French Navy test centre at Fréjus-St Raphaël as "10S-30".

C/n 2   SO-94.   To French Navy.

C/n 3   SO-94.   To French Navy?

C/n 4   SO-94.   To French Navy?

C/n 5   SO-94.   To French Navy?

C/n 6   SO-94.   To French Navy, acceptance flt at Nantes factory 04 Jun 52 as SO-94M and del 05 Jun 52 (possibly re-delivery?).

C/n 7   SO-94.   To French Navy.

C/n 8   SO-94R.   To French Navy, Escadrille 56S.

C/n 9   SO-94.   To French Navy, Escadrille 56S.

C/n 10   SO-94.   To French Navy, del circa May 50.

C/n 11   SO-94.   To French Navy, in service 1950.

C/n 12   SO-94R.   To French Navy, Escadrille 56S.

C/n 13   SO-94.   To French Navy?

C/n 14   SO-94.   To French Navy?

C/n 15   SO-94.   To French Navy, presumed del circa Sep 52 as SO-94R.

Question marks denote aircraft not evidenced in French Navy service.   Exact number of SO-94R conversions not known.

SO-95 Corse II.   Twelve passenger transport with conventional landing gear and (Renault)-SNECMA 12S engines for delivery to French Navy with some aircraft being temporarily used in civil marks for commercial evaluation or operation.   F/f 17 Jul 47 (SO-95M).   Total of 45 built as SO-95M military transport and SO-95R radar trainer.

C/n 1   SO-95.   To French Navy.

C/n 2   SO-95.   To French Navy.

C/n 3   SO-95.   To French Navy.

C/n 4   SO-95.   To French Navy.

C/n 5   SO-95.   To French Navy.

C/n 6   SO-95.   To French Navy.

C/n 7   SO-95.   To French Navy.   WFU circa 1964.

C/n 8   SO-95.   To French Navy, acceptance flt 25 Oct 49.   Damaged by fire on apron of Escadrille 56S, Agadir, 20 Feb 57 but repaired.

C/n 9   SO-95.   To French Navy.

C/n 10   SO-95.   To French Navy.

C/n 11   SO-95.   To French Navy.

C/n 12   SO-95.   To French Navy.

C/n 13   SO-95.   Presumed to French Navy but not traced.

C/n 14   SO-95.   To French Navy.   Accident with Escadrille 4S 10 Jun 61 but repaired.   Flown to Rochefort 05 Oct 62 for use as instructional airframe.

C/n 15   SO-95.   To French Navy.   Ground accident Algiers 23 Jan 62.   To Rochefort 1962 as instructional airframe.

C/n 16   SO-95.   To French Navy.

C/n 17   SO-95.   To French Navy, acceptance flt at Nantes factory 09 Nov 50.

C/n 18   SO-95.   To French Navy.

C/n 19   SO-95.   To French Navy.

C/n 20   SO-95.   Presumed to French Navy but not traced.

C/n 21   SO-95.   To French Navy.

C/n 22   SO-95.   Presumed to French Navy but not traced.

C/n 23   SO-95.   To French Navy.

C/n 24   SO-95.   To French Navy.

C/n 25   SO-95.   To French Navy.

C/n 26   SO-95.   Built at Suresnes 1948.   F-BBIB, to Service des Marchés et de la Production Aéro-nautique (CofA 17 May 49).   Lsd for 500 hrs to Air Maroc 1949 (or Mar 50).   Canc Oct 52.   Presumed to French Navy but not traced.

C/n 27   SO-95.   Completed at Suresnes 1949/50.   F-BBIH, to SGACC (CofA 03 Aug 50).   Canc Apr 54 and transferred to French Navy.

C/n 28   SO-95.   F-BBIF, to SNCASO (Bombay) (CofA 09 Mar 49).   To VT-DBW, del 13 Jun 49 to Air Services of India for 300 hr evaluation.   Retd to France Dec 49.   Retd to Air Services of India (or another example?), del Mar 50.   Retd again to France later in 1950.   To French Navy.

C/n 29   SO-95.   Built at Suresnes 1949.   F-BBIG, to Service des Marchés et de la Production Aéro-nautique (CofA 24 Aug 49).   Canc late 1952 and to French Navy.

C/n 30   SO-95.   Presumed to French Navy but not traced.

C/n 31   SO-95.   To French Navy.

C/n 32   SO-95.   To French Navy.

C/n 33   SO-95.   To French Navy.

C/n 34   SO-95.   Presumed to French Navy but not traced.

C/n 35   SO-95.   To French Navy, del autumn 51.

C/n 36   SO-95.   Presumed to French Navy but not traced.

C/n 37   SO-95.   To French Navy (retd to service Dec 62 after long period of storage).

C/n 38   SO-95.   Built at Nantes-Bouguenais 1951/52. F-BBIQ, to SGACC (CofA 31 Jul 52).   Canc Apr 54 and to French Navy.

C/n 39   SO-95.   Completed 1950/2.   F-BBIR, to SGACC (CofA 29 Jan 53).   Canc Apr 54 and to French Navy.

C/n 40 SO-95.   To French Navy.

C/n 41 SO-95.   To French Navy. Used as trials air-
        craft for ventral ASW radar from Jul 58. To
instructional airframe, Rochefort. Reportedly the
last French Navy SO-95 to survive, wingless fuselage
reported on site Jul 72, before being scrapped.

C/n 42 SO-95.   Built at Suresnes 1948. F-WBIA, loaned
        to Aigle Azur 1949.   Then F-BBIA after overhaul,
to DTI (CofA 31 Aug 50). Canc Oct 52 and to French
Navy. Believed cvtd to VIP version as SO-95V.

C/n 43 SO-95.   (F-WBIC NTU DTI.) To French Navy.
        Landing accident with Escadrille 4S 01 Oct 62
but repaired for service throughout 1963-5.

C/n 44 SO-95.   Built at Suresnes 1949. F-BBIE, to
        Service des Marchés et de la Production Aéro-
nautique (CofA 10 May 49).   Canc Jun 54 and to French
Navy.

C/n 45 SO-95.   F-BBID, to SNCASO (Buc) (CofA 29 Oct 48).
        To VT-DDI Air Services India Mar 50. Retd to
France 1950/51. F-BBID canc Sep 50. To French Navy.
WFU 1965.

Note: Pre-1961 French Navy aircraft sported conspicuous
        unit codes (eg in the case of the SO-95 "50S.9",
"56S.1", "4S-6" or "2S-10").   With new regulations
issued in 1961 the so-called SC Aéro number, ie a
permanent serial number, was allocated by the Navy
Service Central de l'Aéronautique, and this change was
gradually reflected on the aircraft, eg "14", "15", "19"
and "41", replacing the discarded unit codes.

        The SO-95 served with Escadrilles 56S, 11S, 2S,
4S and 50S.

NORD NORATLAS

The Nord 2500 was designed from the outset
as a military transport to meet French Air
Force requirements for a medium-sized cargo
aircraft also suitable for military passenger
and paratroop transport.   Two prototypes were
ordered in 1947 (the unlucky competitor being
the Bréguet 891R Mars), and the first was flown
on 10 September 1949 with SNECMA 14R engines.
The N2500-01 was followed by the N2500-02
powered by Bristol Hercules 739s, which had
been judged the more reliable engines.   Both
aircraft had civil test registrations, and a
handful of production aircraft or prototypes
for new variants also appeared on the French
civil register during the fifties, as recorded
in the following list, either for temporary use
by the manufacturers or, in two instances when
aircraft were on temporary loan to commercial
operators.

Of the three pre-production aircraft which
followed, one became the SNCAN demonstrator
from 1953 as F-BFRG.   This aircraft undertook
a demonstration tour to South America in March
1953, which led to high hopes of sales,
particularly in Brazil where licence-production
was considered.   Aerovias ordered five aircraft, with delivery of the first in June 1954,
but eventually the order was not finalised and
all other expected sales failed to materialise,
causing much disappointment.

In the meantime, production of the military
Nord 2501 for the French Air Force had been
launched, the first aircraft flying in the
autumn of 1952 at Les Mureaux (the final
assembly line later centring on Bourges).   In
1953 the official name of the new aircraft
became the Noratlas, conforming with Nord
practice of including a "Nor-" prefix in the
aircraft names.   A demonstration to Israel of
the fifth pre-production Noratlas with civil
registration F-BBGP led to an order for
military transports known as N2501-IS (IS for
Israel).

Early in 1954 evaluation of a N2501 by Aigle
Azur in Indochina was envisaged, but with the
end of the war this was not undertaken.

Another aspect of commercial operation was
considered as a mail aircraft for use by the
French "Postale", but this was also dropped.

(By contrast, the next military transport to be
ordered by the Armée de l'Air, the Franco-
German Transall, did enter postal service years
later.)

Following the evaluation of a production aircraft in August and September 1954 on its
French West Africa network, UAT ordered two
N2501A, delivered in September 1954 and February
1955.   African operations started with F-BGZA
in December 1954 with a base at Douala (French
Cameroon).   In 1955 a new version appeared as
the N2502, fitted with two auxiliary 400 kgp
Turboméca Marboré II jets mounted at the wing-
tips.   The aircraft was evaluated by UAT, and
five N2502As were delivered to them between 1956
and 1958, the first two N2501A aircraft being
brought up to N2502 standard.

In 1956 an attempt to boost commercial sales
of the type was made by fitting to the
demonstration aircraft power-plants which were
used world-wide, so that possible customers
could standardise on maintenance.   The Pratt &
Whitney R-2800 CB 17 was chosen and fitted to
F-BFRG, which was then redesignated the N2503.
In the following year the aircraft was rebuilt
again, this time with added wing-tip Marborés,
and became the N2508.   The N2508 was promptly
tested in a distant country, India.   A private
Indian airline, Kalinga Airlines, had plans to
establish a route between Calcutta and Lhasa in
Tibet, with government agreement from both India
and China.   The use of the additional jets
developing about 880 lb thrust each was
mandatory for use of the aircraft above highly
mountainous **areas**.   Kalinga was said to have
expressed an **interest** in four to six aircraft,
but no order **was** finally obtained.   Another
N2508 appeared in May 1960 and was sold to
Germany.

The history of the military Noratlases also
forms part of this monograph, including the
German order for 137 aircraft for Luftwaffe use.
Several of these aircraft were later sold to
civil operators.   The order was placed in 1957,
and after initial delivery of 25 (or 20) SNCAN-
built aircraft, production was continued by a
Flugzeugbau-Nord consortium, the aircraft being
designated N2501D (D for Deutschland = Germany).

Nineteen fifty-five was also the year when a

small order for commercial Noratlases was placed, this time by Air Algérie. Two N2501As were obtained from the French Air Force production line (c/n 148 and 149, with two replacement airframes for the military order being again allocated the same numbers), and when a third aircraft was purchased (an N2502B) the first two Air Algerie N2501As were converted to N2502B standard. They were generally similar to UAT's aircraft of the "A" variant, "B" only denoting Air Algérie. The three aircraft were mostly used for oil research support in the Sahara.

Only one Armée de l'Air Noratlas appeared in temporary civil marks (other than for use by the manufacturers). This was c/n 164 in 1972 on loan to the Knights Hospitaller.

On the other hand, unlike the French Air Force the Luftwaffe was not entirely satisfied with its Noratlas fleet and transferred many aircraft to other countries for military use and also had many aircraft put up for sale as they became surplus to requirements. For this reason a large number of former military N2501Ds have entered service with various operators all over the world.

Surprisingly, ex-commercial Noratlases found their way to a military operator when the six surviving UAT aircraft were purchased in 1960 by the Portuguese Air Force, who also bought the three Air Algérie aircraft after the independence of Algeria.

The Noratlas still serves in many countries but must generally be considered as a military transport; commercial use could only be as a freighter, and it was of course much cheaper to buy and operate surplus American aircraft. In 1978, however, several surplus Armée de l'Air aircraft were placed on the French civil register.

Specification

N2501A

High-wing twin-boom twin-engined freighter and transport aircraft.

Span: 33m
Length: 22m
Height: 6m
Wing area: 107 sq m
Empty weight (freighter): 13,150 kg
Maximum weight: 21,000 kg
Payload: 3,730 kg
Power-plants: 2 x Bristol-SNECMA Hercules 758/759,
        1,626 hp at 2,500 rpm
        2,119 hp at 2,800 rpm on take-off
Maximum speed: 440 km/hr
Cruising speed: 324 km/hr at 1,500m
Take-off distance: 660m
Service ceiling: 7,500m
Passenger configuration: 45 passengers

N2501D

German licence-built variant. Specification as per N2501.

Military N2501

Span: 32.50m
Length: 21.96m
Height: 6m
Wing area: 101.2 sq m
Empty weight: 13,300 kg
Loaded weight: 20,600 kg

Maximum weight: 22,000 kg
Power-plants: 2 x Bristol-SNECMA Hercules 758/759
Maximum speed: 405 km/hr at 1,500m
Cruising speed: 320 km/hr
Service ceiling: 7,100m

N2508

Similar to N2502A but with two Pratt & Whitney R2800 CB 17s (2,500 hp) with Hamilton Standard 43E60 three-bladed propellors and Turboméca Marboré II wingtip auxiliary jet (880 lb st). No performance figures are available.

Production

There follows a basic production list; details on Armée de l'Air histories have been limited to acceptance dates and SOC dates only, separated by a slant (/). In a few instances some military codes have nevertheless been given; last code used by aircraft WFU as a general rule, but also a few representative codes of interest for little-known units, in particular those overseas.

When no SOC date is given, the aircraft must be assumed to have been current at October 1978. Most aircraft in the process of being WFU are usually last reported in storage at Entrepôt de l'Armée de l'Air 601 (EAA 601) at Châteaudun (Dept d'Eure-et-Loire), and this must be kept in mind for further reports and updates.

Aircraft with DIT mention have been grounded permanently for instructional purposes, usually at Rochefort-sur-Mer (Dept de la Charente Maritime), a major Armée de l'Air technical school.

Details are briefly mentioned in the case of modified aircraft: VIP is self-explanatory, ECM denotes special electronic countermeasures aircraft known to have been operated by Escadrille Electronique 54 "Dunkerque". Calibration aircraft served with Escadrille de Calibration 57 "Commercy". CIET aircraft are navigation flying classrooms with conspicuous radar nose. But while air-craft of the heavy transport OCU, CIET 340 proper, are coded "340-..", the navigational trainers sport the codes of navigational school Groupement Ecole 316 (hence the "316-.." codes). The N2501SNB (Systeme Navigation Bombardement) incorporates a Mirage IVA navigation and weapons-delivery system for training purposes and is also fitted with a conspicuous radar nose (CIFAS 328, hence the 328- codes).

Regarding codes, French AF Noratlases usually sport a unit code with the last two letters of the call-sign. Frequent changes occur (eg c/n 124 = "61-YD" c/s F-RAYD Nov 56/Apr 60 to "64-IK" F-RAIK up to Apr 62. "61-NP" F-RANP Sep 62/Jun 65..."63-BP" F-RABP late 68 or early 69), and this is the subject of extensive research work. On some occasions only the last two letters of the call-sign are shown with no reference to the unit, and the last three in the case of the navaids calibration air-craft (call-signs F-RCAA/RCAZ), or even no code at all. Giving extensive details on this would have been beyond the scope of this monograph and would mean in any case writing the major part of French military transport history, the Noratlas having b een the backbone of COTAM (French military Transport Command) for a quarter of a century!

Nord 2500

C/n 01 Prototype. Built Issy-les-Moulineaux (SW of
      Paris). F/f 10 Sep 49 from Melun-Réau-
Villaroche (Dept de Seine-et-Marne) as F-WFKL with SNECMA 14R engines and with Claude Chautemps at the controls and Georges Détré as co-pilot. To CEV for official trials 20 Apr 50. Destroyed in unknown circumstances.

Nord 2501

C/n 02 Prototype. F/f 28 Nov 50 as F-WFUN with Bristol

Hercules 739 engines. To CEV for official trials 01 Dec 50. Attended Salon du Bourget 1951. Crashed 06 Jul 52 at Lyon-Bron and destroyed. Famous French aviatrix Maryse Bastie among the six victims.

C/n 03 F/f   Sep 52. Del to French AF 23 Oct 52 but left at disposal of CEV and attended Salon de l'Aviation 1957. F-ZABT with CEV 1957-58 at least; test mods included radar-nose. Officially transferred from French AF to CEV 28 Sep 67. Last reported Jun 70, fate unknown.

C/n 04 Built 1952. F/f   as F-WFRG. To F-BFRG (CofA 20 Feb 53) SNCAN. Demo tour to Latin America 28 Feb-23 May 53. Del 07 Sep 53 to French AF but left at disposal of manufacturers. Lsd to UAT 09 Jan 55-10 Feb 55. Officially handed back to SNCAN 09 Dec 55 as F-BFRG. Became Nord 2503-01 (also mentioned as N2503-04) F-WFRG early 1956 (qv). Canc Jun 57 as Nord 2501. To VT-DKA for a short time in 1958 when lsd to Kalinga A/L. Retd to F-BFRG. Later became N2508 (qv).

C/n 05 Built 1953. F/f   . Accepted by French AF 27 May 53. Civilianised as F-BBGP (CofA 02 Oct 53) SNCAN. Demo tour to Israel. Canc early 1954 and retd to military marks. Oper by CEAM at Mont-de-Marsan (Dept des Landes) (where codes included "118-IG" F-SDIG and "118-IB" F-SDIB). Involved in Franco-German Transall programme as ETB aircraft. Transferred to CEV 04 Jun 73, coded "BS" (F-Z.BS).

C/n   One airframe, identity not known, used for static tests at EAT (Etablissement Aéronautique de Toulouse), Toulouse (Dept de la Haute-Garonne).

In the production aircraft which follow (all of which were French AF 2501s), aircraft c/n 1 to 25 were slightly different from later production models, and maximum take-off weight was limited to 19,600 kg, giving a payload ranging from 1,900 to 3,350 kg.

C/n 1 F/f 24 Oct 52 at Les Mureaux, SW of Paris, with Georges Détré at the controls. Acceptance 12 Jun 53/SOC 12 May 76 French AF.

C/n 2 F/f 22 Nov 52. Acceptance date not traced. Oper by French AF as F-SDAC when crashed at Pic de Costabonne, south of Prades (Dept des Pyrénées Orientales) on the Franco-Spanish border during night of 29/30 Dec 53. SNCAN employees were among fatalities.

C/n 3 26 Jun 53. Crashed 25 Nov 54 at Eschau, nr Strasbourg (Dept du Bas-Rhin), believed F-SDAD of CEAM at the time.

C/n 4 15 Nov 53/26 Sep 69 but also oper for tests with CEV, particularly in connection with development of electronic systems for French-modified DC-7C AMOR programme. Coded "BJ" in Aug 70, with modified radar-nose and displaying large "4" as identity. To DIT Rochefort as instructional airframe. Last reported Jul 75.

C/n 5 23 Mar 54. Transferred to CEV 28 Jan 69 and allocated F-ZJNA. Last reported in service Jun 76 (aircraft in fact oper by CEV since at least Mar 59).

C/n 6 15 Dec 53/02 Aug 76. Derelict at Châteaudun Jun 77 as "312-BJ" F-TEBJ, identity allocated since 07 Jan 70 (not oper by CEV as reported elsewhere).

C/n 7 09 Dec 53/12 Aug 76 (stored at Châteaudun since at least Jul 72 as "340-HL" F-RBHL).

C/n 8 23 Feb 54. Was F-ZXRA, date and circumstances unknown. At CEV Brétigny in 1954. To French AF 23 Feb 54 after del to French AF apparently delayed. SOC 12 Aug 76.

C/n 9 28 Dec 53. Stored at Châteaudun Jul 72 as "118-IC" F-SDIC, SOC 21 Aug 73. Instructional at Rochefort, first reported there Jul 74, uncoded. Last reported Jul 75.

C/n 10 10 Mar 54/25 Sep 75, at Châteaudun as "340-VD" F-RAVD. To DIT Rochefort. Reported 1978.

C/n 11 29 Mar 54/17 Jun 76.

C/n 12 10 Mar 54/12 Aug 76.

C/n 13 27 Feb 54/18 Jun 73, at Châteaudun as "340-VF" F-RAVF. To DIT Rochefort. Last reported there Jul 75, coded "F" only.

C/n 14 26 Mar 54/31 Oct 75, at Châteaudun as "340-VE" F-RAVE. To instructional at Rochefort, reported 1978.

C/n 15 26 Feb 54/21 Jun 76. Used 1972-75 by Châteaudun-based ferry unit Escadrille de Convoyage EC.70 with code "070-MC" F-SDMC.

C/n 16 26 Apr 54/12 Jul 76.

C/n 17 29 Mar 54/16 Aug 76.

C/n 18 26 Apr 54/transferred 20 Apr 72 to DTCA (AIA de Clermont-Ferrand, Dept du Puy-de-Dôme) and allocated F-ZJQR, current May 78. Since reported coded "CR" but not yet confirmed.

C/n 19 14 Apr 54/12 Aug 76.

C/n 20 21 Apr 54/12 Aug 76.

C/n 21 28 Apr 54/12 Aug 76. Stored at Châteaudun by Sep 75 and engineless at Châteaudun Jun 77, still coded "XG" F-RAXG.

C/n 22 04 May 54. Crashed 16 Sep 60 in the Aures Mountains (Algeria) as F-RBHO.

C/n 23 04 May 54.

C/n 24 28 Apr 54. Current Jun 78 as "312-BJ" F-TEBJ named "Le Dinosaure".

C/n 25 13 May 54. To ECM.

From c/n 26 onwards, maximum take-off weight was increased to 20,500 kg, payload range being raised to 2,800-4,300 kg.

C/n 26 26 May 54. Crashed 29 Nov 57 near Tiznit/Ifrane (Morocco).

C/n 27 13 May 54. SOC pending during summer 1978.

C/n 28 23 Jun 54. To ECM.

C/n 29 21 May 54. Transferred to CEV 31 May 72, oper at Istres. Retd to French AF 23 Feb 77 and SOC 09 Jun 77 at Châteaudun uncoded. Gate guardian at BA 123 Orléans-Bricy with marks "61-QP", representing an aircraft of ET 3/61 "Poitou". BA 123 is the major base of French AF Transport Command, located in Dept du Loiret.

C/n 30 21 Jun 54.

C/n 31 27 May 54.

C/n 32 21 Jun 54.

C/n 33 14 Jun 54. To ECM.

C/n 34 Date of acceptance not traced but believed 25 Jun 54 (last aircraft of the first French AF initial order). SOC 12 Oct 65. To DIT Rochefort coded "N". Last reported uncoded at Rochefort Jun 76.

C/n 35 18 Oct 54. Crashed 29 Apr 64 near Foix (Dept de l'Ariège).

C/n 36 27 Jul 54. To French AF/CEAM and not CEV as quoted elsewhere. "118-IG" in Jul 74, c/s F-SDIG.

C/n 37   09 Jul 54.

C/n 38   12 Jul 54/27 Aug 76.   Derelict at Châteaudun
Jun 77 as "64-IA" F-RAIA.

C/n 39   30 Aug 54/26 Apr 78.   To ECM.

C/n 40   To French AF 15 Sep 54.   Retd to manufacturers
(SNCAN) 26 Jan 56 to become conversion N2501E
airframe c/n 40.   F/f as such 09 Aug 56 Melun Réau-
Villaroche.   Experimental variant on original
production N2501, incorporating auxiliary Turboméca
Marboré II jets at the wing-tips, modified flaps and
air-brakes for improved short-field performance and
kneeling undercarriage.   Operational evaluation in
Algeria.   Crashed on landing 26 Apr 57 at Melun-
Villaroche (Dept de Seine-et-Marne), believed oper at
the time on behalf of CEV.   Five killed.   (designation N2506 used for some time as this aircraft had been
earmarked for N2506 development.)

C/n 41   03 Sep 54.   To ECM.

C/n 42   17 Sep 54.   To ECM.

C/n 43   27 Sep 54/27 Nov 76.

C/n 44   05 Oct 54.   Surplus to requirements.   To Cie
Gyrafrance 18 May 78 with civil reg F-BZCK.   One
of 13 Noratlases stored at Le Bourget Jun 78 to at least
Jun 79 pending re-sale, believed to Africa.   (See c/n 52,
55, 56, 58, 60, 64, 65, 72 and 76, F-BZCL/F-BZCT
respectively, and c/n 103, 133 and 144 unregistered.)

C/n 45   06 Oct 54.   Destroyed 15 Feb 63.

C/n 46   13 Oct 54/27 Aug 76.

C/n 47   26 Oct 54/27 Oct 76.

C/n 48   08 Nov 54.   Crashed 25 Jan 56 near Mouzaïaville
(Algeria) as "64-KG" F-RBKG.

C/n 49   28 Oct 54.   Crashed 30 Jul 71 at Pau (Dept de
Pyrenees Atlantiques).   Pau is the main French
AF parachute training centre.

C/n 50   21 Dec 54.   Earmarked for preservation at Musée
de l'Air, Le Bourget, and SOC 04 Jun 76.
Currently on display in the open as "64-BH" F-RABH, its
final service code.

C/n 51   09 Nov 54/16 Feb 77.

C/n 52   Originally N2502 variant.   F/f believed 01 Jun
55 as F-WHHP SNCAN.   Painted in UAT colours.
Attended Salon du Bourget 1955.   Reverted to N2501 (wing
tip jets deleted) and del to French AF.   Accepted 26 Oct
55.   To Gyrafrance as F-BZCL 18 May 78 (see c/n 44).

C/n 53   24 Nov 54.   SOC pending summer 1978.

C/n 54   22 Dec 54.

C/n 55   31 Dec 54.   To Gyrafrance as F-BZCM 18 May 78
(see c/n 44).

C/n 56   28 Sep 54.   To Gyrafrance as F-BZCN 18 May 78
(see c/n 44).

C/n 57   27 Jan 55.

C/n 58   Built 1955.   To French AF 14 Jan 55 but to SNCAN
as F-BAHR (CofA 27 May 55).   Canc Jun 57 and
reverted to French AF.   Surplus to requirements and to
Gyrafrance 18 May 78 as F-BZCO (see c/n 44).

C/n 59   11 Feb 55.   Crashed 07 Dec 66 in Dept de
l'Ariege.

C/n 60   21 Apr 55.   To Gyrafrance 18 May 78 as F-BZCP
(see c/n 44).

C/n 61   31 Mar 55/21 Dec 76.   Stored derelict at
Châteaudun Jun 77 as "340-VJ" F-RBVJ.

C/n 62   09 Feb 55.   Destroyed 25 May 62.

C/n 63   10 Feb 55/22 Feb 78.

C/n 64   08 Mar 55.   To Gyrafrance 18 May 78 as F-BZCQ
(see c/n 44).

C/n 65   23 Feb 55.   To Gyrafrance 18 May 78 as F-BZCR
(see c/n 44).

C/n 66   01 Apr 55.   To ECM.

C/n 67   13 Apr 55.   Destroyed 19 Jul 61 at Bizerte,
Tunisia.

C/n 68   26 May 55.   Crashed 19 Jun 57 near Bouinan, near
Blida (Algeria).

C/n 69   04 Apr 55.   Crashed 19 Oct 71 near Epernay (Dept
de la Marne) as "62-QH".

C/n 70   13 Apr 55.   Crashed 25 Jan 56 at Mouzaiaville
(Algeria) as F-RBKO.

C/n 71   13 Mar 55.   Crashed 10 (or 11) Jan 56 near
Miliana (Algeria) as F-RBKA.

C/n 72   09 May 55.   To Gyrafrance 18 May 78 as F-BZCS
(see c/n 44).

C/n 73   27 Jun 55.   Destroyed 18 Jun 58 as F-RBKC.

C/n 74   02 Jun 55.   Crashed 14 Nov 67 at Dakar (République
du Sénégal) as F-RANZ.

C/n 75   14 Jun 55.

C/n 76   07 Jul 55.   To Gyrafrance 18 May 78 as F-BZCT
(see c/n 44).

C/n 77   01 Jul 55.   Last oper as "312-BG" F-TEBG late
1976.   WFU and stored at Châteaudun with same
code.   Derelict by Jun 77 but not officially SOC until
30 Mar 78.

C/n 78   26 Aug 55.   SOC 18 Oct 77.   Gate guardian at
BA726 Nîmes-Courbessac (Dept du Gard) erected on
concrete blocks.

C/n 79   09 Sep 55.

C/n 80   03 Sep 55.   This was the first N2501 to
incorporate larger dorsal fins.   Side windows of
flight deck were modified and enlarged for improved
visibility, and these are also conspicuous on the so-
called Series 80 aircraft up to last production aircraft.

C/n 81   03 Oct 55.

C/n 82   16 Nov 55.   Crashed 12 Aug 57 near Bizerte
(Tunisia).

C/n 83   09 Nov 55.

C/n 84   19 Dec 55.

C/n 85   03 Jan 56.

C/n 86   24 Jan 56.

C/n 87   18 Jan 56.

C/n 88   27 Jan 56.

C/n 89   20 Jan 56.

C/n 90   24 Feb 56.

C/n 91   01 Feb 56.

C/n 92 05 Mar 56.

C/n 93 27 Feb 56.

C/n 94 05 Mar 56.

C/n 95 23 Mar 56.

C/n 96 16 Mar 56.   Belly-landed 13 Jan 59 as F-RANS but repaired.

C/n 97 11 Apr 56.

C/n 98 12 Apr 56.

C/n 99 12 Apr 56.

C/n 100 07 Apr 56.

C/n 101 27 Apr 56.

C/n 102 30 Apr 56.   Crashed 15 Jan 60 near Ghardaia (French Sahara, Algeria) as "61-QU".

C/n 103 22 May 56.   To Gyrafrance 18 May 78 and stored with others at Le Bourget as "64-IB" F-RAIB (see c/n 44).

C/n 104 25 May 56.

C/n 105 15 May 56.

C/n 106 04 Jun 56.   Loaned to fledgling air element of the République de Djibouti (formerly Territoire Français des Afars et des Issas up to 27 Jun 77) from late 1977.   Based at BA188 Djibouti.   Still with French AF identity up to Apr 78 when reported as F-SCJD.

C/n 107 05 Jun 56.

C/n 108 19 Jun 56.

C/n 109 22 Jun 56.

C/n 110 29 Jun 56.

C/n 111 27 Jun 56.

C/n 112 04 Jul 56.

C/n 113 20 Aug 56.   Calibration aircraft.

C/n 114 28 Aug 56.   To CIET.

C/n 115 24 Aug 56.

C/n 116 03 Sep 56.

C/n 117 04 Sep 56.

C/n 118 10 Sep 56.

C/n 119 24 Sep 56.   VIP conversion.

C/n 120 To French AF 27 Sep 56 but regd F-BAOH (CofA 01 Oct 56) for SNCAN.   Retd to French AF and canc Jun 57.

C/n 121 02 Oct 56.

C/n 122 10 Oct 56.

C/n 123 09 Oct 56.

C/n 124 F/f 12 Oct 56.   Acceptance 31 Oct 56.   Accident (circumstances unknown) in late 1968 or early 1969 and airframe considered BER.   SOC 07 Oct 69.   Instructional at Rochefort with last code "63-BP" F-RABP and named "Le Bûcheron".

C/n 125 22 Oct 56.   Camouflaged, being one of several aircraft using three-tone scheme of dark green,

sand and chocolate brown in 1976.   Initially for use in Djibouti with French AF unit GAMOM/ETOM 88.   Codes were either full unit codes with "88-" and "last two" or, in this case, last two letters of call-sign only, c/n 125 being "JF" F-SCJF 1976-66.

C/n 126 05 Nov 56.   Camouflaged as "JB" F-SCJB for use in Djibouti 1976 (see c/n 125).

C/n 127 21 Dec 56.   Oper with French Polar Expedition to Greenland Apr 57-Jul 58 as "61-YG" F-RAYG.   Cvtd to N2501SNB.

C/n 128 28 Jan 57.   Cvtd to N2501SNB.

C/n 129 23 Nov 56.

C/n 130 28 Dec 56.

C/n 131 15 Jan 57.

C/n 132 10 Dec 56.   Cvtd by mid 1973 with port side windmill generator and modified forward section, as "328-ER" F-UKER.   By 1977 was N2501SNB "328-ED" F-UKED.

C/n 133 07 Feb 57.   To Gyrafrance 18 May 78 (see c/n 44).   At Le Bourget from May 78, unregd and coded "62-KP" F-RBKP.

C/n 134 12 Feb 57.   SOC pending during summer 1978.

C/n 135 13 Mar 57.   N2501SNB.

C/n 136 20 Mar 57.

C/n 137 29 Mar 57.   Stored at Châteaudun Jun 77 as "316-FM" F-TEFM (ex CIET).   SOC pending during summer 1978.

C/n 138 05 Apr 57.   Crashed Oct 57 at El Goléa (French Sahara) and repaired.

C/n 139 24 Apr 57.   Oper in French West Indies as "58-MI" F-RHMI by 1973.

C/n 140 09 May 57.   Camouflaged with full unit code "88-JA" F-SCJA 1976 (see c/n 125).

C/n 141 09 May 57.   Camouflaged (see c/n 125).

C/n 142 17 May 57.

C/n 143 03 Jun 57.   Crashed 06 Sep 74 near Sartène (Island and Dept de Corse).

C/n 144 06 Jun 57.   To Gyrafrance as early as 23 Feb 78.   At Le Bourget with other Gyrafrance aircraft (see c/n 44), unregd and still coded "XA" F-RAXA.

C/n 145 25 Jun 57.

C/n 146 26 Jun 57.   Calibration aircraft.

C/n 147 11 Jul 57.

C/n 148 22 Jul 57.   To French AF (replacement on production line for civil diversion below).

C/n 148/1 (also referred to as 148B)   Built 1957.   F-OBDX (CofA 13 Sep 57), to CGTA/Air Algérie.   In service 08 or 17 Sep 57.   Cvtd to N2502B Sep 58.   Sold to Portugal Oct 62.

C/n 149 30 Jul 57.   To French AF (replacement on production line for civil diversion below).

C/n 149/2 Built 1957.   F-OBDY (CofA 22 Oct 57), to CGTA/Air Algérie.   In service 23 Oct 57.   Cvtd to N2502B Nov 58.   Sold to Portugal Oct 62.

C/n 150 28 Sep 57.

C/n 151 17 Oct 57.   Calibration aircraft.

C/n 152 08 Dec 57.   Crashed 15 Jun 65 at Fort Gouraud
          (now Fderik), Mauretania.

C/n 153 08 Dec 57.

C/n 154 11 Dec 57.

C/n 155 11 Dec 57.   N2501SNB.

C/n 156 06 Mar 58.   Camouflaged (initially as "64-IW"
          F-RAIW by Jun 76, then as "JE" F-SCJE), see c/n
125.

C/n 157 26 Mar 58.   To CIET.   Now reverted to standard
          aircraft, was "62-KS" 1977-78.

C/n 158 11 Mar 58.

C/n 159 05 May 58.

C/n 160 05 Jun 58.

C/n 161 05 Jun 58.

C/n 162 25 Jun 58.

C/n 163 25 Jul 58.   Oper in Pacific as "82-PO" F-RBPO
          alongside c/n 187 1973-76.   Retd to France,
believed cvtd to N2501SNB but unconfirmed.

C/n 164 To French AF 31 Jul 58.   On loan to Oeuvre
          Française de l'Ordre de Malte and regd F-BPXR
(CofA 28 Mar 72).   Retd to French AF and canc May 72.
Stored at Châteaudun Jun 73.

C/n 165 01 Aug 58.

C/n 166 30 Sep 58.   Served overseas (with French AF unit
          ETOM 55 at Dakar, Sénégal, was "55-KD" F-RAKD Jan
78).

C/n 167 24 Sep 58.

C/n 168 31 Dec 58.

C/n 169 03 Nov 8.

C/n 170 26 Nov 58.   Calibration aircraft.

C/n 171 08 Dec 58.   To CIET.

C/n 172 18 Dec 58.   To Ferry Unit (as "M-A", later
          "070-MA" F-SDMA from 1969-75 at least).

C/n 173 31 Dec 58.   Oper by CEV May 72 as F-ZJQZ.
          Retd to French AF.

C/n 174 03 Feb 59.

C/n 175 28 Feb 59.

C/n 176 23 Mar 59.

C/n 177 19 May 59.

C/n 178 11 May 59.   Crashed 21 Apr 64 at Bouar (then
          République Centrafricaine) as "62-WL".

C/n 179 19 May 59.

C/n 180 17 Jun 59.   N2501SNB.

C/n 181 03 Jul 59.   Oper in Indian Ocean mid 1972 as
          "50-WW" F-SDWW.

C/n 182 27 Jul 59.   Crashed 24 Nov 77 as "64-BR" near
          Beziers (Dept de l'Hérault) with heavy loss of
life while on training flight.   SOC 22 Mar 78, TT 8,614
hrs.

C/n 183 03 Aug 59.

C/n 184 23 Sep 59.

C/n 185 12 Oct 59.

C/n 186 07 Jan 60.   N2501SNB.

C/n 187 07 Dec 59.   SOC pending summer 1978 (aircraft
          retd from Pacific use as "82-PN" F-PBPN - see
c/n 163).

C/n 188 19 Jan 60.   To CIET.

C/n 189 02 Mar 60.   To CIET.

C/n 190 07 Mar 60.   Crashed 03 May 65 near Limoges   (Dept
          de la Haute-Vienne) as "63-BV" F-RABV.

C/n 191 30 Mar 60.

C/n 192 05 May 60.   Calibration aircraft.

C/n 193 31 Oct 60.   To CIET.

C/n 194 08 Dec 60.   To CIET.

C/n 195 19 Jan 61.   Transferred 09 Sep 77 to Air Element
          of Niger (5U-M..).   See also N2501D

C/n 196 13 Mar 61.   To CIET.

C/n 197 08 Mar 61/16 Mar 77.   To CIET.   Derelict at
          Chateaudun Jun 77 as "316-FV" F-TEFV.

C/n 198 22 Mar 61.

C/n 199 13 Apr 61.   To Ferry Unit 1972-75 at least as
          "070-MD" F-SDMD and day-glo applied to aircraft.

C/n 200 03 May 61.   N2501SNB.

C/n 201 22 Jun 61.

C/n 202 05 Jul 61.

C/n 203 02 Aug 61.   N2501SNB.

C/n 204 12 Oct 61.   Crashed 20 Mar 62 Col des Deux
          Bassins (Algeria) as "61-ND" F-RAND.

C/n 205 01 Dec 61.

C/n 206 30 Jan 62.   Was camouflaged and uncoded at Reims
          Jun 76 (see c/n 125).

C/n 207 28 Feb 62.

C/n 208 30 Aug 61.

Nord 2501A - Production for Civil Operators

C/n 1   F-BGZA, to UAT del 10 Sep 54.   Cvtd to N2502 Oct
          55.   To Portuguese AF 09 Dec 60 (as FAP 6406?).

C/n 2   F-BGZB, to UAT del 10 Feb 55.   Cvtd to N2502   .
          Written off 29 Mar 59 near Boda, Central African
Republic.

     For N2501A c/n 148/1 and 149/2 see above.

Unidentified

     The fuselage of a French AF Noratlas which crashed in
North Africa was kept for paratroop instructional purposes
near the French military airfield of Bou Sfer (Mers-el-
Kébir, Algeria) as late as Sep 64.   Its fate is not
known after the final French withdrawal from the naval
base which enclosed the airfield.   A three-digit serial
number (unfortunately not recorded) pleads in favour of
102 or 204, but this is not confirmed.

Note: Three Nord 2501s were delivered to Arkia Feb 59/ Mar 59/Apr 59, but no civil registrations were issued. These were probably transferred to the Israeli AF.

Nord 2502

C/n 1 See N2501A c/n 1 above.

C/n 2 See N2501A c/n 2 above.

C/n 3 F-BGZC. To UAT del 24 Apr 56. Sold to Portuguese AF 04 Aug 60 (as FAP 6401?).

C/n 4 F-BGZD. To UAT del 19 Jul 56. Sold to Portuguese AF 04 Aug 60 (as FAP 6402?).

C/n 5 F-BGZE. To UAT del 20 Oct 57. Sold to Portuguese AF 29 Aug 60 (as FAP 6404?).

C/n 6 Nord 2502A F-BGZF. To UAT del 31 Jan 58. Sold to Portuguese AF 20 Oct 60 (as FAP 6405?).

C/n 7 Nord 2502A F-BGZG. To UAT del 07 Mar 58. Sold to Portuguese AF 04 Aug 60 as FAP 6403.

C/n 52 Was a temporary conversion of French AF N2501 c/n 52 (qv).

C/n 2 Nord 2502C. Built 1958. F-BGKG (CofA 20 Mar 58), to Nord Aviation. Canc from civil register late 1960 and transferred to CEV. Was F-Z.NV, believed F-ZJNV, coded "NV" 1971-74 at least.

C/n 3 Nord 2502B. Built 1958. F-OBDZ (CofA 31 Jul 58), to CGTA/Air Algérie, in service 28 Jul 58. Sold to Portuguese AF Oct 62.

Nord 2503

C/n 01 Ex N2501-04 with P&W R2800-CB17 engines and three-bladed propellors. F-WFRG. F/f 31 Jan 56 at Les Mureaux with Marcel Perrin at the controls. To CEV for official trials Jun 56. Later rebuilt as N2508-01 (qv).

Nord 2504

C/n 01 F/f 17 Nov 58 at Melun-Villaroche with Jean Caillaud at the controls. Variant for French Navy (first of 24 originally intended for French Navy, four on 1957 fundings and 20 in 1958). In the event, only c/n 01 was del, the remainder being canc. Civil registration applied together with French Navy roundels initially, but trace has been lost in records and only photographic evidence of F-WIP. remains (possibly F-WIPA?). In service coded "01" at Centre d'Expériences de Fréjus St Raphaël (Dept du Var), initially with Escadrille 10S (c/s in the F-YCJA/Z block with no permanent allocation, later in the F-YDEA/Z range). Following disbandment of 10S, this is now the only aircraft of Section Essais (Test Flight).

Nord 2505

ASW-variant for French Navy. Not built.

Nord 2506

Designation also used for some time for N2501E c/n 40 (see N2501 c/n 40). C/n 01 was second N2506 to fly. Short-field performance.

C/n 01 F/f 22 Aug 57 with two Hercules 759 plus two auxiliary Marboré jets. F-WIEX. To French AF accepted 09 Sep 59. Oper by CEAM as "118-IA" F-SDIA when SOC 12 Sep 69. To DIT Rochefort. Instructional until scrapped Jan 75.

Nord 2507

Not built. Search and Rescue project as SE-161

replacement with 12 hrs' endurance (qv).

Nord 2508

C/n 01 Originally Nord 2501 c/n 04 built 1952. F/f as F-WFRG. To SNCAN F-BFRG 20 Feb 53, touring Latin America on demo 28 Feb-23 May 53. Del 07 Sep 53 to French AF but left at disposal of manufacturers. Lsd to UAT 09 Jan-10 Feb 55. Officially retd to SNCAN 09 Dec 55 as F-BFRG. Cvtd to Nord 2503 c/n 01, f/f as F-WFRG 31 Jan 56. Cvtd to Nord 2508 c/n 01, f/f as F-WFRG 29 May 57 with auxiliary wingtip Marbore II jets. To F-BFRG 06 Nov 57, evaluated by Kalinga in India Nov 57 for planned Calcutta-Lhasa (Tibet) route. Planned order for four-six aircraft not finalised and retd to France. Sold to Germany May 63, YA034 Jun 63 (c/n 01B adopted). Cvtd to Nord 2501D c/n 002A Jun 67, allocated 5359 13 Nov 67 but NTU. B/u Lemwerder Mar 68.

C/n 02 F-WJDZ, f/f May 60. Sold to Germany May 63, YA035 Jun 63 (c/n 01A adopted). Cvtd Jul 67 to Nord 2501D c/n 001A. To 5358 FFS-S Nov 69. To Greek AF 53-258 Nov 70.

German-built Examples

Almost 200 Nord 2501D Noratlases were bought by the West German AF. Originally 20 were ordered from France and 112 (or 117?) from HFB on 15 Aug 56. Later, in 1957, the order was changed to 25 French- and 161 licence-built examples, making a total of 186. The first 25 aircraft were built by Nord Aviation and delivered complete. The remaining 161 aircraft were built by Flugzeugbau Nord GmbH, a combine made up of Hamburger Flugzeugbau GmbH (HFB), Weserflug and Siebel. All aircraft were completed and test-flown by HFB at Hamburg-Finkenwerder. A 187th aircraft was ordered in early 1962 for delivery in 1964. Two N2508s were bought from France in May 63 for Transall equipment evaluation.

The majority of aircraft were withdrawn from service in the years 1970-72, and small numbers were sold to civil operators and the Greek AF. A number of aircraft had been transferred prior to this to the Nigerian and Israeli AFs.

C/n 001 To WGAF. H/o Le Bourget 13 Nov 56, del to Memmingen 17 Dec 56 (uncoded). AS571 Dec 56, 5201, b/u Lemwerder 70.

C/n 002 To WGAF. AS572 Jan 57, 5202, b/u Lemwerder 70.

C/n 003 To WGAF. AS573 Mar 57, 5203, b/u Lemwerder 70.

C/n 004 To WGAF. AS574 Mar 57, GA241 , D.9513 63 (5204 NTU), b/u Finkenwerder Jan 73.

C/n 005 To WGAF. AS575 Mar 57. Written off Berchtesgaden on test flight 15 Jly 58.

C/n 006 To WGAF. GB101 Mar 57, GA231 May 58, AS575 59, 5205. To 5U-MAM Niger AF Aug 69, coded "AM".

C/n 007 To WGAF. GB102 Mar 57, GA232 Jly 58, GR232 Nov 59, AS574 Apr 63, 5206. To 5U-MAN Niger AF Jun 69, coded "AN".

C/n 008 To WGAF. GB103 Sep 57, GA233 Jly 58, GR233 Nov 59, YA110 Oct 61, YA571 65. 5207. To D-ACUS Jan 71, stored at Lübeck, b/u 1976.

C/n 009 To WGAF. GB104 Jun 57, GA234 Jly 58, GR234 Nov 59, GA242 Aug 62, 5208, b/u Finkenwerder Sep 71 (possibly c/n 140).

C/n 010 To WGAF. GB105 Jly 57, GA235 Jly 58, GR235 Nov 59, AS586 Nov 62, 5209, b/u (at Lübeck?) 71.

C/n 011 To WGAF. GB106 Jly 57, GA236 Jly 58, GR236 Nov

112

59, GC236 Jun 62, 5210.   To 5U-MAO Niger AF
Jun 69.

C/n 012 To WGAF.   GB107 Oct 57, GA237 Jly 58, GR237 Nov
59, GC237   62, 5211, b/u Lemwerder 1970.

C/n 013 To WGAF.   GB108 Dec 57, GA238 Jly 58, GR238 Nov
59, GC238   62, 5212, b/u Lemwerder 1970.

C/n 014 To WGAF.   GB109 Dec 57, GA239 Jly 58, GR239 Nov
59, AS588 Nov 62, 5213.   To D-ANAS Dec 71 Nora
Air Services KG.   Cargo services were planned out of
Kassel but only a permit for ferry from Lemwerder to
Bremen was issued 16 Dec 71.   Aircraft moved there since
and in 1974 transferred to airport operator against
outstanding parking fees and used subsequently by fire
brigade.

C/n 015 To WGAF.   GB110 Dec 57, GA240 Jly 58, GR240 Nov
59, GA248   62, D-9512   63 (5214 NTU), b/u
Finkenwerder Dec 72.

C/n 016 To WGAF.   GB111 Dec 57, GA241 Jly 58, GR241 Nov
59, GC241   62, 5215, b/u Lemwerder 1970.

C/n 017 To WGAF.   GB112 Feb 58, GA242 Jly 58, GR242 Nov
59, AS589 Dec 62, BF589 Apr 65, 5216, b/u
Lemwerder 1970.

C/n 018 To WGAF.   GB113 Feb 58, GA243 Jly 58.   Written
off at Bursa, Turkey, 24 Jly 59.

C/n 019 To WGAF.   GB114 Feb 58, GA244 Jly 58, GR244 Nov
59, GA243 Aug 62, 5217, b/u Finkenwerder 1972.

C/n 020 To WGAF.   GB115 Feb 58, GA245 Jly 58, GR245 Nov
59, AS590 Dec 62, 5218, b/u Lemwerder 1970.

C/n 021 To WGAF.   GB116 Apr 58, GA246 Jly 58, GR246 Nov
59, GB120 Jly 62, 5219.   To Greece 52-119 Nov
69.

C/n 022 To WGAF.   GB117 Apr 58, GA247 Jly 58, GR247 Nov
59, GB121 Jly 62, 5220, b/u Finkenwerder 1972.

C/n 023 To WGAF.   GB118 Apr 58, GA248 Jly 58, GR248 Dec
59, AS591 Nov 62, 5221.   To Greece 52-121 Nov
69.

C/n 024 To WGAF.   GB119 Apr 58, GA249 Jly 58, GR249 Nov
59, AS592 Nov 62, 5222.   To Greece 52-122 Nov
69.

C/n 025 To WGAF.   GB101 May 58 LTG61, GA231 Sep 59,
GR231 Nov 59, GC233   62, 5223.   Abroad (to
Greece?).

C/n 026 To WGAF.   GB102, f/f 06 Aug 58 as first German-
built aircraft, h/o to LTG61 09 Sep 58.   GA102
Sep 59, GA239 Dec 59, 5224.   Sold abroad (to Greece?)
1970.

C/n 027 To WGAF.   GB103 Oct 58, GA103 Sep 59, GA233 Nov
59, AS567 Aug 65, 5225.   To Greece 52-125 Nov
69.

C/n 028 To WGAF.   GB104 Dec 58, GA104 Sep 59, GA234 Nov
59, 5226.   Sold abroad (to Greece?).

C/n 029 To WGAF.   GB105, f/f 21 Oct 58.   GA105 Sep 59.
GB105 Nov 59, GA117   , 5227.   Sold abroad (to
Greece?) 1970.

C/n 030 To WGAF.   GB106 Feb 59, GA106 Sep 59, GA246 Nov
59, 5228.   To Greece 52-128 Jan 70.

C/n 031 To WGAF.   GB107 Mar 59, GA107 Sep 59, GA237 Nov
59, GC231   , 5229, b/u Lemwerder 1970.

C/n 032 To WGAF.   GB108 Mar 59, GA108 Sep 59, GB112 Dec
59, GA118 Feb 60, AS568 Aug 65, 5230.   To
Portugal Nov 69.

C/n 033 To WGAF.   GB109 Mar 59, GA109 Sep 59, GB109 Nov
59, GA119 Feb 60, 5231.   To Greece 52-131 Jan 70.

C/n 034 To WGAF.   GB110 Mar 59, GA110 Sep 59, GB110 Nov
59, GA120 Mar 60, GC234 Mar 65, 5232.   To Greece
52-132 Nov 69.

C/n 035 To WGAF.   GB111 Apr 59, GA111 Sep 59, GA247 Dec
59, 5233.   To Greece 52-133 Nov 69.

C/n 036 To WGAF.   GB112 Apr 59, GA112 Sep 59, GA250 Dec
59, AS569 Aug 65, 5234.   To Portugal Nov 69.

C/n 037 To WGAF.   GB113 Apr 59, GA113 Sep 59, GA251 Dec
59, 5235.   To Greece 52-135 Jan 70.

C/n 038 To WGAF.   GB114 Apr 59, GA114 Sep 59, GA244 Dec
59, 5236.   To Greece 52-136   70.

C/n 039 To WGAF.   GB115 Apr 59, GA115 Sep 59, ? Nov 59,
GA245 Feb 60, 5237.   To restaurant at Schwelm
1971.

C/n 040 To WGAF.   GB116 Apr 59, GA116 Sep 59, GB116 Nov
59, GA121 Mar 60, 5238.   To Greece 52-138 Nov 69.

C/n 041 To WGAF.   GB231 Apr 59, GA231 Dec 59, 5239.   To
Greece 52-139 Apr 70.

C/n 042 To WGAF.   GB232 May 59, GA232 Dec 59, 5240.   To
Greece 52-140 Apr 70.

C/n 043 To WGAF.   GB233 Jun 59, GB113 Nov 59, ? Feb 60
(either GB102 or 116), GB101 Aug 60, 5241.
Ferried from Finkenwerder to Hamburg without regn or
serial 23 Nov 70 after purchase by Elbeflug.   D-ACUG
allocated Jan 71, never flown as such.   Stored at Hamburg
until sold to "Pott's Park" and del by rail 09 Jun 72.
Since preserved there on children's playground (not as
restaurant).

C/n 044 To WGAF.   GB234 May 59, GB114 Dec 59, GB231 Aug
60, GC101 Oct 61.   To Tanzanian AF Feb 65 but
not del.   To Portuguese AF as 6413 May 65.   Retd to
LTG63 as GC101 Aug 65.   To Portuguese AF as 6413 again
Nov 65.

C/n 045 To WGAF.   GB235 Jun 59, GA235 Nov 59.   To
Portuguese AF as 6416 Nov 66.

C/n 046 To WGAF.   GB236 Jun 59, GA236 Dec 59, GC246 Dec
62, 5242.   To Greece 52-142 Feb 70.

C/n 047 To WGAF.   GB237 Jun 59, GB107 Nov 59, GB102 Aug
60.   To Tanzanian AF Feb 65 but not del.   To
Portuguese AF as 6415 Nov 65.

C/n 048 To WGAF.   GB238 Jun 59, GA238 Dec 59.   To
Portuguese AF Sep 65.   To GA238 again Nov 66.
To Greece 52-143 Nov 70.

C/n 049 To WGAF.   GB239 Jly 59, GB108 Jan 60, GB103 Aug
60, 5244.   To Greece 52-144 Apr 70.

C/n 050 To WGAF.   GB240 Jly 59, GB120 Nov 59, GB232 Aug
60, GC102 Sep 61, 5245.   To Portuguese AF Nov
69.

C/n 051 To WGAF.   GB241 Jly 59, GA241 Nov 59, GC244 Jly
62, 5246.   To Portuguese AF Nov 69.

C/n 052 To WGAF.   GB242 Aug 59, GA242 Nov 59, GC242 Jun
62.   To Nigerian AF Jan 65.   Back as GA109 Sep
66, 5247.   To Greece 52-147 Aug 70.

C/n 053 To WGAF.   GB243 Aug 59, GA243 Dec 59, GC243 Jly
62, 5248.   Abroad Nov 69 (believed Portugal).

C/n 054 To WGAF.   GB244 Aug 59, GB104 Nov 59, GB233 Jly
60, GC103 Oct 61, 5249.   To Portuguese AF Dec 69.

Top   : MD-312B Flamant development with single fin,
          F-WCZN (J Delmas)

Bottom : The MD-316T, no.01, coded T (J Delmas)

Top   :   MD-320 Hirondelle under construction at Bordeaux-
            Merignac (USIAS via Charles W Cain)

Bottom :   MD-415 Communaute F-WJDN at Le Bourget (Charles
            W Cain)

Top   : Mystere 20 F-BOEF of Air France at Bordeaux-
         Merignac in September 1966 prior to delivery
         (Dassault via J Delmas)

Centre : Mystere 20 N20FE in Federal Express colours
         (J Delmas)

Bottom : SN.601 Corvette of the French regional carrier
         Air Languedoc (Air Languedoc/SNIAS via J-P Dubois)

Top   :   SCAN.30 F-WFHB on display on the banks of the
            Seine in Paris, with Salmson 8AS-00 engines
            (Charles W Cain)

Centre :   SCAN.30 G-ARIX with Gipsy Queen II engines at
            Southampton in May 1967 (Peter J Marson)

Bottom :   Re-engined SCAN.30 N7775C with Lycoming R-680Es
            as 'Pace Gannett' at Whiteman, USA in August
            1966 (John P Stewart)

C/n 055 To WGAF.   GB245 Aug 59, GA245 Oct 59, GB103 Feb 60, GB104   60.   Destined for transfer to Nigerian AF due Jan 65 but was written off at Ahlhorn 24 Nov 64 prior to this date.

C/n 056 To WGAF.   GB246 Sep 59, GB106 Dec 59, GB234 Aug 60, GC104 Oct 61, 5250.   To Portuguese AF Nov 69.

C/n 057 To WGAF.   GC101 Sep 59, GA101 Sep 59, GA240 Dec 59, 5251.   To Greece 52-151 Nov 69.

C/n 058 To WGAF.   GC102 Sep 59, GA117 Sep 59, GA2.. Dec 59, GB1.. Feb 60, GB235? Aug 60, GC105 Oct 61, 5252.   To ES61, b/u Manching 1974/75.

C/n 059 To WGAF.   GC103 Sep 59, GA118 Sep 59, GA248 Dec 59, GC248 May 62.   To Tanzanian AF Feb 65 but not del.   To Portuguese AF 6414 Oct 65.

C/n 060 To WGAF.   GC104 Sep 59, GA119 Oct 59, GA249 Nov 59.   W/o Monte Argentera, Italy, 15 Mar 60.

C/n 061 To WGAF.   GC105 Oct 59, GA120 Nov 59, GB115 Nov 59, GB105 Aug 60.   To Nigerian AF Nov 64.   Back to LTG62 as GB105 1965 or 1966, 5253.   To Greece 52-153 Aug 70.

C/n 062 To WGAF.   GC106.   Written off on f/f near Stade 16 Sep 59, should have become GA121 (NTU).

C/n 063 To WGAF.   GC107 Oct 59, GB117 Nov 59, E861.   Stored as "G117" (less letter B) Jun-Aug 60.   GB106 Sep 60, 5254.   To Greece 52-154 Apr 70.

C/n 064 To WGAF.   GC108 Nov 59, GB118 Dec 59, GB236 Aug 60, GC106 Oct 61, 5255.   To Greece 52-155 Mar 70.

C/n 065 To WGAF.   GC109 Nov 59, GB119? (serial unconfirmed) Dec 59, GB107 Aug 60, YA112 1961/62, YA572 Jly 65, 5256.   To D-ACUT Jan 71.   Stored Lübeck.   To Hermeskeil for preservation 1974.

C/n 066 To WGAF.   GA250 allocated Sep 59 but NTU.   F/f and certification as "G110" (second letter C missing) Dec 59, GB121 Dec 59, GB237 Aug 60, GC107 Oct 61, GA113 Sep 67, 5257.   To Transall unit at Mont-de-Marsan (see previous note) May 68.   To LTG61 Jun 68.   Written off Erding 12 Feb 69.

C/n 067 To WGAF.   (GA251), G111 Dec 59, GB111 Dec 59, GB108 Aug 60, AS589 Aug 65, 5258.   To Greece 52-158 Jan 70.

C/n 068 To WGAF.   G112, GB101 Nov 59, GB238 Aug 60, GC108 Oct 61.   To Nigerian AF Jan 65.   Back to GA233 of FFS-S Sep 66, LTG61 Jun 67, 5259.   To Greece 52-159 Nov 70.

C/n 069 To WGAF.   G113, GB102 Nov 59, GB231 Dec 59, GB109   60, 5260.   To Greece 52-160 Apr 70.

C/n 070 To WGAF.   G114, GB103 Nov 59, GB232 Dec 59, GB239 Aug 60, GC109 Oct 61, GA115 Sep 67, 5261.   To Greece 52-161 Nov 69.

C/n 071 To WGAF.   G115, GB233 Dec 59, GB110 Apr 60, 5262.   To Greece 52-162 Aug 70.

C/n 072 To WGAF.   G116, GB234 Dec 59, GB112? (serial unconfirmed) Mar 60, GB240 Jly 60, GC110 Oct 61, GA118 Sep 67, 5263.   To Greece 52-163 Jan 70.

C/n 073 To WGAF.   G117, GB235? (unconfirmed) Dec 59, GB1.. Feb 60, GB111 Aug 60, YA111   61, YA573 Jly 65, 5264.   To D-ACUV Jan 71.   To Basle Feb 71.   Stolen from there and flown to Kerkyra, Corfu, as N65171 07 Jly 72.   Sold (in public auction) to Sir Atlantique, Kigali, as 9XR-KD Jan 74.   Fate unknown.

C/n 074 To WGAF.   G118, GB236   , GA1.. Mar 60, GB241 Aug 60, GC111 Oct 61, ND111 Jun 67, 5265.   To LTG61 May 68.   To Greece 52-165 Jun 70.

C/n 075 To WGAF.   G119, GB237   , GA... Mar 60, GB112 Aug 60, 5266.   To Greece 52-166 Jun 70.

C/n 076 To WGAF.   G101, GB238   , GB242 Aug 60, GC112 Oct 61, 5267.   To Greece 52-167 Apr 70.

C/n 077 To WGAF.   GB239   , GB113 Mar 60, FFS-S Nov 64, AS598 Aug 65, 5268, LTG61 Mar 69.   To Greece 52-168 Nov 70.

C/n 078 To WGAF.   GB240 Mar 60, GB243 Aug 60, GC113 Oct 61, 5269.   To Greece 52-169 Apr 70.

C/n 079 To WGAF.   GB241 Mar 60, GB114 Jly 60, 5270.   To Greece 52-170 Jun 70.

C/n 080 To WGAF.   GB242 Mar 60, GB244 Aug 60, GC114 Oct 61, 5271.   To Greece 52-171 Feb 70.   Written off at Elefsis 1978.

C/n 081 To WGAF.   GB243 May 60, GB115 Aug 60, 5272.   To Greece 52-172 Jun 70.

C/n 082 To WGAF.   GB244 May 60, GB245 Aug 60, GC115 Oct 61, 5273.   To Greece 52-173 Aug 70.

C/n 083 To WGAF.   GB245 May 60.   To France Sep 60.   To Israel Oct 62.

C/n 084 To WGAF.   GB246 Apr 60, GA249 Apr 60, GC247 Jan 63, 5274, LTG61 Nov 69.   To Greece 52-174 Aug 70.

C/n 085 To WGAF.   GB247 Apr 60, GB117 Aug 60, 5275.   To Greece 52-175 Jun 70.

C/n 086 To WGAF.   GB248 Apr 60, GB246 Aug 60, GC116 Oct 61, 5276.   To Greece 52-176 Apr 70.

C/n 087 To WGAF.   GB249 Apr 60, GB118 Aug 60, 5277.   To Greece 52-177 Jun 70.

C/n 088 To WGAF.   GB250 Apr 60, GB247 Aug 60, GC117 Oct 61.   Via France (Lahr Air Base) to Israel Oct 62.

C/n 089 To WGAF.   GB251 May 60.   To France 1960.   To Israel 1962.

C/n 090 To WGAF.   KA101 May 60.   To France Sep 60.   To Nigerian AF as 301, fate unknown.

C/n 091 To WGAF.   KA102 Jun 60.   To France Sep 60.   To Nigerian AF as 302, fate unknown.

C/n 092 To WGAF.   KA103 Jun 60.   To France Sep 60.   Destined for Nigerian AF as 303 but believed to have found its way to Israel 1962.

C/n 093 To WGAF.   KA104 Jun 60.   To France Sep 60.   Destined for Nigerian AF as 304.   To Israel? (see c/n 092).

C/n 094 To WGAF.   KA105 Jun 60, GB248? (unconfirmed) Aug 60, GC118 Oct 61, 5278.   To FFS-S 1969.   To D-ACUR Jan 71.   Landed at Wunstorf with engine failure on del from Neubiberg to Lübeck 09 Feb 71 and stored Wunstorf until sold to Wolf Air Transport of Frankfurt t/a Portalia Air Cargo as 9XR-KH Jan 71.   Written off in Djibouti 18 Apr 73 due to engine failure on take-off to Asmara with load of arms.   Wreck still lying there.

C/n 095 To WGAF.   KA106 Jun 60, LTG62 Jly 60, GB120 Aug 60, GC250 Jly 62, GA250 Apr 67, 5279.   Written off at Königsdorf 19 Nov 70.

C/n 096 To WGAF.   KA107 Jun 60, GB249 Aug 60, GC119 Oct 61.   Via Lahr to Israel Oct 62.

C/n 097 To WGAF.   KA108 Jun 60, GB121? (unconfirmed) Aug 60, GC251 Jly 62, GA236 Mar 67, 5280.   Temporarily

mispainted as D-ACJB, corrected to D-ACUB and del Jan 71.
Stored at Lübeck until broken up 1975-77.

C/n 098 To WGAF.   KA109 Jun 60, GB250 Aug 60, GC120 Oct
61, GA237 Feb 67, 5281.   To Portuguese AF Oct
70.

C/n 099 To WGAF.   KA110 Jun 60, BF571   60, 5282.   B/u
Fassberg 1973 (with only 239 hrs TT!).

C/n 100 To WGAF.   KA111 Jly 60, GB251? (unconfirmed)
Aug 60, GC121 Oct 61, GB235   66, 5283.   To
D-AMFA Dec 71.

C/n 101 To WGAF.   KA112 Jly 60, GB116 Aug 60, 5284.   To
D-AMFB Dec 71.   Preserved as 5284 at Diepholz.

C/n 102 To WGAF.   KA113 Jly 60, GA101   60, 5285.   To
HC-AXF.

C/n 103 To WGAF.   KA114 Jly 60, GB119 Jly 60.   Written
off at Wahn 23 Jan 61.

C/n 104 To WGAF.   KA115 Jly 60, GA102 Aug 60, 5286.   To
Portuguese AF Nov 70.

C/n 105 To WGAF.   KA116 Aug 60, GA103 Aug 60, (AS570 Aug
65 NTU), 5287 LTG61.   To Greece 52-187 Nov 70.

C/n 106 To WGAF.   KA117 Aug 60, GA104 Aug 60, 5288.   To
Greece 52-188 Nov 70.

C/n 107 To WGAF.   KA118 Aug 60, GA105 Aug 60, 5289.   To
Greece 52-189 Nov 70.

C/n 108 To WGAF.   KA119 Sep 60, GA106 Oct 60, 5290.
B/u Neubiberg 1973.

C/n 109 To WGAF.   KA120 Sep 60, GA107 Oct 60, 5291.   To
D-ACUC Jan 71.   To Lübeck 26 Jan 71, to Basle
Feb 71, to Lübeck Aug 71.   Stored there and b/u 1973-76.

C/n 110 To WGAF.   KA121 Sep 60, GA108 Oct 60, 5292.   To
D-ACUD Jan 71.   Stored at Lübeck and b/u 1973-
76.

C/n 111 To WGAF.   KA122   60, GA109   60, GC249   62,
BD592 1966 or 1967, 5293.   To Deipholz for
storage 1972 or 1973 and b/u there 1974.

C/n 112 To WGAF.   KA123 Sep 60, GA110 Sep 60, 5294.   To
D-ACUH Jan 71.   To Lübeck 30 Jan 71, to Basle
Aug 71.   Sold to scrap dealer at Söllingen/Baden and del
by road 12 Jan 78.

C/n 113 To WGAF.   KA124 Oct 60, GA111 Oct 60, 5295.
B/u Neubiberg 1972/73.

C/n 114 To WGAF.   KA125 Oct 60, GA112 Oct 60, 5296.
B/u Neubiberg 1972/73 but also reported as
Greece 52-196(?).

C/n 115 To WGAF.   KA126 Oct 60, GA113 Oct 60, YA...
62, YA574   65, 5297.   Fate unknown.

C/n 116 To WGAF.   KA127 Oct 60, GA114 Oct 60, GC253 Jan
63, GA253 Apr 67, 5298.   Written off at Alverca,
Portugal, 26 Aug 70 and used for spares there (reported
Mar 73).

C/n 117 To WGAF.   KA128 Nov 60, GA115 Nov 60, GC254?
(unconfirmed)   62, GA114   66, 5299.   To
D-ACUK Jan 71.   Stored at Lübeck.   Preserved at
Oldenburg-Wechloy Feb 77.

C/n 118 To WGAF.   KA129 Nov 60, GA116 Nov 60, 5300.   To
D-ACUF Jan 71.   To Atesa as HC-BDO Apr 77, del
ex Lübeck 20 Apr 77.

C/n 119 To WGAF.   KA130 Nov 60, GB122 Feb 61, 5301.   To
calibration unit FmRg61 Lechfeld   70.   To E861
72.   To Deipholz for storage   74, b/u there.

C/n 120 To WGAF.   KA131 Nov 60, GA122 Nov 60, AS570
65, 5302.   To D-ACUL Jan 71, stored at Lübeck
etc.

C/n 121 To WGAF.   KA132 Jan 61, GA123 Jan 61, 5303.   B/u
Neubiberg 1973.   Scrap now lying at Seifertshafen.

C/n 122 To WGAF.   KA133 Jan 61, GA124 Jan 61, 5304.   To
D-ACUN Jan 71.   Stored at Lübeck (without reg)
and b/u 1973-76.

C/n 123 To WGAF.   KA134 Dec 60, GA125 Jan 61, 5305.   B/u
Manching 1973, parts used for c/n 142.

C/n 124 To WGAF.   KA135 Jan 61, GA252 Jan 61, 5306.   To
D-ACUP Jan 71.   B/u at Lübeck 1973-76.

C/n 125 To WGAF.   KA136 Jan 61, GB119   61, 5307.   To
Greece 53-207   71.

C/n 126 To WGAF.   KA137 Jan 61, GA253? (unconfirmed) Jan
61, AS599 Aug 65, 5308.   To D-AMFC Dec 71.
Disappeared at Diepholz Apr 73, fate unknown.

C/n 127 To WGAF.   KA138 Feb 61, GC252? (unconfirmed)
62, GB107 Apr 64, 5309.   To D-AMFD Dec 71 and b/u
74.

C/n 128 To WGAF.   KA139 Mar 61, GB252? (unconfirmed)
61, GC122 Oct 61, GB231   66, 5310.   To D-AMFE,
Dec 71 and b/u 1974.

C/n 129 To WGAF.   KA140 Feb 61, GA255 Feb 61, 5311.   B/u
Neubiberg 1973.

C/n 130 To WGAF.   KA141 Mar 61, GA254   61, 5312.   To
D-ACUQ Jan 71.   Del to Lübeck 03 Feb 71, to Basle
22 Jun 71 and grounded there.   To scrap-dealer at
Söllingen/Baden by road 12 Dec 77.

C/n 131 To WGAF.   KA142 Mar 61, GB123 Mar 61, 5313.   To
D-AMFF Dec 71.   Disappeared at Diepholz Apr 73,
fate unknown.

C/n 132 To WGAF.   KA143 Mar 61, GB253? (unconfirmed) Mar
61, GC123? (unconfirmed) Oct 61, BD593   66,
5314.   B/u Fürstenfeldbruck 1974/75.

C/n 133 To WGAF.   KA144 Mar 61, GB124 Mar 61, 5315.   To
D-AMFG Dec 71.   B/u 1974.

C/n 134 To WGAF.   KA145 Apr 61, GB254 Apr 61, GC124 Oct
61.   Via Lahr to Israel Oct 62.   To 4X-AOR 1977
but NTU.

C/n 135 To WGAF.   KA146 Mar 61, AS576   61, GA241 Mar 67,
5316.   To D-AMFH Dec 71.   B/u 1974.

C/n 136 To WGAF.   KA147 Apr 61, GB255? (unconfirmed)
61, GC125 Oct 61, GB252   66, 5317.   To D-AMFI
Dec 71.   B/u 1974.

C/n 137 To WGAF.   KA148 May 61, AS577   61, 5318.   To
HC-AXK.

C/n 138 To WGAF.   KA149 May 61, AS578   61, 5319, LTG62,
D9579   72, 9913   76 (still operating on radar
target flts from Nordholz).

C/n 139 To WGAF.   KA150 Jun 61, GB125   61, 5320.   To
Greece 53-220   71.

C/n 140 To WGAF.   KA151 Jun 61, AS579 Jly 61, 5321.   To
5U-MAP   69.   (See note on c/n 008).

C/n 141 To WGAF.   KA152 Jly 61, GB111   61, 5322.   To
Greece 53-222   71.

C/n 142 To WGAF.   KA153 Jly 61, AS580   61, 5323.   To
D-AMFV Dec 71. Not del to Diepholz but remained
at Manching until sold as scrap there Dec 72.   Intended

to become a cafe in Bavaria with parts from c/n 123.

C/n 143 To WGAF. KA154 Aug 61, AS581   61. Written off 17 Jly 67.

C/n 144 To WGAF. KA155 Sep 61, AS583   61, 5324, LTG62. To D-AMFJ Dec 71. B/u 1974.

C/n 145 To WGAF. KA156 Aug 61, AS582   61, 5325. To HC-AXG. Del 03 Nov 71 to ATESA.

C/n 146 To WGAF. KA157 Oct 61, AS584   61, 5326, LTG62. To D-AMFK Dec 71. B/u 1974.

C/n 147 To WGAF. KA158 Oct 61, AS585 Oct 61, 5327. Sold as scrap at Wunstorf Dec 72.

C/n 148 To WGAF. KA159 Oct 61, AS586? Oct 61, GC232 62, GB253   66, 5328. To Greece 53-228 1971.

C/n 149 To WGAF. KA160 Oct 61, AS587   61, 5329, LTG62   70. To D-AMFL Dec 71. B/u 1974.

C/n 150 To WGAF. KA161 Feb 62, GC231   62. Via Lahr to Israel Oct 62.

C/n 151 To WGAF. KA162 Feb 62, GC234   62. Via Lahr to Israel Oct 62.

C/n 152 To WGAF. KA163 Apr 62, GC235 Apr 62, GB104 Sep 66, 5330, D-9580   72, 9914   76. Still active with 138.

C/n 153 To WGAF. KA164 Jly 62, GC239 Jly 62. Via Lahr to Israel Oct 62.

C/n 154 To WGAF. KA165 Jly 62, GC240 Jly 62, GB108 66, 5331, D-9570   71, 9915   76. Still active with 138 and 152.

C/n 155 To WGAF. KA166 Oct 62, GC245 Nov 62, GB113 66, 5332, WaSLw50   71. Sold abroad Oct 74, presumably to Greece.

C/n 156 To WGAF. KA167 May 63, AS593   63, 5333, LTG62   70, WaSLw50   71. WFU Dec 74 and stored at Fürstenfeldbruck. Sold to West African Air Cargo as 9G-ACI 1977 but not del. Still stored.

C/n 157 To WGAF. KA168 May 63, AS594   63, 5334, LTG62   70. Sold abroad 1972 (reportedly to Nigeria).

C/n 158 To WGAF. KA169 May 63, AS595   63, 5335, LTG61   70. To D-AMFM Dec 71. B/u 1974.

C/n 159 To WGAF. KA170 Jun 63, AS596   63, 5336, LTG61   70. Sold as scrap at Neubiberg Dec 72 but reported as sold to Nigerian AF and then as 9G-ACH.

C/n 160 To WGAF. KA171 Jly 63, AS597   63, 5337, LTG61   70. Preserved at Neubiberg since 1972.

C/n 161 To WGAF. KA172 Jly 63, GB231 Aug 63, GB102 Sep 66, 5338. To D-AMFN Dec 71. B/u 1974.

C/n 162 To WGAF. KA173 Jly 63, GB232 Sep 63, 5339, WaSLw50   71. Sold abroad 1975 (Greece?).

C/n 163 To WGAF. KA174 Aug 63, GB233 Sep 63, 5340, WaSLw50   71. To Greece 53-240 later (1975?).

C/n 164 To WGAF. KA175 Aug 63, GB234 Sep 63, 5341. WaSLw50   71. To Greece 53-241   75.

C/n 165 To WGAF. KA176 Aug 63, GB235 Oct 63 for del to France only, destined for Nigerian AF as 305 Nov 63 but went to Israel as 4X-FAG.

C/n 166 To WGAF. KA177 Sep 63, GB236 Oct 63. To France Oct 63. Destined for Nigerian AF 306 Dec 63, but to Israel instead.

C/n 167 To WGAF. KA178 Sep 63, GB237 Oct 63. To France Oct 63. Destined for Nigerian AF 307 Nov 63 but to Israel instead.

C/n 168 To WGAF. KA179 Oct 63. To France Oct 63. Destined for Nigerian AF 308 Dec 63 but to Israel instead.

C/n 169 To WGAF. KA180 Nov 63. Destined for Nigerian AF as 309 but NTU. Via France to Israel instead.

C/n 170 To WGAF. KA181 Nov 63. Destined for Nigerian AF as 310 but NTU. Via France to Israel instead.

C/n 171 To WGAF. KA182 Nov 63, GB235 Dec 63. Written off 15 Jun 65.

C/n 172 To WGAF. KA183 Nov 63, GB236 Dec 63, 5342. To D-AMFO Dec 71. B/u 1974.

C/n 173 To WGAF. KA184 Jan 64, GB237 Feb 64, 5343, LTG61 70. To calibration unit Lechfeld   71. B/u Lechfeld 1973.

C/n 174 To WGAF. KA185 Feb 64, GB238 Mar 64, 5344. To D-AMFP Dec 71. B/u 1974.

C/n 175 To WGAF. KA186 Feb 64, GB239 Mar 64, 5345. B/u Wunstorf 1973.

C/n 176 To WGAF. KA187 Mar 64, GB240 Apr 64, 5346. B/u Wunstorf 1973.

C/n 177 To WGAF. KA188 Feb 64, GB241 Mar 64, 5347. B/u Wunstorf 1973.

C/n 178 To WGAF. KA189 Mar 64, GB242 Apr 64, 5348, WaSLw50   71. To HFB for apprentice training (instructional airframe) 1972.

C/n 179 To WGAF. KA190 Mar 64, GB243 Apr 64, 5349. To D-AMFQ Dec 71. B/u 1974.

C/n 180 To WGAF. KA191 Apr 64, GB244 May 64, 5350. To D-AMFR Dec 71. B/u 1974.

C/n 181 To WGAF. KA192 Apr 64, GB245 May 64, 5351. B/u Wunstorf 1973.

C/n 182 To WGAF. KA193 Apr 64, GB246 May 64, 5352. To Greece 53-252   72.

C/n 183 To WGAF. KA194 Apr 64, GB247 Jun 64, 5353. To D-AMFS Dec 71. B/u 1974.

C/n 184 To WGAF. KA195 May 64, GB248 Jun 64, 5354. To D-AMFT Dec 71. B/u 1974.

C/n 185 To WGAF. KA196 May 64, GB249 Jun 64, 5355. Preserved at Krummenort near Hohn as 5255 since 1972.

C/n 186 To WGAF. KA197 Jun 64, GB250 Jly 64, 5356. Preserved at Ahlhorn since 1972.

C/n 187 To WGAF. KA198 Jly 64, GB251 Aug 64, 5357. To D-AMFU Dec 71. B/u 1974.

Two Nord 2508 aircraft were supplied to the West German AF in 1963 for the evaluation of Transall systems. Full details are given in the preceding section covering French-built aircraft.

NOTES

1   West German AF unit codes except as stated otherwise are given below:

AS567/599 FFS-S = Flugzeugführerrsschule "S" (pilots' school "S") at first at Memmingen, in 1957 to Wunstorf.
GA101/125, 229/255 LTG61 = Lufttransportgeschweder 61

(air transport wing 61) at first at Erding, to Neubiberg 22 Apr 58.

GB101/119 from Mar 57 to Jly 58 Voraus-Kommando LTG61 (preliminary air transport wing 61 command) at Erding, to Neubiberg 22 Apr 58.

GB101/125, 231/255 from Apr 59 LTG62 Celle (in service 01 Dec 59), to Köln-Bonn (Wahn) Mar 60, to Ahlhorn 30 Apr 63.

GC101/109 (119) from Sep 59 to Apr 60 "Reserve" (destined for a not yet established LTG63 and stored at LPR2 = Luftwaffenparkregiment 2 (at depot 2) at Diepholz.

GC101/125, 231/254 from 15 Nov 61 LTG63 Celle, to Hohn Oct 67.

GR231/249 "Reserve", all used by FFS-S at Wunstorf.

ND111 used by the joint German-French Transall troop evaluation unit associated with CEAM at Mont-de-Marsan.

BF571 & 589 TSLw3 = Technische Schule der Luftwaffe 3 (technical school 3) at Fassberg.

BD592 & 593 WaSLw50 = Waffenschule der Luftwaffe 50 (weapons school 50) at Fürstenfeldbruck.

YA034, 035, 110/112, 571/574 ESt61 = Erprobungsstelle 61 (evaluation unit 61) at Manching.

5201 et seq introduced on order dated 13 Nov 67 and applied on aircraft between Jan 68 and Jun 68.

D-9513 et seq semi-military serials allocated by MBL = Musterprüfstelle der Bundeswehr für Luftfahrtgerät, for radar target representation, based at Nordholz.

2   D-ACUA et seq regd to Elbeflug, only received permit to fly from their bases to Lübeck (where the technical base was to have been built) or to Basle (for Jet Aviation conversion).

3   KA101 et seq Test flight serials of HFB.

4   D-AMFA et seq Regd to Stockleigh Holdings Ltd, Dublin, only received permits to fly from their bases to Diepholz for temporary storage until export certificates would be granted.   They were refused Mar 72, and the aircraft were sold as scrap Apr 74.

5   Israeli examples

4X-FAA/FAC/FAE/FAG (c/n 165), 4X-FAH/FAJ/FAL "043" and "045" are known to be ex-German aircraft.   One of them became Uganda "701" in 1972.

6   Portuguese examples

A total of at least six aircraft were abandoned by the Portuguese AF in Angola and Mozambique: C9-ARC/D/E (operated by COMAG) are possibly Nord 2502Fs with jets discarded, and C9-ARH/I/J/K are confirmed as Nord 2501Ds (also operated by COMAG).

TRANSALL

The Franco-German C.160 Transall (for TRANSport ALLiance) was an interesting early example of international co-operation.   The first prototype of this military transport was flown in 1963, and production for both the French Armée de l'Air and the admittedly reluctant German Luftwaffe (the C-130 Hercules reportedly having been preferred) was shared between Nord Aviation at Bourges in central France and two German concerns (VFW/Fokker at Lemwerder and MBB/HFB at Finkenwerder).   There were also hopes of sales to other military customers, but so far only Turkey and South Africa have adopted the type, and the 180th and last Transall was rolled out and flown in 1972.

In March 1967 a civil version was announced as the Transall 161 with redesigned fuselage reminiscent of the Carvair or Argosy.   A large nose-door hinging upwards would have given access to a larger freight load (189 cu m against 140 in the original C.160), but the market for a heavy freighter on short and medium routes offered no great prospects, and the project was discontinued.

It was however clear that sooner or later commercial variants of the Transall would be feasible, as evidenced with the L.100 versions of the somewhat similar Lockheed C-130 Hercules.

Up to 1976 however use of the C.160 under civil registration had been limited to a handful of aircraft only.   Putting aside the original V1 prototype temporarily evaluated in 1966 in civil marks and another German-registered aircraft loaned to the International Red Cross by the Federal Republic in 1969 and re-registered in Switzerland, scheduled commercial use of the freighter has been limited to the French Aéropostale only, with four Armee de l'Air machines on lease, while in the summer of 1976 an early prototype was overhauled by SOGERMA for delivery to Air Affaires Gabon.

From the mid-thirties France had always shown interest in domestic night mail services, an interest dictated by geographical conditions. Following the pre-war experiments with Air Bleu, night mail routes were resumed soon after the Liberation, first with French-built Ju-52s and later with a fleet of Dakotas.   The famous Postale de Nuit is the result of a close agreement between the PTT (French Postal Administration) and Air France, with a large fleet of aircraft operated by Air France on behalf of the Post Office under what is known as the CEPM, or Centre d'Exploitation Postal Métropolitain. Over the years remarkable achievements have been obtained, with an exceptional record of safety and reliability.

The Postale de Nuit currently operates 15 Fokker F.27s and four Transalls on a closely-knit network, where accurate timing is essential. Every night an average of seven to eight million letters (about 200 metric tons of mail) is carried throughout France.   Out of this, nearly 50 per cent is airfreighted by the four Transalls.

The story begins in 1972 when the head office for the Aviation Postale Intérieure, with an urgent need for heavy freighters, obtained the

long-term loan of four French military aircraft
to supplement the existing fleet of the CEPM.
Conversion to civil standards was undertaken by
SOGERMA at Bordeaux-Mérignac, and after a short
spell under provisional test registration for
civil certification the four Transalls entered
service in 1973.   Modifications included
commonality with the F.27 instrument panel, so
that 20 crew teams could fly on both types.

Typical payload is 13 metric tons of mail,
carried in a maximum of 26 containers of three
cubic metres each (roughly 600 kg of mail).
Loading or unloading the whole cargo takes
hardly more than 20 minutes, an essential
feature for the short stop-overs at transfer
points.   Since 1973 the Air France Transalls
have been used mostly over the longest and
busiest route, namely Paris to the island of
Corsica (serving both major airports at Ajaccio
and Bastia) via Lyons, Marseilles and Nice,
flying approximately 1000 hours each year on a
four to six-night schedule.   Average
reliability is extremely high and marginally
above that of the F.27, itself a very reliable
aircraft.

The only snag has been the use of a type
designed as a military transport under
commercial conditions and regulations, but on
the whole night mail operation of the Transall
has been very successful.   By 1976 the need
for a fifth aircraft was already mentioned.
In that same year another C.160 was delivered
to a civil customer fur use in Africa.   In
addition, the French Government had realised by
that time that the capacity of the French Air
Force transport command was somewhat limited,
particularly in the event of an overseas
emergency.   Interest in the Lockheed C-141
for Armée de l'Air use was mentioned, but
proposals for an order were finally rejected.
With a need for at least 25 more aircraft for
COTAM (Commandement des Transports Aériens
Militaires), it was felt necessary to build
more Transalls and take advantage of this to
look for potential military export orders.
Possible commercial sales could also be
envisaged, even knowing that this kind of cargo
aircraft does not sell very well on the civil
side (when the 1500th Lockheed Hercules
appeared early in 1978, only 69 aircraft had
gone to civil operators).

On 29 October 1976, after some preliminary
talks, an agreement was reached between SNIAS,
VFW-Fokker and Messerschmitt-Bolkow-Blohm to
re-start production of the C.160 Transall.   On
the French side the industrial proposals were
submitted in November to both the Prime
Minister and the President of the Republic,
and by late 1977 first deliveries were expected
to take place in 1979 with final assembly
in Toulouse.   The French Air Force
has already announced intentions to deliver 25
new aircraft to the 64eme Escadre de Transport
from 1980, with 'old' aircraft of the first
production batch going to the unit as early as
1979 to replace ageing Noratlases.

Industrial provisions are for the delivery
of pre-equipped fuselage sections, a technique
first innovated with A300 production.   From
early 1977 world-wide demonstration tours were
undertaken both by the French and the Germans
to Africa, the Middle East and the Far East
(including a civil registered Luftwaffe aircraft, D-ACTR).   The Transall also made
international headlines on two recent
occasions: first in April 1977 when the so-
called Verveine operation, conducted with 12

aircraft (plus a commandeered UTA DC-8 and an
Air France Boeing 747F) airlifted Moroccan
troops and supplies to Zaire during the border
clash with neighbouring Angola.   Twelve hundred
hours were flown by the aircraft during the
controversial airlift, which took place between
6 and 18 April 1977, and a thirteenth Transall
was flown in to West Africa with a spare engine
for a quick on-the-spot change of a Rolls Royce
Tyne to a C.160 which had suffered a minor
breakdown.   There is little doubt that the
whole operation began to be used as a convincing
argument for the demonstration tours.

The second much-publicised use occured in May
1978 when French Foreign Legion paratroops were
dropped on Kolwezi following the border clash in
Shaba Province, Zaire.   The use of the aircraft
was tactical only, the troops having been
ferried to Zaire from their base in Corsica by
jet transport.

Ironically the Transall aircraft used during
the Kolwezi operation were operated alongside
Hercules transports - the American aircraft
still being the major competitor of the C.160 on
the international market.

Less publicised for diplomatic reasons is the
use of French military Transalls in 'hot' areas
of Africa, such as the logistic support of
French Jaguars deployed to N'Djamena in Tchad or
operational use in Mauritania.   The year 1978
also saw French Transalls flying to Lebanon in
support of the French United Nations troops
despatched there during the civil war.   This
was again an occasion to demonstrate the somewhat limited long-range capacity of the aircraft
as jet airliners had again to be commandeered.

On the peaceful side, the Transall has also
been used for mercy and rescue operations.
Thus in 1970, when a severe earthquake struck
Peru, four Armee de l'Air Transalls with an airborne field hospital were flown to the disaster
area.   Two hundred hours were flown, 3000
passengers and 500 tonnes were airlifted, and
sixty medevac flights were logged.

In 1973 another C-160 was despatched to the
Sahel area in Africa to help combat the extreme
drought there by cloud-seeding operations, in
which eight out of ten were successful.
Another example of Transall use is in the
emergency delivery of heavy or bulky spares for
Air France when and where needed; for instance,
in 1976 27 breakdown missions were made for the
benefit of the national airline, and 228 hours
were logged.

The Transall remains primarily a military
transport, but undoubtedly efforts will be made
to promote civil sales too.   The French Postal
Administration has already announced that the
purchase of further planes might be anticipated.
In August 1977 a Transall, modified as a water-
bomber, was also demonstrated at Marignane
(home-base of the French fleet of Canadair
CL-215s), but faced strong and cheaper
competition from a DC-6 water-bomber converted
by UTA-Industries.   Introduction of the DC-6
water-bomber alongside the Sécurité Civile's
Canadair CL-215s on 1 September 1978 during
fire-fighting operations in the area of Nice
probably means that ex-Armée de l'Air converted
DC-6s will be preferred.

VFW-Fokker announced in March 1978 that work
on a new batch of Transalls had begun, and the
first order for the "nouvelle série" was signed
on September 29 1979 for three aircraft to be
delivered to Indonesia in 1982 for use in a
Government airlift of 2½m people from Java to
less-populated islands in the archipelago.

Specification

C.160 Transall

High-wing military transport with semi-prepared field surfaces capability.

Span: 40m
Length: 32.05m
Height: 11.7m
Wing area: 160 sq m
Empty weight:
All-up weight: 47,932 kg with an 8,000 kg payload (for 4,500 km strategic mission)
44,320 kg with an 8,000 kg payload (for 2 x 1,200 km tactical mission)
Maximum weight: 49,100 kg with 16,000 kg payload
Power-plants: 2 x Rolls-Royce Tyne 20 turboprop engines (5,665 SHP each on take-off; 6,100 TEHP)
Cruising speed: 500 km/hr at 8,000m
Range: 4,700 km
Take-off run: 700m

Dimensions of hold and rear doors (for loading and paradropping) allow entry of standard European railway loads. Usable volume 140 cu m.

C.161 Transall Civil (1967 project)

Dimensions as for C.160 except overall length (33.05m)
Maximum weight on take-off: 53,100 kg
Empty weight: 28,100 kg
Cruising speed: 475 km/hr
Maximum speed 485 km/hr
Range: 550 nm with 20,000 kg payload
2,500 nm with 10,000 kg payload
Take-off run at maximum weight: 1,780m
Power-plants: 2 x Rolls-Royce Tyne 20 (5,540 SHP on take-off) plus
1 x General Electric CJ-610-1 (1,295 kg/ thrust) auxiliary turbojet

Production

Transall construction numbers are prefixed with a letter "V", denoting Versuchsmuster (evaluation specimen) or "A", denoting pre-production aircraft (with French-style numbers 01, 02 etc). Production aircraft are allocated either "F" or "D" (for France or Germany), and the prefix "Z" was allocated for South African aircraft (Zuid Afrika).

The manufacturers of each individual aircraft are also indicated as follows:

N   - Nord Aviation, Bourges (Dept du Cher), France
VFW - Vereinigte Flugtechnische Werke, Lemwerder
HFB - Hamburger Flugzeugbau, Hamburg-Finkenwerder

Both VFW and HFB are the German partners, HFB later becoming MBB.

Prototypes

__V-1__   N.   Roll-out 09 Nov 62 and f/f 25 Feb 63 at Melun-Villaroche as D-9507 Musterprüfstelle der Bundeswehr für Luftfahrtgerät (MBL). Handed over to German MoD 04 May 63, exhibited at 1963 Paris Salon. D-ABEX allocated and permit issued 01 Dec 65 to MBL at request of Swedish Govt for cold weather evaluation at Vidsel. To France Mar 66 as F-ZWWV "WV". Exhibited as such at 1967 Paris Salon. To CEV as F-ZADK "DK" 28 Feb 68. Loaned to CEAM 1968-69 as F-SDBJ, retaining code "DK". Ret to CEV. To SNIAS Chateauroux 27 Jul 62-18 Apr 73. To CEV 18 Apr 73-23 Jul 74 when grounded at Rochefort (Dept de la Charente Maritime), as instructional airframe still as "DK" with number 373C at least by 1977 and still present mid 1978.

__V-2__   VFW.   F/f 25 May 63. Exhibited at 1963 Paris Salon and 1964 Hannover Show. To CEV as F-ZADH "DH", retaining this identity while on trials with CEAM

from 21 Jun 64-05 Aug 65 at least. Intended for Luftwaffe 13 Nov 67 as 5001 but NTU. Took part in ELD068 airlift to Woomera, Australia. Also reported as F-ZWWY "WY" but not confirmed. To Nord Aviation as F-ZJYD "YD" Oct 68. Ret to Luftwaffe and allocated 5001 but not painted on aircraft. To AIA Clermont-Ferrand as grounded test airframe for hydraulics experiments. Still as "YD" and in natural metal finish.

__V-3__   HFB.   F/f 19 Feb 64. D-9509. On trials with CEAM and CEV as F-ZADI "DI" from 21 May 64-09 Jun 66 at least, during which time aircraft exhibited at 1965 Paris Salon. To HFB as KA200, later to Est 61 (test and evaluation unit) at Manching. To 5002 13 Nov 67. Offered for sale by West German Govt Jan 76 and after modification to C.160G by SOGERMA, Bordeaux. Became TR-LWE Jul 76 for Air Affaires Gabon.

__V-4__   VFW.   Static prototype at Lemwerder.

__V-5__   N.   Dynamic prototype at EAT, Toulouse

__Note__   An unidentified prototype has been reported with French AF CEAM as "118-BJ" but not confirmed.

Pre-Production Aircraft

All aircraft (with the possible exception of A-03) passed through the French CEAM for Franco-German trials, but exact details are not recorded.

__A-01__   N.   F/f 21 May 65 at Melun-Villaroche. Shown at 1965 Paris Salon as D-9524. To WGAF as YA051 of Est 61 Manching 1965 and oper by French CEAM at Mont-de-Marsan (Dept des Landes) from 19 Jul 65. To 5003 13 Nov 67, still with Est 61. WFU, SOC and b/u at Wunstorf 1971. Wreck still present 1979, belonging to 4/TSLw 3.

__A-02__   VFW.   F/f 24 Jun 65 as D-9526. To CEV as F-ZWWS "ZS" Sep 65. To CEAM at BA118 Mont-de-Marsan as "118-BS" F-SDBS from 11 Oct 65-06 Jan 66 at least, during which time ret to CEV. To French AF 15 Nov 68 as "61-MI" F-RAMI May 70, this being the second allocation of code following loss of c/n F-14.

__A-03__   HFB.   F/f 27 Aug 65 as D-9525. To WGAF as YA052 of Est 61 Manching 1965. 5004 allocated 13 Nov 67 but NTU. Remained as YA052 until conversion to D-ABYG 08 Oct 68 for Biafra airlift. Regn canc 17 Oct 68. Re-regd HB-ILN 26 Oct 68 and oper by Balair while on lease to International Red Cross Committee. Reverted to D-ABYG 23 Apr 70 for 1970 Hannover Show. Stored at Lemwerder May 70-73, when transferred as instructional airframe 5004 to LTG63, Hohn. Completely dug into earth at Hohn 1975.

__A-04__   N.   F/f 31 Jan 66 at Melun as D-9527. Wore marks "A04-WT", later F-ZWWT "WT" Feb 66. To CEAM as "118-BT" F-SDBT Jun 66. To French AF 61 Escadre de Transport as "61-ZA" F-RAZA by 1972. Loaned to CEV from 27 Mar 72 as "61-BI" F-ZABI and still coded as such in Jun 78 when flying with test nose probe and MAD-like tail boom. Used for trials in connection with development of Transall Nouvelle Serie up to at least 31 Jan 80.

__A-05__   VFW.   F/f 05 Feb 66 as D-9528. To WGAF Est 61, Manching, as YA053 1966 and based at CEAM Mont-de-Marsan Jun 66. 5005 13 Nov 67. To LTG63 13 Jun 68 for handling instructions. To FFS-S 13 May 69. To 4/TSLw3 as instructional airframe at Wunstorf 1971 and b/u there. Tail to paratroop school Schongau. Fuselage finally b/u Wunstorf 1975.

__A-06__   HFB.   F/f 20 Apr 66 as D-9529. To F-ZWWU "WU" with CEV, shown as such at 1966 Hannover Show. With CEAM by Aug 66 as "118-BU F-SDBU. To French AF 17 Jul 68 as "61-ZB" F-RAZB 1969. Current.

Production

The French AF aircraft are operated on a centralised servicing system, being pooled between three escadrons within the COTAM 61° Escadre de Transport. The call-sign blocks for the "61-MA" and "61-ZA" ranges of unit codes are F-RAMA and F-RAZA respectively.

Early WGAF Transalls were allocated the original post-war Luftwaffe codes, a combination of two letters and three digits, but the current system of four digit identities was soon introduced, starting from 5001 with prototype V-2. Call-signs/serials KA and KM were allocated to HFB and VFW respectively for test flights of production aircraft for the French AF and, in some cases, for WGAF aircraft also. The dates quoted with the units to which aircraft were delivered are the ferry flight dates, rather than the handover dates.

F-1   VFW. F/f 13 Apr 67 as KM101 in Germany. To France as "340-YE" F-RBYE 02 Aug 67. To French AF 61-MA 22 Nov 67.

F-2   HFB. F/f 09 Jun 67 as KA201. To French AF 61-MB 28 Mar 69 after 250 hr test programme.

F-3   N. F/f 07 Jun 67. Del to French AF 19 Oct 67. To CIET as "340-YF" F-ZJYF 14 Nov 67 and then to 61-MC .

F-4   VFW. F/f 10 Sep 67 as KM104. To French AF 61-MD 03 May 68.

F-5   HFB. F/f 05 Sep 67 as KA202. To French AF 61-ME 01 Mar 68.

D-6   HFB. F/f 02 Nov 67. To WGAF LTG63 as 5006 30 Apr 68.

D-7   VFW. F/f 07 Nov 67 as KM102. To WGAF Est 61 Manching as 5007 09 Apr 68. To LTG61 17 May 71.

D-8   N. F/f 11 Jan 68. To WGAF LTG63 as 5008 20 Aug 68.

D-9   N. F/f 27 Feb 68. To WGAF LTG63 as 5009 08 Aug 68.

D-10   VFW. F/f 12 Jan 68 as KM103. Exhibited at 1968 Hannover Show. To WGAF LTG63 as 5010 02 Sep 68.

F-11   HFB. F/f 15 Feb 68 as KA203. To French AF 61-MF 30 May 68.

F-12   N. F/f 12 Apr 68. To French AF 61-MG 12 Sep 68.

F-13   VFW. F/f 06 Feb 68 as KM105. To French AF 61-MH 31 Oct 68. Unofficially adorned with French Navy roundels mid Jun 78.

F-14   HFB. F/f 26 Mar 68 as KA204. To French AF 61-MI 26 Oct 68. DBR on landing at Flores, Azores, 23 May 69 and subsequently SOC. (Code 61-MI c/s F-RAMI has since been re-allocated to A-02.)

F-15   N. F/f 28 Jun 68. To French AF 61-MJ 30 Jan 69.

F-16   VFW. F/f 03 Mar 68 as KM106. To French AF 61-MK 28 Mar 69 but not formally accepted until 29 Dec 69. One of four aircraft loaned by French MoD to French Postal Service (Centre d'Exploitation Postale Métropolitain) and cvtd to C-160P (P for Postal). F/f as such 23 Feb 73 with provisional test reg F-WUFP. Re-regd F-BUFP and to French AF. Lsd to Air France and del 08 Jul 73 (CofA 03 Sep 73).

The loan of the four converted Transalls (see also c/n F-47, F-49 and F-50) which had officially remained a property of the Etat-Major de l'Armée de l'Air (EMMA - French AF HQ) in spite of operation by Air France on behalf of the CEPM, was to end on 01 Jan 78. The French MoD however stated on 25 Sep 78 that the four aircraft would not be returned and would remain in CEPM hands. On the other hand, it is planned to add three more aircraft for delivery to the French AF in addition

to the 25 improved Transalls now launched into production, and the three additional transports will be funded by the French postal service as a compensation.

F-17   HFB. F/f 17 Apr 68 as KA205. To French AF 61-ML 13 Aug 68.

F-18   N. F/f 16 Sep 68. To French AF 61-MM 14 Mar 69. Demonstration tour to Africa and Malagasy Oct/Nov 77 together with French VAB airborne armoured vehicle.

D-19   VFW. F/f 11 Apr 68 as KM107. To WGAF 5011 28 Jan 69. To Turkish AF ETI-019 1971.

D-20   HFB. F/f 08 Jun 68 as 5012. To WGAF Est 61 11 Jun 69. To Turkish AF ETI-020 1971.

D-21   N. F/f 23 Oct 68 as 5013. To WGAF Est 61 22 May 69 but only formally accepted 30 Jun 70. To FFS-S 1970. To Turkish AF ETI-021 1971.

D-22   VFW. F/f 09 May 68 as 5014. To WGAF LTG63 21 Feb 69. To Turkish AF ETI-022 1971.

D-23   HFB. F/f 10 Jul 68 as 5015. To WGAF LTG63 12 Nov 68. To Turkish AF ETI-023 1971.

D-24   N. F/f 20 Dec 68 as 5016. To WGAF LTG63 23 May 69. To Turkish AF ETI-024 1971.

D-25   VFW. F/f 10 Jun 68 as 5017. To WGAF LTG63 14 Apr 69.

D-26   HFB. F/f 05 Aug 68 as 5018. To WGAF FFS-S 25 Mar 69. To Turkish AF ETI-026 1971.

D-27   N. F/f 07 Feb 69 as 5019. To WGAF FFS-S 12 Jun 70. To Turkish AF ETI-027 1971.

D-28   VFW. F/f 14 Jun 68 as 5020. To WGAF LTG63 02 May 69. To Turkish AF ETI-028 1971.

D-29   HFB. F/f 06 Sep 68 as 5021. To WGAF LTG63 22 Apr 69. To Turkish AF ETI-029 17 May 71.

D-30   N. F/f 27 Mar 69 as 5022. To WGAF LTG63 24 Sep 69. To Turkish AF ETI-030 1971.

D-31   VFW. F/f 30 Jul 68 as 5023. To WGAF LTG63 01 Jul 69. To Turkish AF ETI-031 1971.

D-32   HFB. F/f 11 Oct 68 as 5024. To WGAF LTG63 28 May 69. To Turkish AF ETI-032 17 May 71.

D-33   N. F/f 16 Apr 69 as 5025. To WGAF LTG63 10 Oct 69. To Turkish AF ETI-033 1971.

D-34   VFW. F/f 30 Oct 68 as 5026. To WGAF LTG63 24 Jul 69. To Turkish AF ETI-034 1971.

D-35   HFB. F/f 31 Jan 69 as 5027. To WGAF LTG63 01 Aug 69. To Turkish AF ETI-035 1971.

D-36   N. F/f 18 Jun 69 as 5028. To WGAF LTG63 13 Dec 69. To Turkish AF ETI-036 1971.

D-37   VFW. F/f 30 Apr 71 as 5029. Built 1969 but used for ground tests until 1971. To WGAF LTG61 22 Sep 71. To FFS-S Jun 73 in exchange for 5044. To LTG62 01 Oct 78.

D-38   MBB. F/f 23 Jan 69 as 5030. To WGAF LTG63 11 Jul 69. To Turkish AF ETI-038 1971. (Note change of manufacturer from HFB to MBB from this aircraft onwards, reflecting the 1968 Messerschmitt-Bölkow merger and the new name of Messerschmitt-Bölkow-Blohm.)

Z-1   N. F/f 28 Feb 69. To SAAF as 331, h/o 11 Jul 69 and toc by 28 Sqn, Waterkloof 23 Jan 70.

D-39   VFW. F/f 14 Feb 69 as 5031. To WGAF LTG63 15

120

Aug 69.   To Turkish AF ETI-039 1971.

D-40   MBB. F/f 27 Nov 68 as 5032.   To WGAF LTG63 23
        Jun 69.   To Turkish AF ETI-040 1971.   In
service until Feb 72, parked in the open and deteriorated
at Etimesgut for two years.   Repair began at Erkilet Feb
74 but work abandoned autumn 1975.   Eventually rendered
airworthy by VFW mechanics and flown to Lemwerder 25 Nov
75 for complete overhaul.

Z-2   N.   F/f 02 May 69.   To SAAF as 333, h/o 25 Aug
        69 and toc by 28 Sqn, Waterkloof 23 Jan 70.

D-41   VFW. F/f 20 Mar 69 as 5033.   To WGAF LTG63 05
        Mar 69.   Badly damaged in forced wheels-up
landing 19 Mar 70 nr Rendsburg but repaired.

F-42   MBB. F/f 12 Feb 69 as KA206.   To French AF
        61-MN 12 Jul 69.

Z-3   N.   F/f 26 Jun 69.   To SAAF as 335, h/o 22 Sep
        69 and toc by 28 Sqn, Waterkloof 23 Jan 70.

F-43   VFW. F/f 30 Apr 69 as KM108.   To French AF
        61-MO 16 Oct 69.

F-44   MBB. F/f 04 Mar 69 as KA207.   To French AF
        61-MP 14 Aug 69.

F-45   N.   F/f 30 May 69.   To French AF as 61-MQ 03
        Nov 69.

F-46   VFW. F/f 27 May 69 as KM109.   To French AF
        61-MR 08 Oct 69.

F-47   N.   F/f 18 Jul 69.   To French AF 61-MS 20 Nov
        69 but not formally accepted until 28 Aug 70.
Cvtd to C-160P and regd F-WUFQ, then re-regd F-BUFQ
owned by French AF.   Lsd to Air France (see c/n F-16),
del 23 Jun 73 (CofA 28 Aug 73).

F-48   N.   F/f 29 Aug 69.   To French AF 61-MT 13 Feb
        70.

F-49   MBB. F/f 01 Apr 69 as KA208.   To French AF
        61-MU 08 Oct 69.   Cvtd to C-160P.   Regd
F-WUFR.   To F-BUFR French AF and lsd to Air France.
Del 28 Jun 73 (CofA 28 Aug 73).

F-50   VFW. F/f 13 Jun 69 as KM110.   To French AF
        61-MV .   Cvtd to C-160P, regd F-WUFS?   To
F-BUFS French AF.   Lsd to Air France, del 01 Jul 73
(CofA 03 Sep 73).

F-51   N.   F/f 02 Oct 69.   To French AF 61-MW 13 Feb
        70.

F-52   VFW. F/f 15 Jul 69 as KM111.   To French AF
        61-MX 05 Feb 70.

F-53   MMB. F/f 24 Apr 69 as KA209.   To French AF
        61-MY 03 Sep 69.

F-54   N.   F/f 17 Nov 69.   To French AF 61-MZ 07 Apr
        70.

F-55   N.   F/f 30 Jan 70.   To French AF 61-ZC 15 May
        70.

D-56   MBB. F/f 21 Jul 69 as 5034.   To WGAF FFS-S 15
        Jan 70.   To LTG62 01 Oct 78.

D-57   VFW. F/f 12 Aug 69 as 5035.   To WGAF FFS-S 05
        Feb 70.   To LTG62 01 Oct 78.

D-58   MBB. F/f 27 May 69 as 5036.   To WGAF LTG63 28
        Nov 69.

D-59   VFW. F/f 27 Aug 69 as 5037.   To WGAF LTG63 20
        Mar 70.   To LTG61 06 Apr 71.

D-60   MBB.   F/f 27 Jun 69 as 5038.   To WGAF FFS-S 16

Dec 69.   To LTG62 01 Oct 78.

D-61   MBB. F/f 24 Jul 69 as 5039.   To WGAF LTG63 17
        Feb 70.

D-62   VFW. F/f 08 Aug 69 as 5040.   To WGAF FFS-S 03
        Mar 70.   To LTG61 31 Mar 71.

Z-4   N.   F/f 01 Aug 69.   To SAAF as 337, h/o 31 Oct
        69 and toc by 28 Sqn Waterkloof 23 Jan 70.

D-63   VFW. F/f 08 Sep 69 as 5041.   To WGAF LTG63 07
        Apr 70.

D-64   MBB. F/f 13 Nov 69 as 5042.   To WGAF FFS-S 09
        Apr 70.   To LTG62 01 Oct 78.

Z-5   N.   F/f 25 Oct 69.   To SAAF as 339, h/o 23 Jan
        70 and toc by 28 Sqn Waterkloof 24 Feb 70.

D-65   VFW. F/f 15 Nov 69 as 5043.   To WGAF LTG63 22
        May 70.

D-66   MBB. F/f 19 Aug 69 as 5044.   To WGAF FFS-S 09
        Apr 70.   To LTG61 29 Jun 73.

Z-6   N.   F/f 21 Nov 69.   To SAAF as 332, h/o 19 Mar
        70 and toc by 28 Sqn Waterkloof 29 Apr 70.

D-67   MBB. F/f 29 Sep 69 as 5045.   To WGAF FFS-S? 20
        Mar 70.   To 4/TSLw3 1976, used as instructional
airframe at Wunstorf (with LTG63 badge).

D-68   VFW. F/f 04 Dec 69 as 5046.   To WGAF LTG63 26
        May 70.

Z-7   N.   F/f 22 Dec 69.   To SAAF as 334, h/o 23 Apr
        70 and toc by 28 Sqn Waterkloof 03 Jun 70.

D-69   MBB. F/f 03 Oct 69 as 5047.   To WGAF FFS-S 20
        Mar 70.   To LTG61 15 Mar 71.

D-70   VFW. F/f 31 Jan 70 as 5048.   To WGAF FFS-S 05
        May 70.   To LTG62 01 Oct 78.

Z-8   N.   F/f 24 Feb 70.   To SAAF as 336, h/o 17 Jun
        70 and toc by 28 Sqn Waterkloof 07 Jul 70.

D-71   VFW. F/f 07 Mar 70 as 5049.   To WGAF FFS-S 04
        Jun 70.   To LTG61 17 Mar 71.

D-72   MBB. F/f 07 Nov 69 as 5050.   To WGAF LTG63 16
        Jun 70.   To LTG61 Mar 71.   To LTG63 by 1975.

D-73   VFW. F/f 03 Apr 70 as 5051.   To WGAF LTG63 06
        Jul 70.

D-74   VFW. F/f 06 Feb 70 as 5052.   To WGAF LTG63 05
        Jun 70.

D-75   MBB. F/f 30 Dec 69 as 5053.   To WGAF FFS-S 24
        Apr 70.   To LTG62 01 Oct 78.

D-76   N.   F/f 13 Mar 70 as 5054.   To WGAF LTG63 11
        Jun 70.

D-77   VFW. F/f 21 Mar 70 as 5055.   To WGAF LTG63 08
        Jul 70.   To LTG61 07 Apr 71.

D-78   N.   F/f 25 Mar 70 as 5056.   To WGAF LTG63 08 Jul
        70.

D-79   VFW. F/f 28 Apr 70 as 5057.   To WGAF LTG   23
        Jun 70.   To LTG61 06 Apr 71.

D-80   MBB. F/f 20 Jan 70 as 5058.   To WGAF FFS-S 27
        Jul 70.   To LTG61 16 Mar 71.

D-81   VFW. F/f 08 May 70 as 5059.   To WGAF FFS-S 03
        Sep. 70.   To LTG61 12 Mar 74 on exchange with 5095.

D-82   MBB. F/f 17 Feb 70 as 5060.   To WGAF FFS-S 23

Jul 70 but not formally accepted until 08 Jan 71. To LTG63 1972/73.

D-83 N. F/f 24 Apr 70 as 5061. To WGAF FFS-S 23 Jul 70. To LTG62 01 Oct 78.

D-84 N. F/f 05 May 70 as 5062. To WGAF FFS-S 02 Sep 70. To LTG61 29 Jul 71.

D-85 VFW. F/f 25 May 70 as 5063. To LTG63 10 Jul 70. Crashed into mountain and written off 09 Feb 75 on approach to Soudha Bay, Crete. Forty-two killed.

F-86 MBB. F/f 12 Mar 70 as KA210. To French AF "61-ZD" 23 Jul 70.

F-87 N. F/f 10 Jun 70. To French AF "61-ZE" 05 Nov 70.

F-88 MBB. F/f 25 Mar 70 as KA211. To French AF "61-ZF" 07 Aug 70.

F-89 N. F/f 15 Jun 70. To French AF "61-ZG" 09 Oct 70.

F-90 MBB. F/f 27 Apr 70 as KA212. To French AF "61-ZH" 20 Aug 70.

Z-9 N. F/f 21 Sep 70. To SAAF as 338, h/o 07 Dec 70 and toc by 28 Sqn Waterkloof 04 Jan 71.

F-91 VFW. F/f 04 Jun 70 as KM112. To French AF "61-ZI" 25 Sep 70.

F-92 MBB. F/f 29 May 70 as KA213. To French AF "61-ZJ" 23 Oct 70.

F-93 N. F/f 30 Jul 70. To French AF "61-ZK" 10 Dec 70.

F-94 VFW. F/f 08 Jun 70 as KM113. To French AF "61-ZL" 24 Sep 70.

F-95 MBB. F/f 05 Jun 70 as KM214. To French AF "61-ZM" 22 Oct 70.

F-96 N. F/f 25 Aug 70. To French AF "61-ZN" 20 Nov 70.

F-97 VFW. F/f 10 Jul 70 as KM114. To French AF "61-ZO" 30 Oct 70.

F-98 MBB. F/f 17 Aug 70 as KA215. To French AF "61-ZP" 27 Nov 70 but not accepted formally until 23 Jun 71.

F-99 N. F/f 25 Jan 71. To French AF "61-ZQ" 26 May 71.

F-100 VFW. F/f 27 Aug 70 as KM115. To French AF "61-ZR" 15 Dec 70.

D-101 MBB. F/f 30 Jul 70 as 5064. To WGAF LTG63 08 Dec 70. To LTG61 05 Apr 71.

D-102 N. F/f 12 Nov 70 as 5065. To WGAF FFS-S 15 Apr 71. To LTG62 01 Oct 78.

D-103 VFW. F/f 12 Sep 70 as 5066. To WGAF LTG61 16 Apr 71.

D-104 MBB. F/f 26 Aug 70 as 5067. To WGAF LTG63 26 Apr 71.

D-105 N. F/f 11 Dec 70 as 5068. To WGAF LTG63 05 May 71.

D-106 VFW. F/f 22 Oct 70 as 5069. To WGAF FFS-S 06 Apr 71. To LTG62 01 Oct 78.

D-107 MBB. F/f 01 Oct 70 as 5070. To WGAF Est 61 05 May 71.

D-108 N. F/f 23 Dec 70 as 5071. To WGAF LTG61 19 May 71.

D-109 VFW. F/f 16 Nov 70 as 5072. To WGAF LTG61 14 May 71.

D-110 MBB. F/f 29 Oct 70 as 5073. To WGAF LTG63 16 Jun 71.

D-111 N. F/f 18 Feb 71 as 5074. To WGAF LTG61 19 May 71.

D-112 VFW. F/f 12 Dec 70 as 5075. To WGAF Est 61 26 May 71 but not formally accepted until 11 Feb 72.

D-113 MBB. F/f 25 Nov 70 as 5076. To WGAF LTG63 04 Jun 71. Was fire-fighting demonstrator at Finkenwerder 08 May 77.

D-114 N. F/f 03 Mar 71 as 5077. To WGAF LTG61 11 Jun 71.

D-115 VFW. F/f 11 Jan 71 as 5078. To WGAF LTG63 09 Jun 71.

D-116 MBB. F/f 29 Jan 71 as 5079. To WGAF LTG63 29 Jun 71.

D-117 N. F/f 29 Mar 71 as 5080. To WGAF LTG61 30 Jun 71.

D-118 VFW. F/f 22 Feb 71 as 5081. To WGAF LTG63 15 Jul 71.

D-119 MBB. F/f 12 Feb 71 as 5082. To WGAF LTG63 27 Jul 71 and was exhibited at 1978 Hannover Show as fire-fighting demonstrator.

D-120 N. F/f 07 Apr 71 as 5083. Stored at Beja from 06 Aug 71. To WGAF LTG61 Oct 72.

D-121 VFW. F/f 03 Mar 71 as 5084. To WGAF LTG63 15 Jul 71.

D-122 MBB. F/f 15 Mar 71 as 5085. To WGAF LTG63 20 Aug 71.

D-123 N. F/f 29 Apr 71 as 5086. To WGAF LTG61 03 Sep 71.

D-124 VFW. F/f 01 Jun 71 as 5087. To WGAF LTG63 18 Oct 71.

D-125 MBB. F/f 08 Apr 71 as 5088. To WGAF LTG63 25 Aug 71. To LTG61 22 Jan 76.

D-126 N. F/f 22 Jun 71 as 5089. To WGAF FFS-S 26 Jan 72. To LTG61 22 May 75.

D-127 VFW. F/f 25 Jun 71 as 5090. To WGAF LTG63 19 Nov 71.

D-128 MBB. F/f 06 May 71 as 5091. To WGAF LTG63 27 Sep 71.

D-129 N. F/f 23 Jul 71 as 5092. To WGAF LTG63 10 Feb 72. Regd D-ACTR 14 Jul 77 for SNIAS/MBB demonstration tour of Far East in Jul 77. Retd 21 Aug 77 to Finkenwerder and to LTG63 as 5092.

D-130 VFW. F/f 09 Jul 71 as 5093. To WGAF FFS-S 20 Dec 71. To LTG62 01 Oct 78.

D-131 MBB. F/f 01 Jun 71 as 5094. To WGAF LTG63 13 Oct 71.

D-132 N. F/f 07 Sep 71 as 5095. To WGAF LTG61 09 Mar 72. To LTG63 spring 1974.

D-133 VFW. F/f 07 Sep 71 as 5096. To WGAF LTG61 10

Feb 72.

D-134   MBB. F/f 29 Jun 71 as 5097. To WGAF LTG63 15 Nov 71.

D-135   N. F/f 12 Oct 71 as 5098. To WGAF LTG61 13 Jun 72.

D-136   VFW. F/f 22 Sep 71 as 5099. To WGAF LTG63 30 Dec 71.

D-137   MBB. F/f 05 Aug 71 as 5100. To WGAF Est 61 20 Dec 71 but not formally accepted until 04 Sep 72. To LTG61 11 Sep 72.

D-138   VFW. F/f 24 Apr 72 as 5101. To WGAF LTG61 07 Sep 72.

D-139   VFW. F/f 10 Nov 71 as 5102. To WGAF LTG61 07 Apr 72.

D-140   MBB. F/f 20 Aug 71 as 5103. To WGAF LTG61 23 Mar 72. To Est 61 Jan 74. To FFS-S May 75. To LTG62 01 Oct 78.

D-141   MBB. F/f 09 May 72 as 5104. To WGAF LTG61 10 Oct 72.

D-142   VFW. F/f 17 Dec 71 as 5105. To WGAF FFS-S 10 May 72. To LTG62 01 Oct 78.

D-143   MBB. F/f 07 Oct 71 as 5106. To WGAF LTG61 29 Mar 72.

D-144   VFW. F/f 26 Jun 72 as 5107. To WGAF LTG61 18 Oct 72. To FFS-S in exchange for 5114 19 Mar 76? To LTG62 01 Oct 78.

D-145   VFW. F/f 13 Mar 72 as 5108. To WGAF LTG61 18 Jul 72. To LTG63.

D-146   MBB. F/f 08 Dec 71 as 5109. To WGAF LTG61 04 May 72.

D-147   MBB. F/f 09 Oct 72 as 5110. To WGAF LTG61 20 Dec 72.

D-148   VFW. F/f 18 Jan 72 as 5111. To WGAF LTG61 15 Jun 72.

D-149   MBB. F/f 26 Jan 72 as 5112. To WGAF LTG61 20 Jun 72.

D-150   VFW. F/f 17 Aug 72 as 5113. To WGAF LTG61 07 Dec 72.

D-151   VFW. F/f 18 Mar 72 as 5114. To WGAF FFS-S 26 Jul 72. To LTG61 19 Mar 76.

D-152   MBB. F/f 10 Mar 72 as 5115. To WGAF LTG61 11 Aug 72.

F-153   N. F/f 25 Nov 71. To French AF "61-ZS" 01 Mar 72.

F-154   N. F/f 20 Jan 72. To French AF "61-ZT" 18 Aug 72.

F-155   N. F/f 30 Mar 72. To French AF "61-ZU" 06 Jul 72.

F-156   N. F/f 19 Apr 72. To French AF "61-ZV" 04 Aug 72.

F-157   N. F/f 13 Jul 72. To French AF "61-ZW" 20 Sep 72.

F-158   N. F/f 08 Sep 72. To French AF "61-ZX" 09 Nov 72.

F-159   N. F/f 28 Oct 72. To French AF "61-ZY" 29 Dec 72.

F-160   N. F/f 15 Jan 73. To French AF "61-ZZ" 01 Mar 73.

NOTES

1   A Memorandum of Understanding was signed on 2 Nov 76 between SNIAS, VFW and MBB for joint production of the nouvelle serie Transall with a single production line at Toulouse. Twenty-eight aircraft were ordered for the Armée de l'Air, of which the last three were to be paid for by the French Postal Service. Using experience gained with the earlier aircraft, provisions have been made for full two-directional in-flight refuelling capacity and a range improvement from 4,500-5,500 km to 7,500 km.

2   It was intended originally that 138/141/144/147/150 should be built by Nord, but due to lack of available space at Bourges these airframes were completed at VFW and MBB after D-151 and D-152 respectively.

A list is given below indicating the production sequences at the three assembly lines. The original intention was for Nord to have constructed 53, VFW 54 and HFB 53, but because of the nine South African aircraft which were additional to the original 160 ordered the production arrangements were altered as shown below so that by the end of production Nord had built 56 aircraft, VFW 57 and MBB 56.

Nord	VFW	HFB
V-1	V-2	V-3
A-01	A-02	A-03
A-04	A-05	A-06
F-3	F-1	F-2
D-8	F-4	F-5
D-9	D-7	D-6
F-12	D-10	F-11
F-15	F-13	F-14
F-18	F-16	F-17
D-21	D-19	D-20
D-24	D-22	D-23
D-27	D-25	D-26
D-30	D-28	D-29
D-33	D-31	D-32
D-36	D-34	D-35
Z-1	D-37	D-38
Z-2	D-39	D-40
Z-3	D-41	F-42
F-45	F-43	F-44
F-47	F-46	F-49
F-48	F-50	F-53
F-51	F-52	F-56
F-54	D-57	D-58
F-55	D-59	D-60
Z-4	D-62	D-61
Z-5	D-63	D-64
Z-6	D-65	D-66
Z-7	D-68	D-67
Z-8	D-70	D-69
D-76	D-71	D-72
D-78	D-73	D-75
D-83	D-74	D-80
D-84	D-77	D-82
F-87	D-79	F-86
F-89	D-81	F-88
Z-9	D-85	F-90
F-93	F-91	F-92
F-96	F-94	F-95
F-99	F-97	F-98
D-102	F-100	D-101
D-105	D-103	D-104
D-108	D-106	D-107
D-111	D-109	D-110
D-114	D-112	D-113
D-117	D-115	D-116
D-120	D-118	D-119
D-123	D-121	D-122
D-126	D-124	D-125
D-129	D-127	D-128
D-132	D-130	D-131

Nord	VFW	HFB	Nord	VFW	HFB
D-135	D-133	D-134	F-157	D-145	D-146
F-153	D-136	D-137	F-158	D-148	D-149
F-154	D-139	D-140	F-159	D-151	D-152
F-155	D-142	D-143	F-160	D-138	D-141
F-156	D-145	D-146		D-144	D-147
				D-150	

While progress in the preparation of this monograph was reaching a final stage, additional information was received on the Franco-German Transall.

It was thought of interest to include for the benefit of all aviation historians the following data concerning the first flight, delivery, roll-out and acceptance dates given in French and German official records.

In the cases of conflict between dates quoted on different official sources it is not possible to determine at present which (if either) is correct. The available information is given in the following tables and provides a good illustration of the difficulties facing even the contemporary aviation historian.

Columns 1 and 2 give first flight dates, column one being supplied by Nord Aviation (but correct to November 1970 only, with data from c/n D-102 upwards not available at the time of going to press) and column 2 from the German source. Both sources agree with the exception of aircraft F-11, F-18 and D-83, which show differences of a day.

Columns 3 and 4 give delivery dates, French and German sources respectively. Here it should be noted that inconsistency between the two columns is the rule, with an occasional serious discrepancy (see F-16).

Column 5 shows roll-out dates (sorti d'usine) from French sources. Comparison with f/f dates show them to be totally erratic (prior to f/f date, similar to f/f date, or even taking place well after f/f was recorded). It is better to consider such dates entirely separately and simply for the record.

Column 6 shows the official acceptance dates for French aircraft only (date de prise en charge) and while generally close to delivery dates, these still include some inconsistencies (see F-16).

Column		1	2	3	4	5	6
		F/f date		Delivery Date		Roll-out date	Official Acceptance Date
		French	German	French	German	French	French
F-1	VFW	13 Apr 67	13 Apr 67	02 Aug 67	02 Aug 67	02 Aug 67	02 Aug 67
F-2	HFB	09 Jun 67	09 Jun 67	27 Mar 69	28 Mar 69	15 Aug 67	27 Mar 69
F-3	N	07 Jun 67	07 Jun 67	09 Nov 67	19 Oct 67	09 Nov 67	14 Nov 67
F-4	VFW	10 Sep 67	10 Sep 67	02 May 68	03 May 68	26 Apr 68	02 May 68
F-5	HFB	05 Sep 67	05 Sep 67	20 Feb 68	01 Mar 68	20 Feb 68	01 Mar 68
D-6	HFB	02 Nov 67	02 Nov 67	22 Apr 68	30 Apr 68		
D-7	VFW	07 Nov 67	07 Nov 67	05 Apr 68	09 Apr 68		
D-8	N	11 Jan 68	11 Jan 68	14 Aug 68	20 Aug 68		
D-9	N	27 Feb 68	27 Feb 68	31 Jly 68	08 Aug 68		
D-10	VFW	12 Jan 68	12 Jan 68	28 Aug 68	02 Sep 68		
F-11	HFB	15 Feb 68	14 Feb 68	29 May 68	30 May 68	14 Feb 68	29 May 68
F-12	N	12 Apr 68	12 Apr 68	11 Sep 68	12 Sep 68	12 Apr 68	11 Sep 68
F-13	VFW	06 Feb 68	06 Feb 68	07 Oct 68	31 Oct 68	06 Feb 68	30 Oct 68
F-14	HFB	26 Mar 68	26 Mar 68	26 Oct 68	31 Oct 68		
F-15	N	28 Jun 68	28 Jun 68	29 Jan 69	30 Jan 69	28 Jan 69	28 Jan 69
F-16	VFW	03 Mar 68	03 Mar 68	20 Jan 70	28 Mar 69 29 Dec 69	25 Jan 69	05 Jan 70
F-17	HFB	17 Apr 68	17 Apr 68	08 Aug 68	13 Aug 68	11 Apr 68	12 Aug 68
F-18	N	17 Sep 68	16 Sep 68	12 Mar 69	14 Mar 69	07 Mar 69	12 Mar 69
D-19	VFW	11 Apr 68	11 Apr 68	15 Jan 69	28 Jan 69		
D-20	HFB	08 Jun 68	08 Jun 68	18 Sep 69	11 Jun 69		
D-21	N	23 Oct 68	23 Oct 68	30 Jun 70	22 May 69		
D-22	VFW	09 May 68	09 May 68	11 Feb 69	21 Feb 69		
D-23	HFB	10 Jly 68	10 Jly 68	08 Nov 68	12 Nov 68		
D-24	N	20 Dec 68	20 Dec 68	09 May 69	23 May 69		
D-25	VFW	10 Jun 68	10 Jun 68	26 Mar 69	14 Apr 69		
D-26	HFB	05 Aug 68	05 Aug 68	17 Mar 69	25 Mar 69		

		F/f Date		Delivery Date		Roll-out Date	Official Acceptance Date
		French	German	French	German	French	French
D-27	N	07 Feb 69	07 Feb 69	11 Sep 69	12 Jun 70		
D-28	VFW	14 Jun 68	14 Jun 68	19 Apr 69	02 May 69		
D-29	HFB	06 Sep 68	06 Sep 68	15 Apr 69	22 Apr 69		
D-30	N	27 Mar 69	27 Mar 69	18 Aug 69	24 Sep 69		
D-31	VFW	30 Jly 68	30 Jly 68	20 Jun 69	01 Jly 69		
D-32	HFB	11 Oct 68	11 Oct 68	20 May 69	28 May 69		
D-33	N	16 Apr 69	16 Apr 69	24 Sep 69	10 Oct 69		
D-34	VFW	30 Oct 68	30 Oct 68	19 Jly 69	24 Jly 69		
D-35	HFB	31 Jan 69	31 Jan 69	24 Jly 69	01 Aug 69		
D-36	N	18 Jun 69	18 Jun 69	21 Nov 69	13 Dec 69		
D-37	VFW		30 Apr 71		22 Sep 71		
D-38	HFB	23 Jan 69	23 Jan 69	03 Jun 69	11 Jun 69		
Z-1	N		28 Feb 69		11 Jly 69		
D-39	VFW	14 Feb 69	14 Feb 69	07 Aug 69	15 Aug 69		
D-40	HFB	27 Nov 68	27 Nov 68	12 Jun 69	23 Jun 69		
Z-2	N		02 May 69		25 Aug 69		
D-41	VFW	20 Mar 69	20 Mar 69	25 Aug 69	05 Mar 69		
F-42	HFB	12 Feb 69	12 Feb 69	12 Jly 69	12 Jly 69	26 Jun 69	11 Jly 69
F-43	VFW	30 Apr 69	30 Apr 69	16 Oct 69	16 Oct 69	02 Oct 69	15 Oct 69
Z-3			26 Jun 69		22 Sep 69		
F-44	HFB	04 Mar 69	04 Mar 69	14 Aug 69	14 Aug 69	04 Mar 69	13 Aug 69
F-45	N	30 May 69	30 May 69	06 Nov 69	03 Nov 69	23 Oct 69	04 Nov 69
F-46	VFW	27 May 69	27 May 69	08 Oct 69	08 Oct 69	27 May 69	07 Oct 69
F-47	N	18 Jly 69	18 Jly 69	03 Sep 70	20 Nov 69 / 28 Aug 70	06 Nov 69	03 Sep 70
F-48	N	29 Aug 69	29 Aug 69	11 Feb 70	13 Feb 70	23 Dec 69	11 Feb 70
F-49	HFB	01 Apr 69	01 Apr 69	08 Oct 69	08 Oct 69	04 Apr 69	07 Oct 69
F-50	VFW	13 Jun 69	13 Jun 69	06 Nov 69	06 Nov 69	13 Jun 69	05 Nov 69
F-51	N	02 Oct 69	02 Oct 69	09 Feb 70	13 Feb 70	09 Jan 70	09 Feb 70
F-52	VFW	15 Jly 69	15 Jly 69	04 Feb 70	05 Feb 70	15 Jly 69	04 Feb 70
F-53	HFB	24 Apr 69	24 Apr 69	03 Sep 69	03 Sep 69	24 Apr 69	02 Sep 69
F-54	N	17 Nov 69	17 Nov 69	07 Apr 70	07 Apr 70	27 Feb 70	07 Apr 70
F-55	N	30 Jan 70	30 Jan 70	15 May 70	15 May 70	08 Dec 69	15 May 70
D-56	HFB	21 Jly 69	21 Jly 69	12 Jan 70	15 Jan 70		
D-57	VFW	12 Aug 69	12 Aug 69	31 Jan 70	05 Feb 70		
D-58	HFB	27 May 69	27 May 69	14 Nov 69	28 Nov 69		
D-59	VFW	27 Aug 69	27 Aug 69	17 Mar 70	20 Mar 70		
D-60	HFB	27 Jun 69	27 Jun 69	21 Nov 69	16 Dec 69		
D-61	HFB	24 Jly 69	24 Jly 69	06 Feb 70	17 Feb 70		
D-62	VFW	08 Aug 69	08 Aug 69	16 Feb 70	03 Mar 70		
Z-4	N		01 Aug 69		31 Oct 69		
D-63	VFW	08 Sep 69	08 Sep 69	31 Mar 70	07 Apr 70		
D-64	HFB	13 Nov 69	13 Nov 69	03 Apr 70	09 Apr 70		
Z-5	N		25 Oct 69		23 Jan 70		
D-65	VFW	15 Nov 69	15 Nov 69	13 May 70	22 May 70		
D-66	HFB	19 Aug 69	19 Aug 69	26 Mar 70	09 Apr 70		
Z-6	N		21 Nov 69		19 Mar 70		
D-67	HFB	29 Sep 69	29 Sep 69	10 Mar 70	20 Mar 70		
D-68	VFW	04 Dec 69	04 Dec 69	14 May 70	26 May 70		
Z-7	N		22 Dec 69		23 Apr 70		
D-69	HFB	03 Oct 69	03 Oct 69	16 Mar 70	20 Mar 70		
D-70	VFW	31 Jan 70	31 Jan 70	27 Apr 70	05 May 70		
Z-8	N		24 Feb 70		17 Jun 70		

		F/f Date		Delivery Date		Roll-out Date	Official Acceptance Date
		French	German	French	German	French	French
D-71	VFW	07 Mar 70	07 Mar 70	30 May 70	04 Jun 70		
D-72	HFB	07 Nov 69	07 Nov 69	03 Jun 70	16 Jun 70		
D-73	VFW	03 Apr 70	03 Apr 70	06 Jly 70	06 Jly 70		
D-74	VFW	06 Feb 70	06 Feb 70	30 May 70	05 Jun 70		
D-75	HFB	30 Dec 69	30 Dec 69	21 Apr 70	24 Apr 70		
D-76	N	13 Mar 70	13 Mar 70	29 May 70	11 Jun 70		
D-77	VFW	21 Mar 70	21 Mar 70	08 Jly 70	08 Jly 70		
D-78	N	25 Mar 70	25 Mar 70	12 Jun 70	08 Jly 70		
D-79	VFW	28 Apr 70	28 Apr 70	12 Jun 70	23 Jun 70		
D-80	HFB	20 Jan 70	20 Jan 70	17 Jly 70	27 Jly 70		
D-81	VFW	08 May 70	08 May 70	27 Aug 70	03 Sep 70		
D-82	HFB	17 Feb 70	17 Feb 70		23 Jly 70 08 Jan 71		
D-83	N	24 Apr 70	23 Apr 70	13 Jly 70	23 Jly 70		
D-84	N	05 May 70	05 May 70	25 Aug 70	02 Sep 70		
D-85	VFW	25 May 70	25 May 70	10 Jly 70	10 Jly 70		
F-86	HFB	12 Mar 70	12 Mar 70	22 Jly 70	23 Jly 70	02 Jly 70	22 Jly 70
F-87	N	10 Jun 70	10 Jun 70	05 Nov 70	05 Nov 70	26 May 70	05 Nov 70
F-88	HFB	25 Mar 70	25 Mar 70	06 Aug 70	07 Aug 70	30 Jly 70	06 Aug 70
F-89	N	15 Jun 70	15 Jun 70	09 Oct 70	09 Oct 70	15 Jun 70	09 Oct 70
F-90	HFB	27 Apr 70	27 Apr 70	19 Aug 70	20 Aug 70	27 Apr 70	19 Aug 70
Z-9	N		21 Sep 70		07 Dec 70		
F-91	VFW	04 Jun 70	04 Jun 70	24 Sep 70	25 Sep 70	15 Sep 70	24 Sep 70
F-92	HFB	29 May 70	29 May 70	24 Sep 70	23 Oct 70	24 Jly 70	23 Oct 70
F-93	N	30 Jly 70	30 Jly 70	10 Dec 70	10 Dec 70	28 Oct 70	10 Dec 70
F-94	VFW	08 Jun 70	08 Jun 70	23 Sep 70	24 Sep 70	10 Sep 70	23 Sep 70
F-95	HFB	05 Jun 70	05 Jun 70	21 Oct 70	22 Oct 70	22 Jly 70	21 Oct 70
F-96	N	25 Aug 70	25 Aug 70	20 Oct 70	20 Nov 70	15 Oct 70	20 Nov 70
F-97	VFW	10 Jly 70	10 Jly 70	28 Oct 70	30 Oct 70	05 Oct 70	28 Oct 70
F-98	HFB	17 Aug 70	17 Aug 70		27 Nov 70	22 Sep 70	23 Aug 71
F-99	N	25 Jan 71	25 Jan 71		26 May 71	23 Mar 71	25 May 71
F-100	VFW	27 Aug 70	27 Aug 70	14 Dec 70	15 Dec 70	30 Sep 70	14 Dec 70
D-101	HFB	30 Jly 70	30 Jly 70		08 Dec 70		
D-102	N		12 Nov 70		15 Apr 71		
D-103	VFW		12 Sep 70		16 Apr 71		
D-104	HFB		26 Aug 70		26 Apr 71		
D-105	N		11 Dec 70		05 May 71		
D-106	VFW		22 Oct 70		06 Apr 71		
D-107	HFB		01 Oct 70		05 May 71		
D-108	N		23 Dec 70		19 May 71		
D-109	VFW		16 Nov 70		14 May 71		
D-110	HFB		29 Oct 70		16 Jun 71		
D-111	N		18 Feb 71		19 May 71		
D-112	VFW		12 Dec 70		26 May 71 11 Feb 72		
D-113	HFB		25 Nov 70		04 Jun 71		
D-114	N		03 Mar 71		11 Jun 71		
D-115	VFW		11 Jan 71		09 Jun 71		
D-116	HFB		29 Jan 71		29 Jun 71		
D-117	N		29 Mar 71		30 Jun 71		
D-118	VFW		22 Feb 71		15 Jly 71		
D-119	HFB		12 Feb 71		27 Jly 71		

		F/f Date		Delivery Date		Roll-out Date	Official Acceptance Date
		French	German	French	German	French	French
D-120	N		07 Apr 71		06 Aug 71		
D-121	VFW		03 Mar 71		15 Jly 71		
D-122	HFB		15 Mar 71		20 Aug 71		
D-123	N		29 Apr 71		03 Sep 71		
D-124	VFW		01 Jun 71		18 Oct 71		
D-125	HFB		08 Apr 71		25 Aug 71		
D-126	N		22 Jun 71		26 Jan 72		
D-127	VFW		25 Jun 71		19 Nov 71		
D-128	HFB		06 May 71		27 Sep 71		
D-129	N		23 Jly 71		10 Feb 72		
D-130	VFW		09 Jly 71		20 Dec 71		
D-131	HFB		01 Jun 71		13 Oct 71		
D-132	N		07 Sep 71		09 Mar 72		
D-133	VFW		07 Sep 71		10 Feb 72		
D-134	HFB		29 Jun 71		15 Nov 71		
D-135	N		12 Oct 71		13 Jun 72		
D-136	VFW		22 Sep 71		30 Dec 71		
D-137	HFB		05 Aug 71		20 Dec 71 04 Sep 72		
D-138	VFW		24 Apr 72		07 Sep 72		
D-139	VFW		10 Nov 71		07 Apr 72		
D-140	HFB		20 Aug 71		23 Mar 72		
D-141	HFB		09 May 72		10 Oct 72		
D-142	VFW		17 Dec 71		10 May 72		
D-143	HFB		07 Oct 71		29 Mar 72		
D-144	VFW		26 Jun 72		18 Oct 72		
D-145	VFW		13 Mar 72		18 Jly 72		
D-146	HFB		08 Dec 71		04 May 72		
D-147	HFB		09 Oct 72		20 Dec 72		
D-148	VFW		18 Jan 72		15 Jun 72		
D-149	HFB		26 Jan 72		20 Jun 72		
D-150	VFW		17 Aug 72		07 Dec 72		
D-151	VFW		18 Mar 72		26 Jly 72		
D-152	HFB		10 Mar 72		11 Aug 72		
F-153	N		25 Nov 71		01 Mar 72	25 Aug 71	29 Feb 72
F-154	N		20 Jan 72		18 Aug 72	15 May 72	18 Aug 72
F-155	N		30 Mar 72		06 Jly 72	20 Jun 72	05 Jly 72
F-156	N		19 Apr 72		04 Aug 72	28 Jly 72	03 Aug 72
F-157	N		13 Jly 72		20 Sep 72	30 Aug 72	19 Sep 72
F-158	N		08 Sep 72		09 Nov 72	25 Oct 72	07 Nov 72
F-159	N		28 Oct 72		29 Dec 72	18 Dec 72	29 Dec 72
F-160	N		15 Jan 73		01 Mar 73	15 Jan 73	01 Mar 73

DASSAULT MD-303/410

The origin of the Dassault twin-engined light transports goes back to a 1946 project, the Marcel Bloch MB-30. This was a twin-engined communications aircraft for eight passengers to be powered by two 325 ch Béarn engines. In 1947 a derivative known as the MB-303 (later MD-303) was built and flown by Dassault (this being the new name adopted by the already noted pre-war aircraft designer Marcel Bloch) at the factory of Bordeaux-Talence.

A purely military type also with two Béarn 6D engines, the MD-303 had been built against the specifications of the so-called "colonial" programme and as such was designed as a communications aircraft with extra training capacity. A re-engined version was ordered in quantity, and production began at the newly-erected Dassault plant of Bordeaux-Mérignac.

Three main variants were delivered, all powered with SNECMA 12S engines. The MD-135 (c/n 1-136) for the Armée de l'Air, initially for service overseas, followed by the MD-312 six-seat transport and communications aircraft delivered to the Air Force (c/n 137-253), with the end of the production run delivered as utility aircraft to the French Navy (c/n 294-318). The MD-311 (c/n 254-293) with distinctive glazed nose were Armée de l'Air bombing and navigation trainers. A few aircraft were also diverted to the CEV, being used generally for communications.

The Flamant (flamingo), as it was known, was a sleek and elegant aircraft and a familiar sight for many years at most military airfields both at home and overseas. It is of interest to note that a few aircraft were transferred to fledgeling air forces in countries which had previously been linked with France, for example Cameroun and Tunisia in 1963, Cambodia and Madagascar in 1964-65. The type has now been withdrawn from use, apart from a small number of Armée de l'Air trainers which it is intended to keep operational until 1985. In spite of the phasing out of substantial numbers of Flamants, none of them has emerged on the civil market. During the early seventies, however, the Armée de l'Air turned many airframes over to civil aero clubs for display purposes only, to the effect that the Dassault Flamant now graces many civil aerodromes as gate guardian.

Obviously, with the number of MD-311/312/315s procured by the military, there was a possible outlet on the civil market, and commercial variants also appeared. First of these was the MD-316X (SNECMA 14X) flown in 1952, followed in 1953 by the single-fin MD-316T with US engines. Neither type went beyond the prototype stage. A single experimental MD-312B with a maximum auw of 6,300 kg was also flown in 1954 as a private venture and was used for a few years as a 'hack' by the manufacturers.

The Générale Aéronautique Marcel Dassault (GAMD) and later Avions Marcel Dassault (AMD) were engaged in a long and successful career as military jet aircraft builders but also evinced interest in the civil world market as evidenced later by the successful biz-jet programmes and the Mercure venture.

Following the production of the Flamant series (successful by French standards with 325 aircraft having been built), a logical step was a modern twin-turboprop light transport which could benefit both from a military order and from civil sales. Thus in the dawn of the light-turborprop era the GAMD MD-415 Communauté was flown.

By 1959 the large majority of former French colonies was gaining independence under the then infant Fifth Republic, and Communauté was the name adopted for what was roughly similar to the British Commonwealth. The name chosen by Dassault for the new aircraft well reflected the target market. Work on the twin-Bastan eight-ten seat transport and training aircraft had begun in mock-up shape in 1958, and the prototype completed in April 1959 had flown the following month. The MD-415 was a clever

blend of proven features from previous Dassault fighter designs, borrowing much experience and even components from the Ouragan, Mystère IVA, Super Mystère B2, Etendard IVM and Mirage IVA! It was however to remain only a prototype, and the larger and heavier 14-seat Communauté A2 derivative was not followed up.

Military applications had been kept in mind as a 14-seat transport, navigational and bombing trainer, and even close-support in the COIN role with underwing rockets and bombs. An entirely military version had been designed from November 1958 and was flown as the GAMD 410 Spirale in 1960. Intended as a close-support and multi-purpose twin-turboprop for use overseas, the Spirale had 90% commonality with the Communauté. Also a private venture, it appeared at the peak of the Algerian war but was not ordered into production.

A last attempt at the production of a twin-turboprop came in 1968 with the MD-320 Hirondelle (swallow), but only one prototype was built.

There follows a listing of Dassault light transport aircraft.

MD-303   F/f 26 Feb 47 (Béarn 6D power-plants); "Colonial" programme aircraft and trainer prototype

MD-315   F/f 06 Jul 47 "Colonial" programme (SNECMA 12S power-plants); c/n 1-136 for French AF

MD-312   F/f 27 Apr 50 communications aircraft (SNECMA 12S); c/n 137-254 for French AF and end of production line c/n 294-318 for French Navy

MD-311   F/f 23 Mar 48 navigation trainer (SNECMA 12S); c/n 255-293 for French AF

MD-316X   F/f 19 Jul 52 (F-ZWRR coded R) civil transport prototype (SNECMA 14X). To F-WFDB

MD-316T   F/f 07 Jun 53 (F-ZWRT coded T) with Wright C-7BA-1 power-plants and single tail fin and rudder assembly. Flown on a joint Dassault/AVIACO proving route Bordeaux-Bilbao 195?

MD-312B   F/f 20 Feb 54. Experimental version with heavier loaded weight retaining SNECMA 12S engines. Regd F-WCZN. To F-BCZN (CofA 25 Feb 59) (Avions Marcel Dassault). Ended with outsize nose-boom in 1960 in connection with radar development for Mirage IVA bomber. CofA exp Jan 62

MD-415-01   Communauté twin-Bastan turboprop. F/f 10 May 59 as F-WJDN. Civil transport and communications aircraft. Scrapped 1967.

MD-410   Spirale, f/f 08 Apr 60 as military derivative for use overseas

MD-415M   Diplomate; scaled-up Communauté executive transport. Super Communauté military STOL variant not built

MD-320   Hirondelle, f/f 11 Sep 68 as F-WPXB (Astazou XIV). Conversion projected as MD-320 Bi-Larzac twin-jet. The aircraft was still intact but inactive at Bordeaux during summer 1976 together with another prototype not ordered into production, the Jaguar Marine. The Hirondelle was still fitted with turbo-props, indicating that the planned jet conversion was probably not undertaken.

Specification

MD-315/311/312 Flamant

Span: 20.21m
Length: 12.58m
Height: 4.5m
Empty weight: 5,000 kg
Loaded weight: 6,400 kg
Power-plants: 2 x (Renault) SNECMA 12S (460 hp each)
Cruising speed: 260 km/hr
Maximum speed: 310 km/hr
Take-off distance: 900-1,050m

With a crew of three, accommodation as a communications aircraft was for three or four passengers on the MD-315 (or two-three stretchers for Medevac), three passengers on the MD-311 and four on the MD-312.

MD-415

Span: 16.43m
Length: 13.0m
Height: 4.3m
Wing area: 36 sq m
Empty weight: 3,610 kg
Maximum loaded weight: 5,900 kg
Payload: 2,200 kg
Power-plants: 2 x Turboméca Bastan of 1,000 shp + 66 kg s/t
Maximum speed: 500 km/hr at 6,000m
Cruising speed: 450 km/hr at 6,000m
Take-off distance: 380m
Service ceiling: 11,000m
Range: 2,500 km (normal)
       3,200 km (with external tanks)
Passenger configuration: 14

MD-410

No details known.

MD-320 Hirondelle

Span: 14.5m
Length: 12.72m
Wing area: 27 sq m
Empty operating weight: 2,900 kg
Maximum take-off weight: 5,400 kg
Power-plants: 2 x 800 shp Turbomeca Astazou XIV
Maximum speed: 500 km/hr
Range: 3,000 km
Ceiling: 9,000m
Take-off distance: 600m
Payload as cargo aircraft: 1,200 kg

Accommodation for six passengers (executive) or 10/14 (feederliner). Twin-jet proposals (with Turboméca Larzac or Pratt & Whitney JT15D) would have cruised at Mach 0.70 with a range of 2,500 km with eight/ten passengers or 2,000 km with 17 passengers.

Production

MB-303 F/f 26 Feb 47 (Béarn 6D power-plants). Colonial programme aircraft and trainer prototype. Roll-out and ground trials started Nov 46. Later referred to as MD-303, reflecting post-war change of name of Marcel Bloch to Marcel Dassault. Aircraft believed to have been cvtd to MD-315 prototype since fate has not been traced.

MD-315 F/f 13 Jul 47. Colonial programme (SNECMA 12S power-plants). Believed to be former MD-303 but not confirmed. C/n 1-136 for delivery to French AF unless stated otherwise.

The following basic production list outline has been deliberately limited to dates of acceptance and SOC respectively, with fates and including details of those aircraft that have been preserved.

This also applies to subsequent production of MD-311 and MD-312 variants. Information on some MD-315 withdrawals is still missing at the time of going to press.

When only the acceptance date is given, it may be assumed that the aircraft was current by May 79.

C/n 01 See above. Oper by CEV 1951. To French AF and SOC 29 Apr 55.

C/n 02 To CEV. To CEAM in May 49. Was F-ZJOC Apr 51. Fate not traced.

C/n 03 To CEAM c/s F-SDDC circa 1950-51. Fate not traced.

Note 1 Details on c/n 02 and 03 are not clear. No trace exists in Armée de l'Air records, and so aircraft must have been used by Dassault and CEV only. These aircraft may have been production aircraft c/n 2 and 3 on temporary transfer from the French AF and referred to as 02 and 03, but this is merely conjecture. In fact, the compilers believe three prototypes were actually flown, but confirmation is required.

Note 2 DIT (Déclassé Instruction Technique) applies to all French AF aircraft permanently grounded for instructional purposes. No maintenance serials are allocated as is the case with the RAF.

C/n 1 27 Feb 49/04 Oct 56 DIT.

C/n 2 14 Jun 49/14 May 55.

C/n 3 28 Oct 49/04 Jun 51 DIT Rochefort-sur-Mer (Dept de la Charente-Maritime). This is a military airfield where many military aircraft have ended as instructional airframes for the benefit of local technical schools - both French AF (as BA721) and French Navy (as CEAN).

C/n 4 01 Dec 49/04 Jun 51 DIT Rochefort.

C/n 5 25 Jan 50/02 Nov 60 DIT Rochefort. Last reported there 1975 (see c/n 96).

C/n 6 Aircraft was flown by Dassault in 1950 with manufacturer's provisional reg F-WFUX. To French AF 25 Jan 50. Transferred to CEV 06 May 65. Was oper by late 1969 on various antennae trials. WFU and last reported on fire dump at Istres Jun 75, still coded BR (F-Z.BR)

C/n 7 23 Mar 50/destroyed 21 Dec 56 at CEV d'Istres (Dept des Bouches de Rhône).

C/n 8 To French AF 21 Mar 50. Transferred to CEV 06 May 65. Crashed 01 Aug 68 and presumed WFU.

C/n 9 01 Apr 50/06 Jun 51 DIT Rochefort.

C/n 10 06 Apr 50/06 Jun 51 DIT Rochefort.

C/n 11 14 Apr 50/destroyed 12 Dec 52 at Serres (Dept des Hautes Alpes).

C/n 12 23 Jun 50. SOC to Domaines 25 Jun 65 (see preliminary remarks in SNCAC NC-701/702 Martinet production list for explanation).

Licence-built Macchi MB-320, the Vema 51, F-BDHN of Aero
Sud at Setif in April 1960 (B Chenel via J Delmas)

Top   :   SFERMA PD-18.S F-ZWVO at Dijon (via Charles W Cain)

Bottom :   SFERMA PD-146 Marquis F-WJHC at Kidlington, UK (Charles W Cain)

Top   :   Caudron C.449 Goëland No.1368 in Armée de
            l'Air colours (via JMG Gradidge/DAS McKay)

Bottom :   Caudron C.449 Goëland No.1358 of Aeronavale,
            coded 55.S.18 (E.C.P.Armees)

Top   :   Morane-Saulnier MS-700 F-WFDC (Charles W Cain)
Bottom :   SNCASO SO.7010 Pegase No.01 (SNCASO via J
            Delmas)

C/n 13 19 May 50/SOC and to Domaines 31 Mar 66.

C/n 14 04 May 50/destroyed 18 Oct 56.

C/n 15 04 May 50/SOC and to Domaines 20 Jan 66.

C/n 16 23 Jun 50/20 Jan 66 to Domaines.

C/n 17 12 Jun 50/15 Jul 65 DIT Rochefort. One of 13 Dassault Flamants there by 1974-76 for instructional purposes, these gradually being sold for scrap.

C/n 18 22 Jun 50/14 Jan 65 to Domaines.

C/n 19 23 Jan 52/07 Jul 65 to Domaines.

C/n 20 23 Jun 50/28 Oct 65 to Domaines (use for fire practice at Les Mureaux for some time un-confirmed).

C/n 21 23 Jun 50/23 Sep 69.

C/n 22 05 Jul 50/28 Oct 65 to Domaines.

C/n 23 05 Jul 50/31 Jul 58.

C/n 24 17 Jul 50/28 Oct 65 to Domaines.

C/n 25 31 Jul 50. Transferred 11 Oct 63 to Cameroun Air Arm (République Fédérale du Cameroun). Coded N (TJ-..N?).

C/n 26 31 Jul 50/23 Feb 62 to Domaines.

C/n 27 12 Oct 50/28 Oct 65 to Domaines.

C/n 28 11 Aug 50/SOC 1964 to Domaines.

C/n 29 11 Aug 50/20 Apr 65 to Domaines.

C/n 30 11 Aug 50/22 Dec 62.

C/n 31 22 Sep 50. Transferred to Tunisia 05 Sep 63. Reportedly one of three aircraft still current (see c/n 63 and 186).

C/n 32 22 Sep 50. Transferred from French AF to CEV 06 May 65. Was F-ZABB by Mar 72. Fate not known.

C/n 33 12 Oct 50/Jul 64 and disposed of 10 Aug 64.

C/n 34 12 Oct 50/22 Apr 70.

C/n 35 02 Nov 50/SOC and to Domaines 28 Oct 65.

C/n 36 02 Nov 50. Accident 24 Oct 60, repaired. SOC and to Domaines 20 Apr 65.

C/n 37 02 Nov 50/20 Apr 65.

C/n 38 15 Nov 50. Transferred from French AF to CEV 06 May 65. Fate not known.

C/n 39 15 Nov 50/28 Oct 65 to Domaines. Displayed at Sarlat (Dept de la Dordogne) although painted as No 41.

C/n 40 15 Nov 50/SOC and to Domaines 20 Apr 65.

C/n 41 19 Dec 50/05 Sep 72. See notes on aircraft c/n 39 and 194.

C/n 42 19 Dec 50. Transferred 29 May 67 to Cameroun Air Arm.

C/n 43 19 Dec 50/19 Oct 72.

C/n 44 19 Dec 50/29 Apr 65.

C/n 45 04 May 51. Transferred 17 Feb 64 to Air Arm of Cambodia.

C/n 46 03 Jan 51/20 Apr 65.

C/n 47 03 Jan 51/27 Feb 67.

C/n 48 03 Jan 51. Destroyed 06 Jun 54 but wreck still at Salon-de-Provence (French AF Academy) 1967. Scrapped circa 1968.

C/n 49 03 Jan 51. Accident 03 Mar 60 and repaired. Wreck used after WFU for fire practice at Dijon (Dept de la Cote d'Or) and completely burnt out. To Domaines 20 Apr 65 and scrapped 1968 at Dijon.

C/n 50 03 Jan 51/15 Sep 72. Derelict at Scaër-Guiscriff (Dept du Finistere), last reported Jun 77 in poor condition.

C/n 51 26 Jul 51/05 Oct 72 and donated to Aero-Club de Flers (Dept de l'Orne) where on display coded 30-QS ex F-UGQS.

C/n 52 09 Feb 51/28 Oct 65 DIT Rochefort (see note c/n 17 - was coded Q).

C/n 53 09 Feb 51/23 Sep 69.

C/n 54 09 Feb 51/20 Apr 65.

C/n 55 12 Mar 51. Transferred from French AF to CEV 06 May 65. Fate not known.

C/n 56 12 Mar 51. Written off 17 Aug 57 at Fort Dauphin and SOC 24 Oct 57.

C/n 57 12 Mar 51. SOC date not traced. To Domaines 20 Apr 65. Believed scrapped at Les Mureaux (Dept des Yvelines).

C/n 58 28 Feb 51/15 Sep 72 and donated to Aero-Club de Quevilly (Dept de la Seine-Maritime) and on display at Grand Quevilly (nr Rouen) at local model club. Coded 118-IS ex CEAM F-SDIS.

C/n 59 26 Feb 51/SOC and to Domaines 20 Apr 65.

C/n 60 06 Jul 51/05 Sep 72. Donated to Aero Club de Muret for display, still coded 30-QT (F-UGQT). Was derelict for some years at Muret-l'Herm (Dept de la Haute Garonne) but is no longer there (last reported Mar 75).

C/n 61 12 Mar 51/SOC and to Domaines 25 Jun 65.

C/n 62 12 Mar 51. Transferred 10 May 65 to Madagascar (Armée de l'Air Malgache).

C/n 63 12 Mar 51. Transferred 05 Sep 62 to Tunisian Republican AF (Al Quwwat Aljawwiya Al-Djoumhouria Attunisia).

C/n 64 16 Jan 51/25 Nov 57.

C/n 65 10 Apr 51/26 May 65 DIT Rochefort (see c/n 17).

C/n 66 10 Apr 51/14 Sep 72. Donated to Aéro Club de Pézenas (Dept de l'Hérault) where on display. Last reported early 1976.

C/n 67 18 Apr 51/SOC and to Domaines 25 Jun 65.

C/n 68 18 Apr 51. Destroyed date unknown but prior to May 54 when centralised records on the type were set up (believed at Ivato, Madagascar).

C/n 69 18 Apr 51. Transferred 24 Feb 64 to Cambodia.

C/n 70 26 Apr 71/SOC and to Domaines 25 Jun 65.

C/n 71 18 Apr 51/SOC and to Domaines 25 Jun 65
(reportedly scrapped at Les Mureaux).

C/n 72 26 Apr 51/SOC and to Domaines 14 Jan 65.

C/n 73 26 Apr 51/SOC and to Domaines 25 Jun 65.

C/n 74 04 May 51.   Transferred to Madagascar 24 Jan 59.
Was 5R-MPA (confused records on this particular aircraft have been found, but the above date is now assumed to be correct).   Possibly SOC in 1966 - unconfirmed.

C/n 75 04 May 51/23 Nov 63.

C/n 76 04 May 51.   Destroyed 15 May 58.

C/n 77 23 May 51/31 Oct 69 for display purposes at French AF base at Reims.   Then used for fire-fighting practice there and last reported on fire dump prior to 1977.

C/n 78 23 May 51/SOC and to Domaines 25 Jun 65.

C/n 79 23 May 51/12 Oct 72 and donated to Aéro-Club d'Oléron, still coded 30-QX (F-UGQX) and currently on display at Oléron-Bois Fleury (Isle d'Oléron) with all marks removed.

C/n 80 23 May 51.   Transferred 08 Jul 64 to Cameroun Air Arm.

C/n 81 05 Jun 51.   Destroyed 23 Oct 57.

C/n 82 05 Jun 51/06 Nov 69 and donated to Aéro Club de Castelnaudary (Dept de l'Aude) and was displayed there for several years before being scrapped Feb 75.

C/n 83 15 Jun 51/SOC and to Domaines 20 Apr 65.

C/n 84 05 Jun 51.   Crashed 22 May 54 at Ivato, Madagascar.

C/n 85 05 Jun 51.   Transferred from French AF to CEV 13 May 65.   Fate not known.

C/n 86 25 Jun 51.   Destroyed 24 Nov 59.

C/n 87 25 Jun 51.   Transferred 16 Apr 65 to Madagascar, coded A.

C/n 88 25 Jun 51.   Destroyed 30 Jul 57.

C/n 89 25 Jun 51/SOC and to Domaines 20 Apr 65, but wreck still in existence at Luxeuil French AF base in 1969.

C/n 90 06 Jul 51/08 Dec 65 DIT Rochefort, still coded H (see c/n 17).

C/n 91 06 Jul 51/08 Dec 65 DIT Rochefort, still coded P (see c/n 17).

C/n 92 06 Jul 51/06 Nov 69 and donated to Aéro-Club des Alpilles.   Preserved for some years as gate guardian at gliding centre of Romanin-Les-Alpilles or Le Mazet-de-Romanin, nr St Rémy-de-Provence (Dept des Bouches du Rhône), together with an ex French AF T-6. The MD-315 still showed traces of its former c/s F-UIAE on the instrument panel.   In fact it had been F-UIAE coded E on fins DATEF.168 from at least Apr 63 up to at least Dec 64.   Both gate guardians disappeared during the mid seventies.

C/n 93 24 Jul 51/SOC and to Domaines 25 Jun 65.

C/n 94 24 Jul 51/SOC and to Domaines 20 Apr 65.

C/n 95 24 Jul 51.   Accident 19 Aug 63 with DATEF.168. SOC 30 Dec 63.

C/n 96 25 Jul 51.   Destroyed 30 Jun 58.   (C/n 96 was evident on MD-315 c/n 5 at Rochefort Jul 75 as an instructional airframe, possibly being a re-use of components).

C/n 97 25 Jul 51/SOC and to Domaines 20 Apr 65.

C/n 98 25 Jul 51.   SOC date not traced, and to Domaines 14 Jan 65.   Wreck reportedly for scrap Aug 77 at French AF base of Toul (Dept de la Meurthe-et-Moselle).

C/n 99 06 Sep 51.   Destroyed 07 Jul 61 with DATEF.168.

C/n 100 06 Sep 51/SOC and to Domaines 25 Jun 65.

C/n 101 13 Sep 51/SOC and to Domaines 25 Jun 65.

C/n 102 13 Sep 51/SOC and to Domaines 20 Apr 65. Dismantled wreck coded S last reported Aug 67.

C/n 103 13 Sep 51.   Transferred to Cameroun Air Arm 05 Jun 63.

C/n 104 13 Sep 51/SOC and to Domaines 25 Jun 65.

C/n 105 26 Sep 51/SOC and to Domaines 25 Jun 65.

C/n 106 26 Sep 51.   Destroyed 20 Nov 52 at Fort Lamy (Tchad).

C/n 107 26 Sep 51/27 Jun 65 DIT Rochefort, still coded N (see c/n 17).

C/n 108 04 Oct 51/SOC and to Domaines 20 Apr 65.

C/n 109 26 Sep 51/SOC and to Domaines 20 Apr 65.

C/n 110 03 Oct 51/SOC and to Domaines 25 Jun 65.

C/n 111 06 Nov 51/SOC and to Domaines 25 Jun 65.

C/n 112 07 Nov 51/SOC and to Domaines 25 Jun 65.

C/n 113 07 Nov 51/07 Aug 72 and donated to Aéro-Club de Fréjus/St Raphaël (Dept du Var).   Currently on display on civil side of French Navy Experimental Centre at Fréjus/St Raphaël with faded Aéro-Club titles and former French AF code 30-QX (F-UGQX) or 30-QY (F-UGQY) still faintly visible.   The code 30-QY is favoured, being the last code used in service from 13 Nov 69 up to 24 Sep 70.

C/n 114 07 Nov 51/SOC and to Domaines 20 Apr 65.

C/n 115 07 Nov 51.   Transferred 24 Feb 64 to Cambodia.

C/n 116 07 Nov 51/SOC and to Domaines 20 Apr 65.

C/n 117 30 Nov 51/SOC and to Domaines 25 Jun 65.

C/n 118 30 Nov 51.   Transferred 17 Feb 64 to Cambodia.

C/n 119 30 Nov 51/SOC and to Domaines 25 Jun 65.

C/n 120 20 Dec 51/SOC and to Domaines 25 Jun 65.

C/n 121 20 Dec 51.   Destroyed 05 Sep 61.

C/n 122 20 Dec 51.   To DIT Nîmes 07 Feb 66 (Dept du Gard).   Believed scrapped.

C/n 123 20 Dec 51/25 Jun 65 DIT Rochefort, still coded M (see c/n 17).

C/n 124 23 Jan 52/18 Dec 72 and preserved uncoded as gate guardian at main French AF aircraft storage depot, Chateaudun.

C/n 125 23 Jan 52.   Transferred 17 Feb 64 to Cambodia.

C/n 126 13 Feb 52/05 Sep 72 and donated to Aéro-Club
        d'Oloron (Dept des Pyrénées-Atlantiques) for
display.   Current situation unconfirmed.

C/n 127 21 Feb 52.   Transferred from French AF to CEV
        06 May 65.   Was F-Z.BK coded BK when last
reported Jul 71.   Fate not traced.

C/n 128 21 Feb 52/SOC and to Domaines 20 Apr 65
        (reportedly transferred to CEV but very unlikely).

C/n 129 21 Feb 52/03 May 55.

C/n 130 21 Feb 52.   Transferred from French AF to CEV 06
        May 65 and SOC from CEV inventory 15 Oct 71.

C/n 131 21 Feb 52/DIT Nîmes 08 Dec 65 (see also c/n 122).

C/n 132 28 Feb 52/SOC and to Domaines 20 Apr 65.

C/n 133 28 Feb 52/SOC and to Domaines 20 Apr 65.

C/n 134 18 Mar 52/SOC and to Domaines 25 Jun 65.

C/n 135 14 Mar 52/28 Oct 65 DIT Rochefort, coded 721-DO/O
        by Jul 75 (instructional only).   See c/n 17.

C/n 136 16 Apr 52.   Transferred 17 Feb 64 to Cambodia.

MD-312 F/f 27 Apr 50.   Communications aircraft (SNECMA
        12S).   C/n 137-253 for French AF - see also
additional batch c/n 294-318 for French Navy.   In the
following list where no SOC date is given the aircraft
must be assumed to be current (by mid October 1978).
Exceptionally recent (presumed current) military codes
with date tie-ups are mentioned.

C/n 137 15 May 51/SOC prior to May 54 when centralised
        type records were set up, but confirmed to have
taken place at Mont de Marsan (Dept des Landes).
Probably crashed 29 Jun 51 with CEAM.

C/n 138 22 Jun 51.   Transferred from French AF to CEV 31
        Aug 65.   Was coded AH (F-Z.AH) by Sep 74.   Fate
not traced.

C/n 139 19 Sep 51.   Transferred 05 Sep 66 to Madagascar.

C/n 140 01 Aug 51/12 Mar 73 and donated to Aéro-Club de
        Narbonne.   On display at Lézignan-Corbières, 20
km W of Narbonne (Dept de l'Aude), still coded 7-JE
(F-UHJE).   Current status not known.

C/n 141 13 Sep 51/12 Mar 73.

C/n 142 24 Mar 52.

C/n 143 03 Apr 52.   Rare code 332-IM at CEAM branch at
        Brétigny Jun 72 deserves mention.   By mid 1978
was recoded 319-CL (F-TECL) in the same 319 range as
most surviving French AF Flamants being used for twin-
engine conversion at the Base Aérienne 702, Avord (Dept
du Cher).

C/n 144 03 Apr 52/21 Jan 69 DIT Rochefort.   Coded 721-
        EO/O.   (See c/n 17).

C/n 145 23 Apr 52.

C/n 146 16 Apr 52/05 Dec 75 and preserved at BA 702
        Avord, still coded 319-CM (F-TECM).   Still on
location 1978.

C/n 147 12 May 52/05 Sep 74.

C/n 148 06 May 52 (was 319-DE F-TEDE May 78).

C/n 149 29 Apr 52/30 Nov 72 and donated to Aéro Club de
        Falaise (Dept du Calvados) where preserved,
still coded 315-SW (F-TESW).

C/n 150 16 May 52/23 Mar 72 and wreck dumped at Cambrai
        French AF Base (Dept du Nord) 1976-77.

C/n 151 Neither acceptance date nor SOC date has been
        traced for this particular French AF aircraft,
but it is confirmed to have been transferred to DIT
Rochefort for instructional purposes.   (No aircraft file
exists in Central Records, therefore WFU prior to May 54;
see c/n 68).

C/n 152 16 May 52/15 Nov 77.   Was 319-DO at least during
        last two years' service and still so coded as a
wreck at Avord May 78.

C/n 153 29 May 52 (was 319-DV F-TEDV May 76).

C/n 154 French AF.   As for c/n 151, but fate not traced
        after SOC.   No aircraft file in Central Records
or at Historical Section of French AF, therefore WFU and
SOC prior to May 54.   (At Base Ecole 702, Avord, from 25
Feb 52 until 26 Nov 52.)

C/n 155 17 Jun 52.   Was in process of being SOC on 31
        Jul 78.

C/n 156 23 Jun 52.   Reportedly oper by CEV in Jun 63 at
        least but not officially confirmed.   Was SOC
(French AF) 01 Aug 72.

C/n 157 23 Jun 52/23 Mar 72.

C/n 158 29 Sep 52.

C/n 159 24 Jun 52/14 Jun 72.

C/n 160 24 Jun 52.   Was 319-DA F-TEDA May 78.

C/n 161 Acceptance Jul 52/01 Jan 72 and donated to Aéro-
        Club de Caen (Dept du Calvados) where preserved
uncoded.

C/n 162 Acceptance date not traced.   Transferred to
        Madagascar Sep 64 and SOC (in Malagasy service)
1966.   (Some reports state del Oct 64 with official date
of transfer 10 May 65!)

C/n 163 01 Jul 52.   Was 319-CN F-TECN May 78.

C/n 164 01 Jul 52.   Was 319-DB F-TEDB May 78.

C/n 165 Acceptance date not traced.   SOC 03 Aug 71 and
        donated to Aéro-Club de Blois (Dept du Loir-et-
Cher) where preserved still coded 44-GI (F-RHGI).   Last
reported Apr 77 and believed still there.

C/n 166 23 Jul 52.

C/n 167 Acceptance Jul 52.   Transferred 28 Dec 71 to
        DTCA and oper by AIA Clermont-Ferrand (Dept du
Puy-de-Dôme) as F-Z...?   Transferred to Armée de l'Air
du Niger but confirmation awaited.

C/n 168 Acceptance date not traced.   Donated 03 Mar 71
        for preservation (uncoded) in the Jardin
d'Acclimatation, Bois de Boulogne, on western outskirts
of Paris.   Still there Nov 77.

C/n 169 Acceptance Jul 52/24 Nov 66.   Wreck still at BA
        702 Avord May 78 coded 4-WA (F-UGWA).

C/n 170 30 Jul 52/24 Jan 74 and donated to Aéro-Club de
        Cherbourg (Dept de la Manche) where on display
by Aug 76.   Current situation not known.

C/n 171 Acceptance date not traced.   Destroyed 31 Aug
        67, believed at Avord (Dept du Cher).

C/n 172 30 Jul 52.   Was 319-DD F-TEDD May 78.

C/n 173 01 Aug 52.   Was 319-CR F-TECR May 78.

C/n 174 01 Aug 52/21 Sep 76.

C/n 175 Acceptance Sep 52/23 Mar 72, reportedly preserved

at Saintes (Dept de Charente Maritime) 1978, still coded 41-GB (F-UGGB).   41-GV also reported.

C/n 176 Acceptance Sep 52/23 Mar 72.   Reportedly preserved.

C/n 177 18 Sep 52.   Was 319-DU F-TEDU Jun 77.

C/n 178 Acceptance Nov 52/23 Mar 72.

C/n 179 02 Sep 52/21 Sep 76.

C/n 180 22 Sep 52.

C/n 181 08 Oct 52/09 May 77.

C/n 182 Acceptance date not traced.   SOC 10 Mar 70.

C/n 183 Acceptance Sep 52/23 Mar 72.

C/n 184 Acceptance Oct 52/23 Mar 72.

C/n 185 06 Oct 52/19 Nov 73 and donated to Aéro-Club de Gap, code 319-CO being painted out.   On display at Gap-Tallard, 14 km south of Gap (Dept des Hautes-Alpes).

C/n 186 Acceptance date not traced.   Transferred 05 Sep 63 to Tunisia (see c/n 31 and 63).   Reportedly still current.

C/n 187 10 Oct 52.   Current as 319-CK (F-TECK).   Cannot therefore be aircraft reported as a Flamant WFU and derelict since Aug 76 at Quimper (Dept du Finistère Sud, Brittany) and quoted as c/n 187.

C/n 188 Acceptance Oct 52/05 Sep 72.   Preserved at Angoulême (Dept de la Charente) with incomplete code 1-GD (being 41-GD F-UJGD).   Last reported there Sep 77.

C/n 189 14 Oct 52.   Was 319-CE F-TECE May 78.

C/n 190 Acceptance date not traced.   SOC and to Domaines 09 Sep 65.

C/n 191 Acceptance Nov 52/25 Nov 71 and donated to Aéro-Club de Vierzon (Dept du Cher) where preserved, still coded XA (F-UIXA).

C/n 192 14 Nov 52/24 Oct 74.

C/n 193 Acceptance date not recorded (see c/n 68).   Since crashed in Luxembourg 01 Apr 54 (with GTLA 2/60) and SOC 07 Sep 54.

C/n 194 Acceptance Dec 52/23 May 72 and donated to Aéro-Club d'Aubigny-sur-Nère (Dept du Cher), uncoded. Traces for former code 30-QL (F-UGQL) faintly visible. The fins sport the identity of MD-315R c/n 41 (qv).

C/n 195 27 Jan 53/21 Sep 76.

C/n 196 28 Nov 52.   Was 5-MK F-UGMK Jun 74.

C/n 197 Acceptance Nov 52/23 Mar 72 and donated to Aéro-Club de Cognac (Dept de la Charente).   Probably not put on display, and last reported in scrapyard at Cognac 1974, coded 332-IL (ex CEAM Brétigny).

C/n 198 Neither acceptance nor SOC date traced, but transfer from French AF to Madagascar 10 Jul 68 confirmed.   Fate unknown.

C/n 199 Date of acceptance not traced.   SOC 23 Sep 69. Gate guardian at Chartres (Dept de l'Eure et Loire).   Current situation not clear.

C/n 200 Acceptance Dec 52/14 Dec 71 and donated to Aéro-Club de Barcelonnette (Dept des Alpes de Haute-Provence) where preserved with former code 13-TA (F-UHTA)

painted out.   On location Aug 78 at airfield of Barcellonnette-St Pons.

C/n 201 01 Dec 52.   Was 319-DJ F-TEDJ May 78.

C/n 202 18 Dec 52.   Was 319-CS F-TECS May 78.

C/n 203 08 Jan 53.   Was 319-CY F-TECY May 78.

C/n 204 Acceptance Dec 52/23 Mar 72.

C/n 205 08 Jan 53/21 Sep 76.

C/n 206 Acceptance Jan 53/23 Mar 72, still coded 118-IP (F-SDIP) ex CEAM on scrapyard at Châteaudun Jun 77.

C/n 207 Acceptance Jan 53/06 Jul 73 when 319-CV (F-TECV) and preserved still coded at Saintes (Dept de la Charente-Maritime) at French AF technical school.   Last reported May 76.

C/n 208 08 Jan 53.   Believed 319-CC.

C/n 209 Acceptance date not recorded.   Crashed 15 Jul 69 at Tournon St Martin (Dept de l'Indre) as 319-CL (F-TECL).   Four killed.

C/n 210 05 Feb 53.   Believed 319-CP or 319-CE.

C/n 211 Acceptance date not traced.   Destroyed 07 Dec 60 at BE702 Avord (Dept du Cher).   SOC 02 Oct 61.

C/n 212 27 Jan 53/31 Mar 76.   Reportedly an instructional airframe at Rochefort but not confirmed.

C/n 213 Acceptance Jan 53/23 Mar 72 and scrapped at Salon-de-Provence May 73.

C/n 214 Acceptance Feb 53/23 Mar 72.   Last reported coded 319-DQ at Dijon prior to Jun 74 and believed scrapped there.

C/n 215 16 Feb 53.   Was 319-DP F-TEDP May 78.

C/n 216 16 Feb 53.

C/n 217 16 Feb 53.

C/n 218 25 Feb 53.   Was 319-CA F-TECA May 78.

C/n 219 Acceptance date not traced.   SOC 30 Jun 70.

C/n 220 20 Mar 53/02 Feb 74 and donated to Aéro-Club de Castres (Dept du Tarn) as 319-CK (F-TECK) and preserved Castelnau-Magnoac.

C/n 221 09 Apr 53.   Was 319-CM F-TECM May 78.

C/n 222 Acceptance Apr 53/14 Dec 72 and donated to Aéro-Club d'Issoudun (Dept de l'Indre).   Preserved at Issoudun-Le Fay in derelict condition with undercarriage collapsed and with previous code 319-DY (F-TEDY) removed but faintly visible.

C/n 223 Acceptance May 53/23 Mar 72.

C/n 224 Date of acceptance not traced.   SOC 13 Mar 69. Last reported in private scrapyard at Issoudun (Dept de l'Indre) Apr 78, still coded 319-DM (F-TEDM).

C/n 225 Date of acceptance not traced.   SOC 04 Feb 71 and to Domaines.

C/n 226 16 Apr 53.   Was 319-CG F-TECG May 78.

C/n 227 29 Apr 53.

C/n 228 05 May 53.

C/n 229 06 Jun 53.   Was 319-DW F-TEDW May 78.

C/n 230 Acceptance May 53/23 Mar 72.

C/n 231 Acceptance May 53/24 Oct 74.

C/n 232 28 May 53.   Was 319-CB F-TECB May 78.

C/n 233 Date of acceptance not traced.   SOC 09 Apr 70.
To fire practice and remains present completely
burnt out at BA200 Apr-St Christol (Dept du Vaucluse) 01
Oct 78.

C/n 234 16 Jun 53.   Was 319-CW F-TECW May 78.

C/n 235 19 Jun 53.

C/n 236 Acceptance Jun 53.   Donated 13 Aug 73 to
Association des Parents des Enfants Inadaptés de
Bergerac at Bergerac (Dept de la Dordogne) and must be
assumed to have been for handicapped childrens' playground.   Not an instructional airframe at Rochefort as
mentioned in other sources.

C/n 237 24 Jun 53.   Was 314-DC or 319-DC Jun 77.

C/n 238 24 Jun 53/24 Oct 74 and derelict at Aix-les-
Milles (Dept des Bouches du Rhône) for about a
year on civil side of airfield together with H-34 c/n
SA-57.   Still coded 91-CW F-UKCW when probably scrapped
late 1975 or early 1976 with vertical fins having at
some time been re-assembled inverted.

C/n 239 07 Dec 53.

C/n 240 27 Jul 53.   Was 319-DY F-TEDY May 78.

C/n 241 29 Jul 53 until WFU and donated to French
national air museum 1977.   In storage at Le
Musée de l'Air, Le Bourget, Sep 77 awaiting future
display, still coded 319-CV F-TECV.

C/n 242 30 Jul 53/04 Oct 76.

C/n 243 31 Jul 53/04 Oct 76.   Wreck still at Cambrai
(Dept du Nord) Jun 78.

C/n 244 17 Sep 53.   Was 319-CO F-TECO May 78.

C/n 245 Acceptance Sep 53/25 Apr 75.

C/n 246 25 Sep 53.

C/n 247 Acceptance Sep 53/23 Jan 74.

C/n 248 23 Oct 53.

C/n 249 Acceptance Oct 53/23 Mar 72.

C/n 250 02 Nov 53.

C/n 251 30 Nov 53.   Was 319-CZ F-TECZ May 78.

C/n 252 29 Dec 53.   Was 319-DQ F-TEDQ 1978.

C/n 253 12 Jan 54.   Was 319-CQ F-TECQ 1968.

MD-311 C/n 254-293.   Navigation trainer (SNECMA 12S
power-plants), with glazed nose as conspicuous
feature.

C/n 254 Judging from French AF Central Records, c/n 254
was the first MD-311 and not the last MD-312 as
generally quoted.   There is however no individual aircraft file for c/n 254, and this may have been retained
by the manufacturers; or else the aircraft may have been
transferred to the French AF and was SOC prior to May 54.
Photographic evidence exists of an aircraft with MD311-01
painted on fins.   Further details are awaited.

C/n 255 15 Nov 51/31 Oct 69 and preserved on display at
Sarre-Union (Dept du Bas Rhin) ex 316-KN F-TEKN.

C/n 256 20 Feb 52.   Destroyed at Aubiers (Dept du Gers)
with c/n 292 13 Aug 69.

C/n 257 28 Feb 52.

C/n 258 20 Mar 52.

C/n 259 08 Oct 52.   Destroyed 20 Aug 59.

C/n 260 05 Jan 53.

C/n 261 08 Jan 53/06 Nov 69.   Preserved for some years
at Nancy-Essey (Dept de Meurthe-et-Moselle) coded
A, but no longer there.

C/n 262 27 Jan 53/12 Apr 78.

C/n 263 27 Jan 53.

C/n 264 Acceptance Feb 53/19 Feb 74 and scrapped at
Tours circa 1974?

C/n 265 06 Feb 53/28 Jan 77.   Believed scrapped at Nancy
1977, still coded 2-HD F-UGHD.

C/n 266 16 Feb 53.   Was 2-HD F-UGHD Jun 78, a code
previously used by c/n 265.

C/n 267 06 Mar 53.   Destroyed 11 Mar 57 and SOC 22 Nov
57.   TT 1,200 hrs.

C/n 268 20 Mar 53.   Destroyed 25 Jan 66.

C/n 269 13 Mar 53/06 Nov 69.   Was derelict at Orléans-
Saran (Dept du Loiret), coded 316-KD F-TEKD and
removed during motorway construction.   Not reported since
1973.

C/n 270 31 Mar 53/25 Apr 78.

C/n 271 04 Apr 53/06 Nov 69.   Preserved at Amboise (Dept
de l'Indre-et-Loire).

C/n 272 09 Apr 53/06 Nov 69.   Preserved at La Baule
(Dept de la Loire-Atlantique), coded 316-KJ F-TEKJ
and scrapped circa Jul 75.

C/n 273 13 Apr 53.   Accident 27 Jul 65 and consequently
DIT 09 Jul 66.

C/n 274 16 Apr 53.

C/n 275 16 Apr 53/06 Nov 69.   Preserved on display at
L'Aigle (Dept de l'Orne), still coded 316-KC
F-TEKC with traces of previous code L still visible on
fins.

C/n 276 29 Apr 53.   Transferred as 316-KQ F-TEKQ to EAA
601 Châteaudun 30 Aug 77 for storage.

C/n 277 08 Apr 53/09 May 72.

C/n 278 23 Apr 53.   Current with French AF navigation
school GE 316 at Toulouse (Dept de Haute-Garonne)
since 06 Sep 77.

C/n 279 30 Apr 53/09 May 72.   Scrapped Reims circa 1973-
1974.

C/n 280 12 May 53.   Stored at EAA 601 since 31 May 77
(not scrapped St Raphaël as other sources
suggest).

C/n 281 23 Apr 53/14 Oct 69 and donated to Aéro-Club d'Aix-
Marseille (Dept des Bouches du Rhône), where
preserved on concrete pylons.   Still coded 316-KK F-TEKK
with additional titling "Aéro-Club d'Aix Marseille" and
"Sgt J P VIALA" on left side of nose, presumably in
memory of French AF NCO.   A marble slab also states that
the aircraft was donated by French AF 1970.

C/n 282 05 Jun 53.   Current with GE.316 since 13 Oct 77.

C/n 283 03 Jun 53.   Destroyed 06 May 56 at Salon-de-
          Provence (Dept des Bouches du Rhône).

C/n 284 03 Jun 53/14 Oct 69.   Was preserved on display
          at Perpignan-Llabanère (Dept des Pyrénées
Orientales) and last reported Nov 73.

C/n 285 08 Jun 53.   Accident at French AF Academy Salon-
          de-Provence (Dept des Bouches du Rhône) 17 May 54
as F-TEBN.   SOC 31 Jul 54.   (Mention in scrapyard at
St-Raphaël circa 1973-75 appears to be uncertain.)

C/n 286 19 Jun 53.

C/n 287 18 Jun 53/17 May 78.

C/n 288 09 Jul 53/25 May 78.

C/n 289 27 Jul 53/09 May 72.

C/n 290 03 Jul 53.

C/n 291 30 Jul 53.

C/n 292 14 Sep 53.   Destroyed with c/n 256 at Aubiers
          (Dept du Gers) 13 Aug 69.

C/n 293 15 Sep 53/25 Oct 73.   Scrapped Toulouse-
          Francazal (Dept de la Haute Garonne).

Note   An unidentified Flamant adorned with Norwegian
          military roundels briefly appeared in the final
scene of a French film "Si Tous les Gars du Monde", but
this was only for filming purposes circa 1955 in a scene
where the Flamant was supposed to be a Norwegian aircraft rescuing a trawler in a hopeless situation.

          While on the subject of Dassault Flamants in
foreign colours, 16 were reportedly delivered to Vietnam,
but apparently the transfer never took place.   The
number of aircraft transferred to Cambodia has been
reported to be six, but only the aircraft mentioned in
the production list should be taken into account.

MD-312   C/n 294-318, batch for French Navy.   The
          Flamants were taken on strength by the acceptance
and ferry unit ERC (Escadrille de Réception et de
Convoyage) before being delivered to communications
squadrons.   The main operator was initially Escadrille
11S (at Les Mureaux and later Dugny-Le Bourget), but the
MD-312 also served with Escadrilles 2S, 3S, 10S
(experimental unit) and various station flights, as well
as the ERC for some time.

          Up to 1961 French Navy aircraft had unit codes
allocated when delivered (eg c/n 304 10s-11 with c/s
F-YCJK;   c/n 306 HY-3 c/s F-YAFC).   From 1961 however
the Service Central de l'Aéronautique (French Naval
Bureau of Aeronautics) decided to issue permanent
identities to French Navy aircraft, now known as S.C.Aéro
numbers - being usually the construction numbers or
corrupted c/n.   In the case of the MD-312, all surviving
aircraft adopted their three-digit c/ns as a permanent
identity.   Unlike French AF call signs usually tied up
to the aircraft unit code, the French Navy have
gradually introduced the use of call-signs allocated to
the pilot or crew proper and not to the aircraft, but
this is not a general rule.

          To ensure identification of unit, a list is
given below of the relevant call-sign ranges:

Escadrille de Réception et de Convoyage: F-YFLA,
                                  now F-YELA
Escadrille 2S: F-YCAA
                                  and F-YCBA
                                  now F-YDAA
                                  and F-YDBA
Escadrille 3S: F-YCCA
                                  and F-YCEA
                                  now F-YDCA
                                  and F-YDDA

Escadrille 10S: F-YCJA
          changed   early seventies F-YDEA
Escadrille 11S: disbanded Feb 69 F-YCKA

          Station and regional flights also had call-signs
in the F-YA.. range (eg F-YAFA Hyères, F-YAJA French West
Africa etc).

          SOC dates for French Navy MD-312s are not yet
available, but they have all be WFU, being replaced early
in 1973 by ten (later 12) Piper PA-31 Navajos.

C/n 294 25 Apr 52.   Kept by ERC for training purposes
          for some time before del to Escadrille 11S.
Instructional airframe at CEAN Rochefort (Centre-Ecole de
l'Aéronautique Navale, the main French Navy technical
school for non-flying personnel).   Last reported there
Jul 76.

C/n 295 Acceptance at Bordeaux 08 Oct 52, but official
          date quoted as 09 Oct 52, when ferried to Toussus-
le-Noble.   Believed scrapped St-Raphaël circa 1973-75.

C/n 296 30 Oct 52.   Last reported late 1969 as F-YCJQ and
          fate not traced.

C/n 297 20 Dec 52.   Instructional for fire-fighting
          practice at Centre d'Instruction Naval de
Querqueville, Cherbourg (Dept de la Manche) circa 1973.
Completely burnt up and sold for scrap 03 Sep 76.

C/n 298 18 Dec 52.   Flew into hills and written off in
          poor visibility 05 Dec 61 nr Solliès-Toucas (Dept
du Var).   Three killed.

C/n 299 24 Dec 52.   Crashed and written off at Massif
          St Martin, nr Roquebrune sur Argens (Dept du Var)
26 Jan 54 (was 10S.10 c/s F-YCJJ).   Six killed.

C/n 300 30 Jan 53.   SOC 1971 when with Escadrille 2S.   At
          scrapyard Lann-Bihoué (Dept du Morbihan) 1972 and
on fire dump when last reported Aug 73.

C/n 301 30 Jan 53.   Last reported in service Feb 72 and
          fate not clear.   Mentioned elsewhere as coded
721-ES, probably indicating instructional use at
Rochefort, but with rare unexplained French AF code on an
ex French Navy aircraft.   Apparently disposed of and
later put on display at civil airport of Vichy Charmeil
(Dept de l'Allier).   Aircraft gradually became derelict,
then totally wrecked.   Last reported on location as
completely broken up.

C/n 302 24 Feb 53.   SOC 1971 when with Escadrille 10S.
          Instructional at Rochefort 1972-75 at least (last
reported Jul 76).

C/n 303 06 Mar 53.   SOC 1971.   Last reported in service
          Jul 69 as F-YCJT.

C/n 304 26 Mar 53.   Last reported in service with
          Escadrille 10S Oct 71.   Fuselage derelict in
scrapyard by road N88 at Bizac, S of Tarreyres, 12 km S
of Le Puy-en-Velay (Dept de la Haute-Loire) Jul 79.

C/n 305 01 Apr 53.   Last reported in service Dec 69 as
          F-YCBV.

C/n 306 16 Apr 53.   Hyères station flight and SOC 02 Jul
          57, circumstances unknown.

C/n 307 06 May 53.   Hyères station flight and other
          squadrons later.   SOC 1971 when with Escadrille
2S.   At scrapyard Lann-Bihoué 1972.   Last reported on
fire dump Aug 73.

C/n 308 21 May 53.   Last reported in service Dec 71 as
          F-YDCJ.   At scrapyard Lann-Bihoué Jul 72 when
last reported.

C/n 309 26 May 53.   Last reported in service Escadrille
          3S Apr 71.   Believed scrapped at Hyères, where

last seen 1976.

C/n 310 11 Jun 53.   Crashed Cannes-Mandelieu (Dept des Alpes-Maritimes) 07 Oct 57, believed 11S-12 F-YCKL.   After repeatedly delayed departure pilot inadvertently took off with controls locked, wooden chocks not having been removed.   Five killed, including an Admiral.   Wreck remained on location for some time.

C/n 311 16 Jun 53.   Last reported in service 1972.   WFU and believed scrapped at St-Raphaël where last seen Aug 73.

C/n 312 25 Jun 53.   Destroyed 01 Feb 54, circumstances unknown (but possibly after hangar collapse due to weight of unexpected now at Le Luc-en-Provence (Dept du Var).   Confirmation wanted.)

C/n 313 30 Jun 53.   SOC 1971.   Fate not traced.

C/n 314 17 Jul 53.   Last reported in service Dec 72 as F-YDEA and F-YDEF.   WFU and believed scrapped at St-Raphaël where last seen Aug 73.

C/n 315 21 Jul 53.   Last reported in service late sixties.   Fate not traced.

C/n 316 20 Jul 53.   Last reported in service Sep 72 as F-YDEN.   Believed scrapped at St-Raphaël circa 1973-75.

C/n 317 21 Jul 53.   Last transferred to Escadrille 10S 25 Jul 69.   Went into inverted spin 19 Dec 69 against strong headwind over Mont Ventoux (Dept du Vaucluse).   Crew managed to recover and land at nearby Orange AF base with aircraft considerably overstressed, twisted wing etc.   SOC on location Jan 70, ferried by road to CEAN Rochefort for instructional purposes. Believed to be the aircraft auctioned there by Domaines Jun 75 but not confirmed.

C/n 318 29 Jul 53.   Only one evidence in service, as F-YCKX, believed in 1968.   Also quoted as 721-EW at Rochefort (see c/n 301).

<u>BUSINESS JETS WITH COMMERCIAL OPERATORS</u>

FALCON

As recorded elsewhere, Marcel Dassault's ventures into the air transport field have not been very successful, but on the other hand a tremendous success was achieved with a business jet, giving the manufacturers yet another string to their bow - to augment the long-established success of the company's military aircraft. The Mystère 20 was designed as a private venture initially, work on the project starting in 1962 with some co-operation from Sud Aviation.   The new aircraft drew much from previous jet fighter experience, and the result was an outstandingly beautiful and clean aircraft, well in Dassault's tradition.   The prototype, F-WLKB, made its maiden flight from Bordeaux-Mérignac on 4 May 1963.   On the same day an American evaluation team led by Charles Lindbergh was on the spot, and by August of the same year a large order had been obtained from the Business Jet Division of Pan American, which was to become sole dealer for the type in America.

First fitted with two Pratt & Whitney JT12 A-6 engines, the prototype was later fitted with General Electric CF700 2B power-plants.   The first three production aircraft (with a longer fuselage) were used for tests, including certification trials (both SGAC and FAA certification for transport operation being granted), and the fourth aircraft - F-WMKF - was delivered to Pan Am as N801F on 12 April 1965.

This was the first of a long line, as American orders were gradually increased.   The export name Fan Jet Falcon was adopted.   It is currently marketed in the western hemisphere by Falcon Jet Corporation, a subsidiary of Pan American and Avions Marcel Dassault, with over 250 aircraft sold in this area at the time of writing.   As expected, more orders for various customers were dealt with by Dassault proper, with a double-barrelled construction number involving the general production line c/n and an extra number starting from 401.   Small military contracts were also obtained.

Over the years the Falcon jet has been improved, and by mid 1977 some 450 had been sold, the aircraft being flown in some 30 different countries with an additional 20 being used by air forces or government agencies.   Out of this regularly increasing total, many have entered airline use, either on scheduled service or within the business jet division, not to mention aircrew training, hence the mention of this twin-turbofan executive aircraft in a monograph dealing with commercial air transports.

Of particular interest is the use of a large fleet of fast freighters by Federal Express in the United States;   thirty-three have been fitted with a freight door and an average 2,200 hours is flown yearly by each aircraft in the company.   Use as a 12-14 seater has also been undertaken by Air France and Touraine Air Transport (TAT), while VIP accommodation is for eight to ten passengers.   One of the latest successes (though not for a commercial variant) was the order for 41 Mystère-Falcon Guardians by the US Coast Guard.

Dassault's range of business jets is now developing quickly with new aircraft (the Falcon 10 and Falcon 50 particularly), but the Mystère 20/Fan Jet Falcon will stay for many years to come, representing the major breakthrough of France's aircraft industry into this world market, in a field where competition remains intense.

<u>Specification</u>

Executive/business jet with 8-14 seat accommodation.

Span: 16.3m
Length: 17.3m
Height: 5.32m
Empty weight/maximum take-off weight/maximum payload are

mentioned in that order under different variants.
Initial production aircraft (Falcon 66) had a wing span
of 14.47m and were powered by 2 x 1,870 kgp GE CF-700 2C
(weights: 7,000/12,000/1,460 kg).

Falcon C

Improved variant.   Optional Saphir I APU.

Power-plants: 2 x 1,870 kgp GE CF-700 2C
Weights: 7,060/12,000/1,401 kg
Maximum range with eight passengers and luggage: 3,480
                                                    km
Cruising speed: 855 km/hr (Mach 0.79)

A single aircraft was designated CC (Cross Country)
with special landing gear for grass field operation (see
c/n 73/419 VH-BIZ).

Falcon D

(Initially Falcon 68.)   From the 172nd production
aircraft.

Power-plants: 2 x 1,930 kgp GE CF-700 2D
Weights: 7,075/12,400/1,363 kg
Maximum Mach number raised to 0.88
Range: 3,540 km

Falcon E

Improved Falcon D.

Power-plants: 2 x 1,960 kgp GE CF700 2D2
Weights: 7,160/13,000/1,632 kg
Range: 3,520 km

Freighter conversion with minimal empty weight
increased by 43 kg.   Payload of 1,500 kg, hold capacity
6.65 cu m.

Falcon F

(Initially Falcon 70.)   Much-modified version
(prototype c/n 173 F-WLCU).

Power-plants: 2 x 2,040 kgp GE CF700 2D2 with optional
                                            reverse
Weights: 7,245/13,000/1,505 kg initially
Range: 3,580 km
Maximum cruise speed: TAS 862 km/hr, 465 kt

As the Fan-Jet Falcon is adequately covered in
several other specialist publications, details are
included only of aircraft used by airlines, in keeping
with the theme of the monograph as a whole.

1   Operated on scheduled passenger lines

TAT (for Air France)

F-BTMF c/n 184/462 owned by Locaero, bought from Europe
     Falcon Service 1974.   Sold to Fokker-VFW Jul 76
as D-COMF (later F-BTMF/F-GAPC).

F-BUFG c/n 175 owned by Locaero, bought from Avions
     Marcel Dassault (ex F-WMKF/N4373F/N866MM) 01 Aug
73.   Sold to Fokker-VFW Jul 76 as D-COFG (later F-BUFG/
F-ODHA).

2   Operated on scheduled freight lines

Federal Express

N1FE   c/n 84 "Karen".   Purchased Mar 72.   Ex F-WJMK/
        N975F/N530L.

N2FE   c/n 132 "Cheryl".   Purchased May 72.   Ex F-WMKG/
        N4348F/N560L.

N3FE   c/n 151 "Shannon".   Ex F-WMKI/N4360F/N810F/
N810PA.

N4FE   c/n 108/430 "Miss Chi".   Purchased Jun 72.   Ex
        D-CBAT/N5CA.

N5FE   c/n 20 "Traci".   Purchased May 72.   Ex F-WMKJ/
        N842F/N367G/N367GA.

N6FE   c/n 50 "Michelle".   Purchased May 72.   Ex F-WNGO/
        N879F/N804F/N6565A.

N7FE   c/n 46 "Lisa".   Purchased Aug 72.   Ex F-WMKG/
        CF-ESO/N23555.

N8FE   c/n 199 "Wendy".   Purchased May 72.   Ex F-WMKH/
        N4388F.

N9FE   c/n 216 "Laurie".   Purchased Jun 72.   Ex F-WLCT/
        N4402F.

N10FE   c/n 16 "Audrey".   Purchased May 72.   Ex F-WNGL/
        N807F/N354H.

N14FE   c/n 227 "Ann Marie".   Purchased Oct 72.   Ex
        F-WMKG/N4410F.

N15FE   c/n 229 "Donna".   Purchased 1973.   Ex F-WJMJ/
        N4411F.

N16FE   c/n 230 "Laura Jane".   Purchased Oct 72.   Ex
        F-WJML/N4412F.

N17FE   c/n 232 "Sonya".   Ex F-WJMN/N4413F.

N18FE   c/n 233 "Polly".   Purchased Apr 73.   Ex F-WLCV/
        N4414F.

N20FE   c/n 235 "Sal".   Purchased Apr 73.   Ex F-WPXJ/
        N4415F.

N21FE   c/n 226 "Angela".   Purchased Oct 72.   Ex F-WPXI/
        N4409F.

N22FE   c/n 223 "Holly".   Purchased Apr 73.   Ex F-WPUX/
        N4407F.

N23FE   c/n 224 "Kristine".   Purchased Apr 73.   Ex
        F-WPUY/N4408F.

N24FE   c/n 220 "Tammy Elaine".   Purchased Apr 73.   Ex
        F-WPUU/N4404F.

N25FE   c/n 221 "Patricia".   Purchased Apr 73.   Ex
        F-WPUV/N4406F.   DBR 27 Sep 75 at Green Airport,
RI.

N26FE   c/n 204 "Hope".   Purchased Apr 73.   Ex F-WMKI/
        N4392F.

N27FE   c/n 207 "Jennifer Jay".   Purchased May 73.   Ex
        F-WMKF/N4395F.

N28FE   c/n 209 "Janet".   Purchased Apr 73.   Ex F-WLCX/
        N4396F.

N29FE   c/n 210 "Adina".   Purchased Apr 73.   Ex F-WPXF/
        N4397F.

N30FE   c/n 211 "Marianne".   Purchased Apr 73.   Ex
        F-WJMK/N4398F.

N31FE   c/n 212 "Colleen".   Purchased Apr 73.   Ex F-WPXG/
        N4399F.

N32FE   c/n 213 "Sarah".   Purchased May 73.   Ex F-WJMM/
        N4390F.

N33FE   c/n 214 "Stacy".   Purchased Apr 73.   Ex F-WNGO/
        N4400F.

N34FE   c/n 215 "Kellie".   Purchased Apr 73.   Ex F-WLCS/
        N4401F.

N35FE c/n 217 "Melanie".   Purchased Apr 73.   Ex
F-WLCY/N4403F.

N36FE c/n 218 "Becky".   Purchased Apr 73.   Ex F-WMKJ/
N4372F.

N37FE c/n 270 "Theresa".   Purchased Apr 73.   Ex
F-WPUZ/N4435F.

N30JM c/n 24.   Lsd Sep 72 from Arkansas Aviation.   Re-
regd N2255Q, retd Jun 73.   Ex F-WNGM/N845F/
N297AR).   Allocated N13FE but NTU.   Later N738RH/N6OSM.

FEC-14 c/n 198, del Oct 72.   Ex F-WNGO/VR-BDK.   Sold
Oct 72 to Arkansas Aviation as N74196.   Later
XC-BIN.

3   Operated by the business division of scheduled air-lines

Iraqi Airways

YI-AHH c/n 337/529.   Ex F-WRQR.

YI-AHI c/n 342/532.   Ex F-WRQP.

YI-AHJ c/n 343/533.   Ex F-WRQR.

Japan Air Lines (training division)

N131JA c/n 282.   Ex F-WMKG/N4436F.   Sold to IASCO 1977.

N132JA c/n 284.   Ex F-WPXM/N4437F.   Sold to IASCO 1977.

N133JA c/n 290.   Ex F-WMKH/N4440F.   Sold to IASCO 1977.

Kingdom of Libya Airlines/Libyan Arab Airlines

5A-DAF c/n 128/436, del 20 Feb 68.

5A-DAG c/n 143/442, del 15 May 68.

5A-DAH c/n 190/465, oper by military as 002, c/s 5A-DAH.

Air Nauru

VH-BIZ c/n 73/419.   Lsd 1967/72 from Business Jet Pty.
Ex F-WJML.   Sold as 9Q-CKZ.   The only Falcon CC
built.

4   Other airlines

Aero Leasing

HB-VBL c/n 126/438, del 21 Apr 68.   Oper by B Cornfeld.

HB-VBO c/n 150/445, del 29 Jan 69.   Sold to C Plane SA.

HB-VDG c/n 58.   Purchased Jul 74.   Ex F-WNGL/N884F/
N600KC/F-BTQZ.

HB-VED c/n 162/451.   Purchased 1975.   Ex F-WNGO/OO-WTB/
D-CBBT.

Air Taxi

EP-AHV c/n 320/519.

Balair

HB-VAV c/n 3/403.   Ex F-WMKG/VR-BCG.   Lsd 197? for UNO
service.

Euralair

F-BOXV c/n 104/454.   Joint operation with EFS since
1967.

F-BRPK c/n 188/464.   Ex F-WJMK/F-BRPF.   Lsd for a while
from EFS.

F-BTQZ c/n 58.   Ex F-WNGL/N884F/N600KC.   Purchased from
EFS 1973.   Sold to Aero Leasing 1974 as HB-VDG.

Fred Olsen

LN-FOD c/n 53/417.   Ex F-BNRE.   Del 26 May 69.   Sold
to Royal Norwegian AF 15 Dec 72 as 053.

LN-FOE(1) c/n 62/409.   Ex F-BOLX.   Purchased Jan 73.
Damaged 12 Dec 73 at Norwich and sold to
California Airmotive Corp for repair as N17401.

LN-FOE(2) c/n 125.   Ex F-WJMN/N4344F/N6810J/N812PA.
Regd 07 May 74.

LN-FOI c/n 41/407.   Ex F-WNGL/SAAF431/F-BOED.   Del 12
Jan 70.   Sold to Royal Norwegian AF 15 Dec 72 as
041.

SATA

HB-VCO c/n 25/405.   Ex F-WNGN/F-BOON.   Purchased May 72.
Sold 23 Dec 75 to IGN as F-BSYF.

CORVETTE

In 1968 Sud Aviation was designing a multi-purpose rear-engined twin-jet aircraft in collaboration with Nord Aviation, and the Sud-Nord prefix SN was adopted.   The prototype of the new aircraft, called the SN-600 Diplomate, was flown in July 1970 as F-WRSN.   The aircraft was destroyed in 1971 during stalling trials at the CEV.

Meanwhile, extensive redesign took place, giving birth to the production SN-601 Corvette.

which had a much longer fuselage with accommodation planned for up to 12 passengers.   In view of prospective sales, long leadtime items were ordered for production at St-Nazaire even before the first flight of aircraft no 1 (F-WUAS) had taken place.   The new business jet, with as many passengers for half the weight of a Mystère 20, seemed particularly attractive in the manufacturer's view, and sales to third level airlines were anticipated.   Another major point was low fuel consumption, some 40% less than on comparative aircraft such as the Mystère 20, HS.125 or Lear-Jet.

The first production Corvette was rolled out in October 1972 and was flown on 20 December, being described by Aérospatiale as an SSE, these three letters standing for Silencieux-Spacieux-Economique (quiet, capacious and economical) - doubtless a private allusion to the SST Concorde.   The registration F-WSSE was reserved but not taken up.

As had been the case with Dassault's Falcon jet, American sales were immediately sought. In 1973 an optimistic order for 70 aircraft by the Atlanta-based US Corvette was announced, while Aérospatiale could only boast six firm orders (with a few more on option).   At the Salon du Bourget the aircraft was renamed Corvette 100 and also offered as a 14-seater for third-level operation.   A proposed 16 to 19-seater stretched version, the Corvette 200, was also announced (with an option for eight from Air Alpes).   Later in the year the American distributor for the type became Air Center at Wiley Post, near Bethany, Oklahoma, six aircraft reportedly being sold from the start.   In the event, the American demonstrators had to be returned to France later with no sales being concluded.

Demonstration tours were organised; to Africa in June and July 1974 (with c/n 4), and to Latin America in July 1975 (c/n 13), but as more and more aircraft were being produced, it became clear that the situation regarding sales was worsening.   In July 1975 the contract for American distribution was amended, but the new organisation, Aérospatiale Aircraft Corporation, also failed to obtain any sales.

Reluctantly the lack of sales success was admitted, and production was terminated at 40 aircraft.   By February 1977, three years after manufacture had started, only four Corvettes had been sold, and fourteen were operated by leasing companies (with some penalty to SNIAS). Thirteen more were being leased by Aérospatiale under a somewhat uncertain scheme, and the situation was hardly improved by the end of 1977.   An analysis of the Corvette failure published in July 1977 by the French administr-ation stated that it accounted for 66% of the total SNIAS losses for 1972-75 (more than Concorde therefore), some 85% of the expenditure programme being lost (900 million French Francs).

Apparently this dramatic blunder was attributed to early misjudgment of a highly competitive but uncertain market and miscal-culation of initial market research.   It can be said however that the Corvette was a fine aircraft for the third-level airlines, as exemplified by its chequered operation in France by Air Languedoc, Air Alpes and Air Alsace, including its use as a replacement for the Mystère 20.   Corvettes were also flown on international routes such as Marseilles-Milan and Marseilles-Barcelona by Air France/Air Alpes.   Air Alsace flew from Strasbourg to a few major European cities including London.

The future will determine whether the Corvette really was an unlucky aircraft when operation of the present fleet still under SNIAS ownership can be considered in perspec-tive.   A further attempt at promoting US sales was made during 1979 with the appointment of Midwest Air Charter and Air National as US distributors, and one demonstrator was flown to the US.

Specification

Span: 12.8m (13.24m with optional wing-tip tanks)
Length: 13.82m
Height: 4.23m
Wing area: 22 sq m
Empty weight: 3,622 kg
Basic operational weight: 3,812 kg
Maximum weight: 6,100 kg
Power-plants: 2 x 1,048 kgp UACL JT15 D-4 turbofans
Maximum cruising speed: 780-800 km/hr depending on
                     weight and altitude
Economical cruising speed: 620 km/hr at 11,000m
Range with maximum payload: 1,465 km or 2,125 km (six
                     passengers) with wing-tip
                     tanks

Accommodation was for crew of one or two on flight deck.   Basic executive version is for six passengers. Third-level arrangement provides for 8-12 passnegers. An improved version was planned as the Corvette 150 with a slightly stretched fuselage, resulting in an increase in baggage hold volume and up to 14 seats in third-level airliner configuration, but development did not proceed.

Production

C/n 01 SN-600 F-WRSN.   F/f 16 Jul 70.   Written off 23 Mar 71 at Istres.

C/n 1 F-WUAS.   F/f 20 Dec 72.   To SNIAS.

C/n 2 F-WRNZ.   F/f 07 Mar 73.   To F-BRNZ SNIAS (CofA 24 Dec 75).

C/n 3 F-WUQN.   F/f 09 Nov 73.   To F-BUQN SNIAS (CofA 31 May 74).   Retd to F-WUQN circa 1975.

C/n 4 F-WUQP.   F/f 12 Jan 74.   To F-BUQP SNIAS (CofA 30 Apr 74).   To Slibail/Air Alsace, del 11 Dec 74, named "Riesling".

C/n 5 F-BVPA.   F/f 30 Aug 74.   To Air Alpes, del 10 Sep 74.   To SNIAS May 75.   To COGESAT/CAS Jul 76.   To SNIAS 1977.   To F-ODJX 1979.

C/n 6 F-WUQR.   F/f 22 May 74.   To F-BVPB Air Alpes (Air France colours), del 16 Sep 74.   To SNIAS May 75.   To COGESAT/CAS Jul 76.   Lsd to TAT 1977 (still in Air France colours Feb 77).   To SNIAS 1977.   To Uni Air, del 05 Apr 78.

C/n 7 F-OBZR.   F/f 24 Jul 74.   To Air Center, del 15 Aug 74.   Re-regd N611AC 1974/75 (23 Jan 75?).   To F-BVPK SNIAS (CofA 29 Jun 76).   To SFACT, del 14 Dec 77.

C/n 8 F-WPTT.   F/f 17 Oct 74.   To 6V-AEA Africair/ASECNA, del 20 Dec 74.

C/n 9 F-WRQK.   F/f 11 Dec 74.   To F-BRQK SNIAS (CofA 20 Jan 75).   To N612AC Air Center Apr 75.   To F-BTTR SNIAS (CofA 18 Nov 76).   Lsd to Nile Valley Airlines, del 15 Jan 77.   To F-OCRN Aero Service, del 21 Oct 77.   Re-regd TN-ADI soon after.

C/n 10 F-BVPO.   F/f 20 Dec 74.   To SNIAS/Air Alsace, del 31 Dec 74.   To Coopamat/still Air Alsace May 75, named "Gewurztraminer".

C/n 11 F-WIFU or F-BIFU.   F/f 15 Apr 75.   To N613AC Air Center.   To F-BTTS(1) SNIAS but regn NTU. To TR-LWY Air Inter Gabon, del 21 Jan 77.

C/n 12 F-BVPC.   F/f 11 Jan 75.   To SNIAS/Air Alpes, del 04 Feb 75.   Lsd to Air Entreprise Jun 75.   Lsd to Air France 27 Jul 75.   Lsd to Air Languedoc 1976?   To COGESAT/CAS Jul 76.   To SNIAS 1977.   To   del 06 Jan 78.

C/n 13 F-BVPD.   F/f 24 Jan 75.   To SNIAS/(Air Alpes NTU).   To COGESAT/Air Languedoc, del 29 Nov 75. Lsd to Air Inter Gulf Nov 75 (or 10 Feb 76?).   To COGESAT/CAS 30 Nov 76?

C/n 14 F-BVPS. F/f 26 Feb 75. To Air Languedoc, del
01 Apr 75. To SNIAS/still Air Languedoc Aug 75.
To COGESAT/still Air Languedoc Nov 75. To COGESAT/CAS.
To SNIAS.

C/n 15 F-WIFA. F/f 19 Mar 75. To SE-DEN Malmros
Aviation, del 04 Apr 75. To SNIAS Apr 78.

C/n 16 F-BVPT. F/f 24 Apr 75. To Air Languedoc (CofA
16 May 75). To COGESAT/still Air Languedoc Nov
75. Lsd to Aero Service International, del 31 Jan 76.
To COGESAT/CAS 1976 (still in ASI colours Oct 77). To
SNIAS Feb 78.

C/n 17 F-WNGQ. F/f 21 May 75. To N614AC Air Center,
del 03 Jun 75. To F-BTTM SNIAS, del 14 Jun 76.
To Sécurité Civile, del 20 Jan 78.

C/n 18 F-WNGR. F/f 18 Jun 75. To N615AC Air Center,
del 29 Jun 75. To F-BTTO SNIAS (CofA 05 Oct
76). Lsd to TAT, del 20 Apr 77.

C/n 19 F-BVPL. F/f 03 Jul 75. To SNIAS/Air Languedoc.
To COGESAT/still Air Languedoc Nov 75. To
COGESAT/CAS. To F-OCJL SNIAS/Air Algerie (CofA 11 Jul
77). To F-BVPL SNIAS late 1977. Lsd to Jetstar, del
28 Apr 78.

C/n 20 F-WNGS. F/f 22 Aug 75. To N616AC Air Center,
del 30 Sep 75. To F-BTTN SNIAS, del 14 Jun 76.
Lsd to TAT, del 02 Apr 77.

C/n 21 F-BVPE. F/f 03 Oct 75. To Air Alsace, del 30
Oct 75 (lsd to Air France). To SNIAS? To
Sterling as OY-SBS 1978.

C/n 22 F-WNGT. F/f 18 Nov 75. To N617AC Air Center
del   . To F-BTTU(1) SNIAS, del 12 Jun 76,
regn NTU. To F-ODFE SNIAS/Aero Service, del 12 Feb 77.
Re-regd TN-ADB late 1977.

C/n 23 F-BVPF. F/f 04 Dec 75. To Air Alsace, del 23
Dec 75 (lsd to Air France). To Air Bail/still
Air Alsace Aug 76.

C/n 24 F-BVPI. F/f 01 Apr 76. To SNIAS. Lsd to
DLT, del 19 Aug 76. To Uni Air, del 10 May 78.

C/n 25 F-WNGU. F/f 10 Jan 76. To F-BVPG SNIAS (CofA
16 Jan 76). Lsd to Air Inter Gulf 1976. To
COGESAT/still Air Inter Gulf Jul 76. To SNIAS Feb 78.
To F-OBZV reserved early 78 but NTU. To Uni Air, del
28 Apr 78.

C/n 26 F-WNGV. F/f 26 Jan 76. To N618AC Air Center,
del 09 Feb 76. To SNIAS, del 12 Jun 76. To
F-ODFQ Jetstar, del 01 Apr 77. Re-regd PH-JSB Jul 77.

C/n 27 F-BVPH. F/f 26 Feb 76. To SNIAS. Lsd to
Aero Service International, del 09 Apr 76. To
COGESAT/CAS Aug 76. To SNIAS 1978.

C/n 28 F-WNGX. F/f 15 Mar 76. To F-BTTL SNIAS (CofA
03 Sep 76). Lsd to Uni Air, del 20 Jan 77.

C/n 29 F-WNGY. F/f 23 Apr 76. To F-BVPJ SNIAS (CofA
10 May 76). Lsd to Air Inter Gulf 1976. To
F-OBZP COGESAT/still Air Inter Gulf (CofA 28 Oct 76).
Later lsd to Air Algérie. To F-BVPJ SNIAS late 1977.
Retd to F-OBZP (CofA 08 Jan 79), still owned by SNIAS,
oper by Air Benin.

C/n 30 F-WNGQ. F/f 30 Aug 76. To F-BTTP SNIAS (CofA
08 Nov 76). Lsd to DLT, del 01 Apr 77. To
OO-MRC Hessenatie, del 02 Jun 78.

C/n 31 F-WNGZ. F/f 04 Jun 76. To F-BTTK SNIAS (CofA
17 Jun 76). Lsd to TAT, del 16 Jan 77.

C/n 32 F-WNGR. F/f 06 Oct 76. To F-BTTQ SNIAS (CofA
14 Oct 76). Lsd to Scanfly, del 02 Jun 77. Re-
regd OY-ARA Jun 77. To SE-DED Corvette K/S Jul 78.

C/n 33 F-BTTT. F/f 14 Dec 76. To SNIAS. Lsd to TAT,
del 07 Jan 77.

C/n 34 F-WNGS. F/f 18 Feb 77. To F-BYCR CAS/Scanfly,
del 06 Oct 77. Re-regd OY-ARB. To SE-DEE
Corvette K/S Jul 78.

C/n 35 F/f 02 Jun 77. To PH-JSC Jetstar, del 16 Jun 77.

C/n 36 F-BTTS(2). F/f 08 Nov 77. To SNIAS (CofA 24
Apr 78). Canc late 1978.

C/n 37 F-BTTU(2). F/f 24 Oct 77. To SFACT, del 29 Nov
77.

C/n 38 F/f 01 Feb 78. To 5A-DCK Libyan Govt, del 05 Jun
78.

C/n 39 F-WNGY. F/f   . To F-OBYG Oct 78. To
Emperor Bokassa 1er as TL-SMI (CofA 24 Jan 79).

C/n 40 F/f   .

This chapter deals with four types of special interest. One was the French licence-built Grumman Widgeon - an unorthodox private venture of the late forties - and this was soon followed by another attempt at French manufacture of a foreign design, involving the VEMA 51 version of the Italian Macchi MB-320.

The other two ventures were closely connected with the development of the successful Turboméca range of turboprop engines, and this aspect of French aviation development deserves a short introduction of its own.

Shortly after the end of World War II Turboméca, a small firm established at Bordes in the Pyrénées mountains, began pioneering small gas turbines. Gradually development and sales became increasingly successful, and a wide variety of turbojet and turboprop power-plants have since been delivered all over the world. The range of aircraft and helicopters currently being powered by Turboméca or foreign licence-built engines is very considerable, and statistics from late 1977 gave 13,595 Turboméca engines in use all over the world, of which 14% were fitted to foreign designs.

Back in the mid-fifties it had been found necessary to test-fly the Turboméca turboprop engines. A subsidiary of Sud-Aviation, the Bordeaux-Mérignac based SFERMA, began conversions of piston-engined aircraft, working closely with Turboméca. First, in 1958, came the Beech-SFERMA PD-18S with the co-operation of Beechcraft. Despite hopes of attracting large orders, the re-engined Beech 18 remained in prototype form only.

Other SFERMA conversions followed. In 1959 the Max Holste MH-153 and Nord 1110 appeared (these were Astazou-powered MH-152 and Nord 1101 airframes respectively), and these were followed in 1961 by a Dornier 27 and Pilatus PC-6A Porter also converted to Astazou turboprop operation. The year 1963 saw the participation of the firm in the power-plant development of the Short Skyvan. But, as far as this monograph is concerned, the most interesting project was the Beech-SFERMA PD-146 Turbo Travel Air/Marquis conversion which led to a small production batch of Astazou-powered Barons in St-Nazaire. There were hopes that the Beech-SFERMA agreements would materialise into large production orders, but again, as with the PD-18S, development of the production variant was soon stopped. The Marquis may be considered to be more of an executive aircraft than a small transport air-craft for regular commercial operation, but its inclusion in this survey of French transport aircraft serves to illustrate a notable facet of post-war development.

SFERMA was disbanded in January 1965, being taken over by Sud-Aviation and later re-emerging as SOGERMA. While on the subject of SFERMA/SOGERMA conversions, mention must be made of a single SE-2010 Armagnac transport converted into the SE-2060 flying test-bed for use by the CEV (1962-65) and also of the Caravelle engine test-bed first fitted with a SNECMA M53 jet engine (in 1973) and with a CFM-56 in 1977.

SCAN-30

When the Grumman G-44 Widgeon made its maiden flight on 28 June 1940 it had been designed to some extent as a scaled-down Goose amphibian mostly for commercial use. But, understandably, production until 1945 was for military use. Then, after the end of the war, the hull was improved and as the G-44A 76 amphibians were manufactured by Grumman for civil customers.

At the same time a small French firm at La Rochelle, the Société des Constructions Aéro-navales (SCAN), was producing some 30 small, pusher flying-boat trainers for the French Navy called the SCAN-20, which had been designed for use during the war. With that limited flying-boat experience, SCAN embarked upon the manufacture under licence of the Grumman Widgeon, a production batch of some 40 being launched (the exact number has not been ascertained but was between 37 and 42 airframes) again demonstrating the considerable optimistm customary in the French post-war aeronautical industry.

With deliveries of the power-plants delayed, the whole batch had been manufactured by the time the SCAN-30 prototype was flown. Three versions were offered, differing in the engines installed: the SCAN-30 and SCAN-30A were fitted with French Salmson 8 AS-00 power-plants. The SCAN-30/1 (also referred to as SCAN-30G) first flew on 14 May 1953 with Gipsy Queen II engines, and the SCAN-30L had Lycoming GO-435 C2 power-plants. As early as 1947 there were talks with the French Navy for an order of SCAN-30s fitted with French Mathis G8R engines. Evaluation of the type was in fact undertaken with a G-44 (c/n 1297) registered F-BENK (later F-BDAE with Air Monaco) but retaining the original US power-plants.

Soon after its appearance in prototype form in 1949 the SCAN-30 hit the headlines when Mme Jacqueline Auriol, the daughter-in-law of the then President of the Republic, was seriously injured in a crash at Les Mureaux on the River Seine. (She was later to recover and become a popular figure during the fifties as the only woman test pilot and broke several world records in jet fighters.) Late in 1950 a demonstration tour to French West Africa was set up for the SCAN-30 (with two other light transports, the Morane-Saulnier MS-700 and the Castel-Mauboussin CM-100), but no orders were obtained. As there were little hopes of selling an amphibian in continental France, efforts were made to sell the SCAN-30 overseas, mostly in Africa, French Indochina and French Guiana.

By 1952 planned sales for the still unsold airframes were as follows: twelve to French Guiana, where conditions for operating such a type were ideal, ten to a commercial operator in Africa, three to France Hydro (see the Latécoère 631 story), five to various flying schools overseas and seven for off-the-shelf customers.

But it was soon learned the hard way that there were no prospective firm customers. Only a few SCAN-30s were sold from the large stock produced, all supported by the Government for administration use overseas. By February 1952 33 airframes were put up for sale by the Public Auction of the French Administration at La Rochelle, a number reduced to 21 in 1954 (consisting of 20 Lycoming-powered SCAN-30s for USA sales and a single Gipsy Queen II-powered SCAN-30G which was soon to appear on the UK Register). In that same year only six SCAN-30 amphibians were current, three in Guiana and three in Indochina (including one presented to Emperor Bao-Dai of Vietnam in 1950/51 and reportedly not taken up - see c/n 7).

Unexpectedly, a solution to the problem was found during the mid 1950s when the unused airframes were purchased in the United States. In February 1952 15 airframes were sold to a French-American company named Intercontinental Aviation, but no further details of this company have been traced. Then, a few years later, one or possibly more companies were involved in converting SCAN-30s to Super Widgeons or Super SCANs. The main company involved in the conversion was the Pacific Aircraft Engineering Corp (PACE) who, in 1958-59, engineered the conversion of about 15 SCANs to 300 hp R-680E radial engines and dubbed the resulting aircraft the PACE Gannett. Lee Mansdorf, who was also connected with PACE at one time, would appear to have carried out some conversion work prior to this in about 1955-56. It is likely that the US-registered aircraft with the -LM suffix were in fact SCAN-30s converted by his company. In addition, there may be a connection between Lee Mansdorf and the reported purchase of 24 airframes by a Gannett Co.

The modifications involved in the PACE Gannett SCAN-30 conversion included the re-building of the hull with five watertight compartments, a new undercarriage uplock mechanism, escape hatch, new hydraulic and electrical systems, the removal of the trailing edges of the wings and their replacement by new ribs, new all-metal, flush-riveted wings with 25 US gallon auxiliary fuel tanks in the wings. In 1959 the PACE Gannett conversion cost $89,950, including a standard instrument panel and electronics.

Further conversion work was undertaken on SCAN-30s during the 1950s, particularly with different engines such as the 270 hp Lycoming GO-480-B1 and the 220 hp Continental W-670 radial. At least two SCAN-30s powered by Lycoming GO-435-C2Bs were fitted with Hartzell three-blade propellors to improve take-off performance.

It is of interest that a note in the FAA type certificate for the Grumman-built G-44A series contains the following specific reference to the imported SCAN-30 aircraft:

"French SCAN-30 aircraft, serial numbers 2, 3, 4, 9 through 16, 20 through 23, 25 through 37, and 41, manufactured by Société des Constructions Aéronavales, under license to Grumman Aircraft Engineering Corp, are eligible for certification when accompanied by a certificate from the French Bureau Veritas.... Aircraft of this type, other than the serial numbers listed above, must be certificated and imported under CAR 10."

This would suggest that at least 29 aircraft were originally imported to the USA.

The major differences between the original SCAN-30 and the Grumman G-44A Widgeon were that the French-built examples had non-anodised hulls (resulting in much corrosion), non-aircraft quality control cables and, of course, metric dimensions and fasteners.

There have been reports in the last few years of one or more SCAN-30s still in crates in the USA, but unfortunately confirmation of this has not been received.

Specification

High-wing cantilever twin-engined amphibian. Operation with one pilot and three passengers. Figures in brackets refer to PACE conversion. SCAN-30 with Salmson engines.

Span: 13.15m
Length: 9.52m
Height: 3.77m
Wing area: 22.3 sq m
Empty weight: 1,600 kg (unknown)
Loaded weight: 2,350 kg (2,495 kg)
Disposable load: 280 kg (unknown)
Power-plants: 2 x 240 hp Salmson 8 AS-00 (300 hp Lycoming R-680E)
See also text for SCAN-30G with 2 x 200 hp Gipsy Queen II and SCAN-30L with 2 x 260 hp Lycoming GO-435-C2 variants
Maximum speed at sea level: 258 km/hr (304 km/hr)
Cruising speed: 230 km/hr (270 km/hr at 66% power at 2,450m)
Range: 750 km (1,600 km)
Service ceiling: unknown (2,450m)
Rate of climb at sea level: unknown (565 m/min)

Production

C/n 01   F-WFDM.   Probably written off 11 Jul 49 at Les Mureaux.

C/n 1   F-WFHA.   Believed f/f 30 Jul 49.

C/n 2   F-WFHB.   To Intercontinental Aviation Feb 52. Later N63LM. Sold as AP-AMW, regd Feb 62 to Govt of Pakistan. No longer current.

C/n 3   F-BFHC (reservation).   To Intercontinental Aviation Feb 52. Later N62L (mfg "1954", Lycoming R-680E) Lee County Land & Title Co (CAR 6/63). To Lehigh Acres Development Corp (CAR 7/64). To Ned Rice (CAR 1/66). Current.

C/n 4   F-BFHD. (Mfg 1949) SCAN (CofA 21 Jul 50). To Intercontinental Aviation Feb 52. Cvtd to SCAN-30L and restored (CofA 09 Feb 55) to Mrs M T Sanstadt (wife of the Intercontinental manager). Accident end May 55 in Bieler See, Switzerland, while operated by Mr Farner. To N4732V, later N57LM, described as a "Super SCAN - one of the first re-engined SCANs with Continental W-670 radials" in Jun 56. Re-regd N115WB to Bestone Inc (CAR 6/63 - mfg 1954, Lycoming R-680E). To Shebago Lake Shores Inc (CAR 1/70). To Air Journeys Inc (CAR 7/71). To Incon Corp (CAR 7/73). To Stanley C Hewitt (CAR 1/77).

C/n 5   F-BFHE (built 1949).   To SCAN (CofA 05 Oct 50). Based at Saigon? Canc Nov 51.

C/n 6 F-BFHF (built 1950). To SCAN (CofA 17 Oct 50).
       Lsd to Service de Santé (Health Dept) in Indo-
china. Written off 25 Jul 51 at Cap St Jacques.

C/n 7 F-BFHG (built 1950). To SCAN (CofA 13 Jan 51).
       Based Saigon. Oper by Mr Boyeaux (probably for
Emperor Bao-Dai). Canc Oct 54. Wreck still at Tan
Son Nhut Aug 57.

C/n 8 F-BFHH (reservation).

C/n 9 To Intercontinental Aviation Feb 52. Regd N62G.
       Canc prior CAR 6/63.

C/n 10 Imported to USA and regd N3923. Canc prior CAR
        6/63.

C/n 11 Imported to USA and regd N7918C. Canc prior CAR
        6/63.

C/n 12 Imported to USA and regd N7911C (mfg "1951",
        Lycoming R-680E). Believed to be first Gannett
conversion. To James Martin (CAR 6/63). Current.

C/n 13 Imported to USA and regd N3924. Canc prior CAR
        6/63.

C/n 14 Imported to USA and regd N58LM. Sold to Canada
        1959 as CF-LFQ (Lycoming R-680E). Regd to
Irenée Garant, Montmagny, PQ (CCAR 3/66). To Marine
Industries Ltd (CCAR 3/68). To Riverton Airways Ltd
(CCAR 6/72). Resold to USA 1975 and regd N48011 to
Allen Enterprises, Alexandria, MN (CAR 7/75).

C/n 15 Imported to USA and regd N60LM. Sold to Canada
        1959 and regd CF-MLC. Regd to Airexec Services
Ltd, Fort St John, BC (CCAR 3/66). Retd to USA as
N7913L. No trace in CAR or CCAR after 1966.

C/n 16 Imported to USA and regd N7912C (mfg "1954",
        Lycoming R-680E). To PACE (CAR 6/63). Sold
to N C Swanson (CAR 1/64). To National City Bank of
Denver (CAR 1/65). To Ohio Valley Aviation (CAR 7/66).
To Newhall Enterprises Inc (CAR 7/68). To Cincinnati
Air Taxi (CAR 1/70). To Marion Burke (CAR 1/72).
Current.

C/n 17 No trace.

C/n 18 No trace.

C/n 19 F-BGTD (mfg 1949). To Aéro Club de France, Buc
        (CofA 23 Jun 53) after conversion to SCAN-30G by
Paul Albert & Legastellois at Buc. Flight tested in UK
by de Havilland. To Soc Technicoptere, Toussus, 1955.
Sold to Lake Air Charters Ltd, Mwanza, as VP-KNV. Sold
to Irish Helicopters Ltd 30 Sep 59 as EI-ALE. Resold
to Bruce Campbell Ltd, Hamble, 01 Feb 60 as G-ARIX.
Written off landing in Southampton Water 19 May 61.

C/n 20 To Intercontinental Aviation Feb 52. Regd
        N7917C and subsequently N68596. Fate unknown.
Canc prior CAR 6/63.

C/n 21 To Intercontinental Aviation Feb 72. Regd
        N7775C (mfg "1964"!, Lycoming R-680E). Regd to
Lee Mansdorf (CAR 7/64). Struck submerged object on
take-off on training flight nr Lake Havasu, AZ, 03 Jul
66. Repaired. Sold to Marvin Tait (CAR 1/71) and to
John Gibbs (CAR 1/75). Current.

C/n 22 To Intercontinental Aviation Feb 52. Imported
        into USA but no details known.

C/n 23 To Intercontinental Aviation Feb 52. Imported
        into USA but no details known.

C/n 24 To Intercontinental Aviation Feb 52. No details
        known.

C/n 25 To Intercontinental Aviation Feb 52. Regd

N7913C (Fairchild 6-440). Regd to PACE (CAR
1/64). Sold to Y S de Yoreo (CAR 7/64). Current.

C/n 26 Imported to USA and regd N2810D (Lycoming GO-480).
        To Reading Eagle Co (CAR 6/63). Sold in Canada
1965 as CF-SPA. Regd to Wright Industrial Equipment Ltd
(CCAR 3/66). Struck sand bar on take-off nr Coppermine,
NWT, 24 Jun 66. Substantial damage sustained and
probably written off, as missing from subsequent CCARs.

C/n 27 Imported to USA and regd N2811D. Reregd N50G
        prior 6/63. Regd N58Q to Rochester Enterprises
Inc (CAR 6/63, Lycoming GO-480). Sold to Eaton Lumber Co
(CAR 7/68) and to Valley Aircraft Co (CAR 1/70). Written
off 13 Jul 69 when aircraft porpoised on glassy water nr
Barrow, AK.

C/n 28 Imported to USA and regd N2812D. Sold in Canada
        1956 as CF-ODR. Regd to Ontario Dept of Lands &
Forests (CCAR 3/66). Sold to Lorne Corley (CCAR 6/72).
Current.

C/n 29 Imported to USA and regd N2813D (mfg "1955",
        Lycoming GO-480). Regd to Graubart Aviation Inc
(CAR 6/63). To North County Management Co (CAR 1/65).
Current.

C/n 30 Imported to USA and regd N2814D. Owned by
        Louisiana Dept of Wildlife & Fisheries from about
1956 until sold to Ted Voorhees (CAR 1/77).

C/n 31 To Intercontinental Aviation Feb 52. Imported
        to USA and regd N7921C. To N4453 Lee Mansdorf
(mfg "1967"!, Lycoming R-680E) (CAR 7/68). Sold to
Executive Aero Inc (CAR 1/70). To Business Aviation
Services Corp (CAR 1/71). To Creature Enterprises Inc
(CAR 7/73). To J N Langley (CAR 1/77). Used in TV
series "Fantasy Island" under Creature Enterprises ownership.

C/n 32 To Intercontinental Aviation Feb 52. Regd
        N7916C (mfg "1954", Lycoming R-680E). To
Honorbuilt Trailer Co (CAR 6/63). To Arthur Eisele (CAR
7/64). To Morse Aero Inc (CAR 1/70). Accident at Miami
Seaplane base 08 Aug 71. Repaired. To Standard Bank &
Trust Co (CAR 7/73) and retd to Morse Aero Inc (CAR 1/77).
For sale in Jul 78 by Aircraft Ltd, now fitted with 300 hp
IO-520E engines, reversible two-blade propellors and
retractable floats. Only 100 hrs TT on airframe.

C/n 33 To Intercontinental Aviation Feb 52. Regd N4451
        (mfg "1967"!). To Great Circle Corp (CAR 7/68).
To Steve Jones (CAR 1/75). Current.

C/n 34 To Intercontinental Aviation Feb 52. Regd N4452.
        To Lee Mansdorf (CAR 7/68, mfg 1949). To
Executive Aero Inc (CAR 1/70). Accident at Ebeye,
Marshall Islands, 22 Jan 71 and again at Majuro, Marshall
Islands, 04 Apr 71. Written off 31 Jul 72 when it sank
off an unknown location in the Marshall Islands while oper
by Lagoon Aviation, still as N4452.

C/n 35 Imported to USA and regd N4120A (mgf 1956,
        Lycoming GO-480). To Anthony Stinis (CAR 6/63).
Re-regd N10BR to Robert Rehbaum.

C/n 36 F-OABR (mfg 1950). To Commissariat Général à la
        Productivité (CofA 27 Oct 55) after conversion to
SCAN-30G. Based Cayenne. Canc Sep 56. Sold in USA as
N4121A (probably 1956). Regd to Anthony Stinis (CAR
6/63), Lycoming R-680E. To Agnes Ilardi (CAR 7/67). To
Joseph Ilardi (CAR 1/71). Canc prior CAR 7/73.

C/n 37 To Intercontinental Aviation Feb 52. Imported
        into USA. No further details known.

C/n 38 F-OALL (build 1952, SCAN-30A). To Department de
        la Guyane (CofA 03 Oct 52). Canc late 1957.

C/n 39 F-OALM (built 1952, SCAN-30A). To SCAN (CofA 02
        Sep 52). Based Cayenne. Canc late 1956.

C/n 40 F-OALN (built 1952, SCAN-30A).   To Department de
la Guyane (CofA 01 Oct 52).   Canc late 1956.

C/n 41 Imported into USA probably 1956 and regd N4122A
(mfg "1956", Lycoming R-680E).   M Baumgartner
Dodge Inc (CAR 6/63).   To Gayle Aviation Inc (CAR 1/65).
To Lambros Seaplane Base (CAR 1/67).   To Gordon Newell
(CAR 7/67).   To Modern Furniture Co (CAR 1/69).   To
Flying Sportsman Inc (CAR 1/77).

Aircraft not identified: F-WFII (photograph in
"Flight" 06 Oct 49, page 489 - misprint for F-WH..?

Notes

1   The SCAN company, now involved in marine small boat
construction only, auctioned "two flying boats" for
scrap as late as 1976.   These were believed to be
SCAN-30 airframes, but details are still missing.

2   Farner SA was a major aircraft concern in 1946/47
based at Granges.   Mr Farner died in a SCAN-30
crash, believed to be that of c/n 4.

VEMA 51

The VEMA 51 was the 1951 French version of
the Italian-designed Macchi MB-320.   Six aircraft had been built in Italy, three of which
were to be operated by East African Airlines.
It was a small twin-engined transport with
accommodation for six.   Power-plants were two
185 hp Continental E-185s.

Jean Lignel, then manager of Air Algérie,
evinced interest in the type and, following
trials of an Italian-built machine at Brétigny,
home of the CEV, rights for licence manufacture
were obtained from Aer Macchi in 1951.   French
production was to be undertaken by the Société
Française de Construction Aéronautique (SFCA),
a small firm created in 1934 which had built
various Lignel-designed light aircraft.
Eventually the single VEMA 51 remained in
prototype form and was used for many years in
Algeria.   It was probably too small for
regular commercial use in numbers, and
demonstration flights, including some at Le
Bourget as late as 1955, failed to bring in any
customers.

Specification

This refers to the Macchi MB-320; the VEMA 51 was
similar.

Span: 13.0m
Length: 8.65m
Height: 3.19m
Wing area: not quoted
Empty weight: 1,490 kg
Loaded weight: 2,250 kg
Payload: 420 kg (including 45 kg luggage and 5
        passengers)
Maximum speed (at height): 322 km/hr (presume at sea
                           level)
Cruising speed (at height, 70% power): 285 km/hr at
                           2,000m

Take-off distance: 260m (unqualified)
Service ceiling: 5,200m
Range with maximum load at 260 km/hr at 2,000m: 1,000 km
Passenger configuration: five passengers plus pilot or
                         four passengers plus two crew

Production

C/n 01 Built 1950 (actually a Macchi-built MB-320, either
        c/n 5874 or 5875 ex I-RAIA or I-MACH).   Regd
F-OAHN.   To Air Algérie (CofA 09 Mar 51).   To F-BDHN
SNCASO (CofA 23 Apr 52).   To Soc de Liaisons et Transport
Aérien, Algiers, Aug 57.   To Aéro-Sud, Bone, Oct 59.
WFU Jun 63 at Algiers.

Two other Macchi MB-320s appeared on the French civil
register and are also worthy of mention:

C/n 5907 F-BBIM of Soc Jas Hennessy, Cognac (CofA 12 Feb
          54), ex I-DACA and VP-KIT, built 1951.   This was
also sold to Aéro-Sud Oct 59 for use with the single VEMA
51 in the transport role until WFU Dec 60 at Hassi Leila,
wreck being derelict at Hassi Messaoud (Sahara).

C/n 5908 F-OAJL, del 28 Oct 51 to Office des Bois du
          Gabon.   Sold to Air Kivu (?), Belgian Congo,
circa 1953.

SFERMA PD-18S

The origin of the SFERMA PD-18S goes back to
October 1957 when, during a visit to France,
the then president of the Beech Aircraft
Corporation evinced interest in the new Turboméca Bastan turboprop.   A re-engined version
of the Beech 18 was planned, but preliminary
studies in the USA soon proved to be too
costly; undertaking the conversion in France
was however found feasible, and a Beech-SFERMA
agreement was reached in December 1957.

The well-known French test-pilot and engineer J
Lecarme spent a few days in Wichita working with
Beech on the many technical problems involved.
Then, in January 1958, a small team under the
guidance of Lecarme and Anthonyssen began work
on re-equipping a Beech 18 with Bastans.   Six
months later the much-modified airframe was
ready;   discernible features included larger
fins, wing fences, new undercarriage and tailwheel and of course new engine nacelles, with

Ratier three-bladed propellors.   The Turboméca light gas turbines were fitted and tested during the summer of 1958, and the Beech-SFERMA PD-18S (PD referring to Project Design) was soon flown.

Costs were shared between Beech, SFERMA, Turbomeca and Ratier (with engines and propellors from a French Government order), and the aim of the whole programme was ostensibly the conversion of military piston-engined Beech 18s to Turboméca turboprops.   Firms in Canada, the UK and Italy showed interest in acquiring rights to re-engine the ubiquitous Beech aircraft on the same pattern, but no conversion programme was ever undertaken.   By mid 1960 three PD-18S aircraft had been flown; one ordered on behalf of the CEV for Bastan development, one as a SFERMA private venture to investigate French civil certification and commercial feasibility (with Sperry-SECAL), in addition to the original conversion prototype.

No significant achievements other than that of gaining experience on Bastan operation was obtained from the Franco-American programme.

Specifications

A twin turboprop low-wing light transport aircraft.

Span: 14.5m
Length: 10.75m
Height: 2.74m
Wing area: 32.4 sq m
Empty weight: 3,075 kg
Maximum take-off weight: 4,449 kg

Maximum speed: 425 km/hr
Cruising speed: 410 km/hr
Take-off distance: 280m
Service ceiling: 9,000m
Power-plants: 2 x 700 shp Turboméca Bastan Turboprops
               with Ratier-Figeac variable pitch air-
               screws
Accommodation: 5-7 passengers

PD-18S Conversions

C/n 01 F-ZWVO.   F/f 19 Sep 58 at Bordeaux-Merignac, previous identity uncertain, presumably CEV or DTI.   To SFERMA.   To CEV 03 Dec 58 for evaluation.   To CGTM late Dec 58.   Aircraft did not revert to C-45 standards and after WFU still as F-ZWVO was dumped at Cazaux gunnery training centre (Dept de la Gironde). Last seen there summer 1978.   Scrapped Sep 78.

C/n 02 F-WHMM.   F/f 18 Nov 59, most probably ex OO-GET (c/n AF-465, USAF 51-11908).   Reverted to C-45 standards as F-BHMM (CofA 03 Feb 62) Sud Aviation.

C/n 03 F-Z...   Built Jun 60, most probably ex OO-GEX (c/n AF-329, USAF 51-11772) which had been sold to SFERMA 29 Apr 60.   Reverted to C-45 standards.   To F-SEBA with CNET, later to F-BIEK and WFU.

MARQUIS

Following the Beech/SFERMA PD-18S turboprop conversion, another Franco-American venture was the SFERMA 60, actually a Beech Travel Air fitted with two Turbomeca Astazou   This promising turboprop engine had first been test-flown on two SFERMA-modified test-beds, the Max Holste-SFERMA 153 (F-WGGC) and the Nord-SFERMA 1110 (F-WJDQ), both conversions flying by mid 1959.

Late in the same year conversion began on a green Beech 95 airframe supplied to SFERMA by Beech Aircraft.   Now referred to as the Beech-SFERMA PD-146 Turbo Travel Air, the aircraft took to the air in July 1960.   This first prototype in its shining natural metal finish was following in 1961 by a second conversion ordered by Mr Szydlowski, the enterprising manager of Turbomeca.   It was planned to use this machine for further flight-test development by the power-plant manufacturers.   A third SFERMA 60 followed, with backing from the French Government.   Five more conversions followed, four being sold to German customers.

The original SFERMA 60 (fitted with Astazou IID turbines for single pilot operation and four passenger) led to a further conversion known as the SFERMA 60A.   Powered by two Astazou IIJs, production was undertaken at St-Nazaire from Beech Baron airframes, allowing five passengers to be carried, still with single pilot operation.   The French name Marquis was adopted.   For some time it was hoped that this SFERMA-Beech co-operation would develop into large-scale production of the type.

Some impact of this could even be traced in Beech project designs of 1961-63 for low-wing monoplanes to be powered by Turboméca Bastans or Astazous.   But the Marquis did not sell very well, and by 1965 fewer than twenty SFERMA conversions had flown out of a total of 27 Beech airframes reportedly imported.

In January 1965 all SFERMA activities and facilities were taken over by the parent concern, Sud Aviation.   Several Marquis are still flying in Europe.

Specification

Beech-SFERMA PD-146 Turbo Travel Air

Bracketed figures for the Beech B.95 Travel Air enable comparisons to be made with the original pre-conversion design.

Span: 11.53m (11.53m)
Length: 7.73m (7.74m)
Height: 3.26m (2.9m)
Wing area: 18 sq m
Empty weight: 1,325 kg (1,165 kg)
Loaded weight: 2,310 kg (1,861 kg)
Power-plants: 2 x 440 shp Turboméca Astazou
               (2 x 180 hp Lycoming O-360-ALA)
Maximum speed at sea level: 460 km/hr (338 km/hr)
Maximum cruising speed: 380 km/hr at 3,000m (309 km/hr)
Take-off distance to 15m: 153m (312m)
Maximum range: 2,130 km (2,270 km)

CM.100 F-WFAV

Top   :   Breguet 890H Mercure F-WFRF at Bangui circa
            1950/52 (R Leclercq via J Delmas)

Bottom :   Breguet 892S Mercure F-WFDO (J Cuny via J
            Delmas)

The Breguet 500 Colmar (SCA via J Delmas)

Top   : Breguet 730 (E.C.P.Armees)

Bottom : Breguet 730 - note the unusual position and
          shape of the floats (E.C.P.Armees)

Production

C/n 01 PD-146 F-WJHC.   Airframe del Oct 59 at Bordeaux.
        F/f 12 Jul 60 after conversion.

C/n 2   PD-146/60 F-WJSI.   Built 1961.   To Turboméca,
        del Jul 61.   Re-regd F-BJSI (CofA 12 Aug 63).
Used as engine test-bed with various types of Astazou.
To Ets Bedochaux Jun 73.

C/n 3   PD-146/60 F-WJSJ.   Built 1963?   To F-BJSJ
        SFERMA (CofA 08 Feb 63).   Lsd to Escadrille
Mercure?   Lsd to Europe Assistance Jun 63, purchased 04
Jan 65 (for ambulance work).   WFU (or DBR?) Oct 66 at Le
Bourget.

C/n 4   PD-146/60 F-WJSO.   Canc 11 Jul 62.   D-ILFA
        allocated 16 Jul 62, permit issued 16 Jul 62,
regd 23 Jul 62.   Crashed 08 Nov 62 during single pilot
night IFR flight from Hanover.   On pull-up after
approach at Bremen, failed to undertake published
missed-approach procedure and hit support cables of
radio transmitter mast at Steinkimmen.   Pilot killed.
Canc 13 Nov 62.

C/n 5   PD-146/60 D-ILFE allocated 30 Jul 62.   Permit for
        ferry St-Nazaire-Bremen 30 Jul 62 (Travelair
GmbH).   Regd 03 Aug 62 (CofA issued 03 Aug 62) for
Travelair.   Canc as sold to Switzerland 21 Aug 62, CofA
for Export as HB-GBB Sep 62.   Restored as D-ILFE with
new permit 08 Nov 62 and regd 09 Nov 62 (CofA 09 Nov 62),
this time to Altenburger Maschinenfabrik Jäckering & Co,
Hamm.   To Gebr Battenfeld Maschinenfabrik GmbH,
Meinerzhagen, 22 Oct 63.   En route from Barcelona in bad
visibility on ILS approach at night to Stuttgart, touched
down far from runway centre-line and swung off runway 17
Mar 64.   Badly damaged but no casualties.   Wreck sold
to Travelair 29 Jul 64.   Repair began in Mar 65 but was
abandoned due to high costs.   Canc 14 Nov 66.   Wreck
now preserved near DC-6 D-ABAH Preussisch-Oldendorf

C/n 6   PD-146/60 D-ILFO.   Built 1963.   Regn allocated
        09 Nov 62, permit issued 01 Apr 63 to Travelair
GmbH & Co KG (CofA 05 Jul 63).   To HB-GBF International
Musical Establishment 26 Jul 63.   To A Ostermann 1965
(oper by Fredair).   To Roethel AG, Basle, 1966.   To
Dr E Ring t/a Ring-Hotel-Finanz AG, Basle, 1969.   To
Travelair AG 1970.   To Fredair AG, Basle, 1972 or 1973.

C/n 7   PD-146/60 F-BLKU reservation reported but NTU.
        To D-ILFI.   Built 1962.   Allocated 03 Sep 62,
permit same day.   Regd 11 Sep 62 to Travelair GmbH & Co
KG, Bremen (CofA 11 Sep 62).   To Battenfeld GmbH 03 Dec
62.   Undercarriage damaged landing at Bremen 19 Jan 63
but repaired.   Swung off runway landing at Meinerzhagen
05 Oct 63, DBR and canc 10 Dec 63.   On request of owner
21 Jan 64 D-ILFI re-allocated 23 Jan 64 but not used
since (still reserved).

C/n 8   PD-146/60 F-BKOO.   To Turboméca (CofA 16 Jan 63.
        To SFERMA Dec 63.   To Sud Aviation Jun 65.   To
France Aviation Aug 65.   Retd to Sud Aviation Jul 66.
To SERIMA 1968.   To SNIAS Nov 69.   WFU (or DBR) Dec 68
at Strasbourg.

C/n 9   PD-146/60A F-BKOP.   To SFERMA (CofA 12 Nov 63).
        To Sud Aviation Jun 65.   Lsd to Air Bourgogne
and Escadrille Mercure 1965.   To France Aviation Jul 66.
WFU (or DBR) Feb 68 at St-Nazaire.   Wreck at Chavenay
1976.

C/n 10 PD-146/60A F-BLLP.   To Turboméca (CofA 23 Nov
        63).

C/n 11 PD-146-60A F-BLLQ.   To SFERMA (CofA 14 Jan 64).
        To Sud Aviation Jun 65.   Lsd to Escadrille
Mercure?   To SERIMA 1968.   To SNIAS Nov 69.   To
Pyrenair Gaspe Mar 71.

C/n 12 PD-146/60A F-BLLR.   To SFERMA (CofA 30 Nov 64).
        To Sud Aviation Jun 65.   To Air Affaires Dec 65.
Lsd to Air Bourgogne and to Air Alpes?   To SERIMA 1968.
To SNIAS Nov 69.   To A Pukacz Jun 70.   To Air Lorient
Aug 73.

C/n 13 PD-146/60A F-BLLS, reserved but NTU.   Completed
        as c/n 26.

C/n 14 PD-146/60A F-WLLT.   Built 1964.   To F-BLLT Sud
        Aviation (CofA 10 Jun 65).   Lsd to Air Bourgogne
and Escadrille Mercure?   To SERIMA 1968.   To SNIAS Nov
69.   To A Pukacz 1970.   To Aero Impex Jan 74.

C/n 15 PD-146/60A F-WLLU.   Built 1964.   To F-BLLU Sud
        Aviation (CofA 17 Jun 65).   Lsd to Escadrille
Mercure?   To SERIMA 1968.   To SNIAS Nov 69.   To Ets
Juster May 70.   To Omnium de Prospective Industrielle
Jan 72.   To Air Hainaut?   To Aviafair   .   To France
Bail Jun 77.   Sold to Line SA Jul 78.

C/n 16 PD-146/60A F-WLLV.   Built 1964.   To F-BLLV Sud
        Aviation (CofA 23 Jul 65).   Lsd to Escadrille
Mercure?   Lsd to Air Dauphine, del 30 Jun 67.   To SERIMA
1968.   To SNIAS Nov 69.   To A Pukacz Jun 70.   WFU (or
DBR) Jul 70 at Gap.

C/n 17 PD-146/60A F-BLLX.   Built 1967.   To Sud Aviation
        (CofA 26 Jul 67).   To SERIMA 1968.   To SNIAS Nov
69.   To A Pukacz Dec 69.   To G Levailland May 72.   To
Lebocey Transport Service May 73.   to Aero Impex Apr 74.

C/n 18 PD-146/60A F-BLLY.   To Sud Aviation (CofA 17 Dec
        65).   To SNECMA 1968.   To Aviation Culture
Loisirs Aug 73.   To Aero Impex 1974.

C/n 19 PD-146/60A F-BLLZ reserved but NTU.   Was nearing
        completion in 1965.

C/n 26 PD-146/60A.   Initially allocated c/n 13 (but not
        mentioned in German register files) D-ILCA.   To
Travelair GmbH & Co KG, Bremen (CofA 05 Jun 64).   To
Battenfeld Maschinenfabriken GmbH, Meinerzhagen, Aug 64.
To Müllers Air-Food-Service GmbH (Issum) Mar 76.   Made
belly-landing on business flight Braunschweig-Marl after
diversion to Düsseldorf 03 Jun 76.   No casualties but DBR
and wreck sold to France!   Canc Oct 76.

After the Liberation of France in 1944-45 there was a pre-war light transport ready for use in quantities by both military and commercial operators, the Caudron Goëland. The type was used and was joined by the NC-702 Martinet, a derivative of the German Siebel 204. The small transport operators however were more likely to use the cheap surplus aircraft of more recent design, such as the Beech 18 and similar British or US-built types. Nevertheless, some undaunted French manufacturers attempted the design and sale of new indigenous aircraft of a similar or smaller size, aircraft which were generally on the borderline of the air-taxi class (the word 'executive' in relation to aircraft not yet having been coined). Thus were born the MS-700, SO-7010 and SIPA 70 as well as French adaptations of foreign designs such as the SCAN-30 or VEMA 51 (both of which are dealt with in the previous chapter, together with the two Beech-SFERMA turboprop conversions).

It seems that all the French-designed aircraft were doomed from the start, mostly due to their too small   capacity for viable commercial operation - while use as private business aircraft had still to be developed in France and elsewhere. There was apparently little, if any, use for such light aircraft, and in any case, as far as the medium-sized transport was concerned, the ubiquitous Douglas DC-3 soon widely replaced the heavier models in this category.

Another aggravating circumstance for immediate post-war (1945-47) air transport in France was the fact that French operators were compelled to buy French (even if it were French-built German types such as the AAC-1). Both French-designed airframes and power-plants were in need of improvement at that time, operational profit-making was extremely rare, and this of course caused many small companies to go bankrupt.

The mysterious Mauboussin M-300 (of which very little is known) is also included in this section, even if it appears that such an aircraft had absolutely no hope of entering commercial service.

GOËLAND

The origin of the Caudron Goëland is not exactly within the scope of this monograph, as the C.440 prototype was flown as early as 1934. This light transport aircraft was built in large quantities both before and during the war for civil and military customers at home and overseas. Several versions appeared, but the most prolific was the C.445M military trainer for the Armée de l'Air.   During the occupation of France the Goëland was one of the few French aircraft to be kept in production under German control.   Apart from a small number of civil aircraft in Vichy French service throughout 1941-42 (and confiscated by DLH in 1943-44) and ex-Armée de l'Air trainers taken over by the Luftwaffe, more aircraft were built to fulfil German orders for use by the Luftwaffe as multi-purpose twin-engined trainers. Ten of those found their way to pro-German Slovakia, and the Regia Aeronautica also took some Goëlands on strength.

As early as 1944, when air services were being resumed over liberated France, the Caudron was considered for the domestic air mail routes then under re-organisation. But what had been a fine aircraft ten years earlier was now judged to be too slow and inadequate for the job, and the NC-701 and AAC.1 (qv) were preferred.

When the war ended SNCAN took over the production line.   The original Caudron works double c/n numbering system was maintained for more batches of Goëlands (with intermediate blocks allocated to large numbers of Caudron C.800 and Nord 2000 gliders).

In addition to a few machines which had survived the war and were kept in service with the Armée de l'Air during the late forties, no fewer than 325 more Goëlands were rolled out in two main variants, a C.445 post-war sub-type and the C.449, both using Renault 6Q power-plants.   These were government orders mostly for military use, but nearly one third of this post-war production was transferred to civil operators under a somewhat complicated scheme, many being later returned to the military services.   Obviously the national airline was the first operator involved.   Allocation to Air France of blocks on the civil register was becoming usual practice, and most Goëlands in the F-BAP. range issued in late 1945 were assigned to the airline in 1945/46.

In 1946 more aircraft of the type were diverted to government agencies and to various small operators later described collectively as third level airlines.   Many Goëlands were thus used during the 1946-50 boom when many air transport companies were launched with what can only be described as premature enthusiasm. Half a dozen aircraft also appeared on the civil register of Belgium. They were registered in the OO-C.. block reserved for aircraft being used in the Belgian Congo.

By the early fifties however the large number of civil Caudrons had dwindled to a handful of old fashioned yet attractive aircraft and, with the exception of military aircraft (including French Navy Goëlands used as trainers), the C.445 and C.449 soon disappeared from the French skies during the mid and late fifties. The name of the famous Caudron pre-war manufacturer now survives only in the few remaining Caudron C.800 gliders.

Specification

Twin-engined light transport aircraft, cantilever wing, of wooden construction with plywood and fabric covering. Two crew members plus six passengers.

Span: 17.6m
Length: 13.65m
Height: 3.4m
Power-plants:
  C.445/3 2 x Renault 6Q-10/6Q-11 (220 hp at 2,500 rpm)
  C.449   2 x Renault 6Q-02/6Q-03 (240 hp at 2,500 rpm/
                                     1,900m)
  C.449/1 2 x Renault 6Q-20/6Q-21 (240 hp at 2,500 rpm/
                                     2,200m; 220 ch/2,500
                                     rpm on ground, 300 ch
                                     on take-off

Empty weight: 2,697 kg (C.449/1)
Maximum weight: 3,700 kg (3,900 kg on some aircraft)
Maximum weight: 3,500 kg or 3,650 kg with restrictions
                 (C.445)
Payload: 515 kg
Cruising speed: 261 km/hr at 1,500m
Range at maximum weight: 1,000 km

Production

To begin, an explanatory note on Caudron aircraft construction numbers. A double-entry system was used; the initial four or five digit number was the general Caudron series, applied consecutively to all Caudron designs built. This was followed by the individual Goëland aircraft c/n (in this case Caudron Goëland C.440 through C.449 sub-types). The last Goëland ever built was therefore c/n 10308/1422, delivered to the French AF on 13 May 48.

Total production of 325 is believed to run from c/n 8902/1098 to 8926/1122 and from 10009/1123 to 10308/1422. With military aircraft, only aircraft where any details are known are included. All aircraft are C.449 unless stated otherwise.

C/n 8902/1098 F-BAPA? NTU.

C/n 8903/1099 F-BAPB. To Air France, del 20 Aug 45. To SNCAN at Meaulte 07 Jul 49. To French Navy Sep 50.

C/n 8904/1100 F-BAPC. To Air France, del 08 Oct 45. To SNCAN at Meaulte 07 Jul 49. To French Navy Jul 50.

C/n 8905/1101 F-BAPD. To Air France, del 30 Aug 45. To SGACC 14 Oct 48. Retd 16 Feb 49 to SNCAN Meaulte for overhaul. Intended for French Navy but withdrawn prior transfer.

C/n 8906/1102 F-BAPE. To Air France, del 21 Sep 45. Probably DBR Jun 56 at Cormeilles.

C/n 8907/1103 F-BAPF. To Air France, del 08 Oct 45. To Soc TARRAN/Soc Tananarivienne de

Pecheries, Ivato, 14 Nov 49. Canc Aug 52.

C/n 8908/1104 F-BAPG. To Air France, del 01 Oct 45. WFU Mar 59 at Cormeilles.

C/n 8909/1105 F-BAPH(1). Regd Oct 45 to French Govt. Canc Nov 45.

C/n 8910/1106 F-BAPI. To Air France, del 10 Nov 45. DBR 21 Dec 57 at Cormeilles.

C/n 8911/1107 F-BAPJ(1). Regd Oct 45 to French Govt. Canc Nov 45.

C/n 8912/1108 F-BAPK(1). Regd Oct 45 to French Govt. Canc Nov 45.

C/n 8913/1109 F-BAPL. To Air France, del 07 Nov 45. Retd to Govt 09 Oct 47. To DAC Tunisia. To CN St-Yan. Canc Sep 52.

C/n 8914/1110 F-BAPM. Intended for Air France. Regd early 1946. Canc Feb 46 as sold abroad. Restored to CN St-Yan (CofA 18 Sep 47). WFU May 56.

C/n 8915/1111 F-BAPN. To Air France, del 27 Oct 45. Regd to SGACC Feb 49. Back to Air France May 52. WFU Mar 59 at Cormeilles.

C/n 8916/1112 F-BAPO. To Air France, del 14 Nov 45. WFU Mar 59 at Cormeilles.

C/n 8917/1113 F-BAPP(1). Intended for Air France but retained by Service du Materiel. Regd Oct 45, canc Nov 45.

C/n 8918/1114 F-BAPQ. To Air France, del 26 Dec 45. DBR 25 Nov 58 at Cormeilles.

C/n 8919/1115 F-BAPR. Intended for Air France. Regd 1946. Retd to Govt 13 Nov 47 for French AF/ECMA.

C/n 8920/1116 F-BAPS. To Air France, del 27 Nov 45. WFU Mar 59 at Cormeilles.

C/n 8921/1117 F-BAPT. To Air France, del 27 Nov 45. To Soc TARRAN/Soc Tananarivienne de Pecheries, Ivato, 11 Jan 50. Canc Aug 52.

C/n 8922/1118 F-BAPU. To Air France, del 04 Dec 45. WFU Mar 59 at Cormeilles.

C/n 8923/1119 F-BAPV(1). Regd Nov 45 to French Govt. Canc Feb 46.

C/n 8924/1120 F-BAPX(1). Regd Nov 45 to French Govt. Canc Feb 46.

C/n 8925/1121 F-BAPY(1). Regd Nov 45 to French Govt. Canc Feb 46. Restored as F-BFAA OFEMA (CofA 08 Jul 48). To OO-CEA Aeromas (CofA 24 Sep 48). To Sabena 11 May 49. Retd to French Air Ministry 19 Dec 50.

C/n 8926/1122 F-BAPZ(1)? NTU. To F-BFAB OFEMA (CofA 08 Jul 48). To OO-CEB Aeromas (CofA 25 Sep 48). To Sabena 11 May 49. Retd to French Air Ministry 19 Dec 50.

C/n 10011/1125 F-BFAC. To OFEMA (CofA 02 Jul 48) (ex military?). To OO-CEC Aeromas (CofA 02 Oct 48). To Sabena 11 May 49. Retd to French Air Ministry 19 Dec 50.

C/n 10016/1130 F-BAPP(2). To SNCAN (CofA 10 Sep 46). To Louis Dolfuss, Tlemcen. WFU May 63.

C/n 10021/1135 F-SCFD. ELA.47 Aug 49.

C/n 10024/1138 F-BAQP. To SGACC (CofA 27 Jul 46).

148

Scrapped at St-Yan 1950.   Canc Feb 51.

C/n 10027/1141 F-BAPH(2).   To Air France, del 19 Feb 46.
WFU Dec 58 at Cormeilles.

C/n 10028/1142 F-BAPJ(2).   To Air France, del 25 Feb 46.
Written off 22 May 47 at Persan-Beaumont.

C/n 10029/1143 F-BAPK(2).   To Air France, del 25 Feb 46.
Retd to Govt 09 Oct 47.   To DAC Morocco.
To CN St-Yan.   Canc Sep 52.

C/n 10030/1144 F-BAQU.   To SGACC (CofA 04 Dec 46).
Canc late 1948.

C/n 10031/1145 F-BAQZ.   To SGACC (CofA 12 Dec 46).
Canc late 1948.

C/n 10032/1146 F-BAPV(2).   To Air France, del 12 Mar 46
(regd to SGACC).   Retd to Govt 13 Nov 47
for French AF/ECMA.

C/n 10033/1147 F-BCCA.   To SGACC (CofA 06 Dec 46).
Canc May 49.

C/n 10034/1148 F-BAPZ(2).   To SGACC (CofA 06 Jun 46).
Canc Apr 49.   To French Navy 1950.

C/n 10035/1149 F-BAQA?   NTU.

C/n 10036/1150 F-BAQB.   To SGACC (CofA 05 Jun 46).
Canc 03 Dec 49.   To French Navy 1950.

C/n 10037/1151 F-BAQC?   NTU.

C/n 10038/1152 F-BAQD.   To SGACC (CofA 29 May 46).   WFU
Dec 49 and canc 28 Dec 49.

C/n 10039/1153 F-BAQE.   To SGACC (CofA 11 Sep 46).
Canc May 51.

C/n 10040/1154 F-BAQF.   To SGACC (CofA 01 Jul 46).   WFU
Dec 49 and canc 28 Dec 49.

C/n 10041/1155 F-BAQG.   To SNCAN (CofA 12 Jun 46).
Sold abroad 13 Jan 47.   To OO-CCJ Congo
Motor, Elizabethville, del 02 Jul 46.   Written off 24
Aug 48.

C/n 11042/1156 F-SCCD.   To ELA.44 1952.

C/n 10045/1159 F-BAQV.   To Air Sud (CofA 04 Oct 46).
Canc Dec 48.   To military markings.

C/n 10051/1165 C.447.   F-RAIM.   GAEL Nov 51.

C/n 10055/1169 C.447.   F-TEBF.   BE.701 1952.

C/n 10057/1171 F-BAQT.   To Air Azur (CofA 05 Aug 46).
Lost at sea 08 Jul 47 between Tunis and
Bizerte.

C/n 10058/1172 F-BAQN.   To CADAF (CofA 01 Aug 46).   To
Aigle Azur.   Canc Sep 49.

C/n 10059/1173 F-BAQX.   To LASO France (CofA 06 Aug 46).
Canc Sep 49.

C/n 10061/1175 F-BAQO.   To Air Transport (Cof A 30 Jul
46).   Canc Sep 49.

C/n 10062/1176 F-BAQM.   To STARO (CofA 02 Jul 46).
Canc Nov 47.   To French AF ELA.41 as
F-SCAU 1950.   To F-SCDB ELA.45 and F-UGBD EC.7.

C/n 10063/1177 F-BAQQ.   To Air Transport (CofA 01 Aug
46).   Canc Sep 49.

C/n 10064/1178 F-BAQR.   To CADAF (CofA 31 Jul 46).   To
Aigle Azur.   Canc Sep 49.

C/n 10067/1181 F-BAQH.   To SNCAN (CofA 22 Jul 46).
Canc 13 Jan 47.   To OO-CCK Congo Motor,
Elizabethville, del 22 Aug 46.   Written off 27 Nov 58.

C/n 10068/1182 F-BAQK.   To TAM (CofA 26 Jun 46).
Transferred to military marks and replaced
by c/n 1130.   Canc Sep 49.

C/n 10069/1183 F-BAQI.   To Escadrille Mercure, del 05
Sep 46.   To Alexandre Dody, Nice, 1947?
Canc Aug 53.

C/n 10071/1185 F-UGAB.   EC.7 1951.

C/n 10074/1188 C.445/3.   Military, then to F-BAVM Aero
Club de la Loire Atlantique, Nantes (CofA
12 Aug 54).   To CN St-Yan 1956/57.   Canc Mar 60.

C/n 10075/1189 F-TFHA.

C/n 10076/1190 F-BAQJ.   To French Govt, possibly for IGN
(CofA 31 May 46).   Written off 11 May 47
at Chatillon-sous-Colmont, Mayenne.   Canc 23 Apr 48.

C/n 10077/1191 F-BCCD(1) NTU.   To F-BAPX(2) Air France,
del 30 Nov 46 (regd to SGACC).   Retd to
Air France May 52.   WFU Mar 59 at Cormeilles.

C/n 10081/1195 F-BAPY(2).   To French Govt (CofA 09 May
46).   Canc late 1946.   Left in Brazil
after demonstration tour when flown Paris-Casablanca-
Dakar-Natal-Rio de Janeiro 09-14 May 46.   To LV-FGI
Enrique Decumex.

C/n 10090/1204 F-SCCF.   ELA.44 1952.

C/n 10095/1209 F-BAQL.   To TAM (CofA 22 Jul 46).   Canc
Sep 49.   To military markings.

C/n 10098/1212 F-BAQY.   To LASO France (CofA 02 Aug 46).
Canc Sep 49.

C/n 10099/1213 F-BAQS.   To Rapides Cote d'Azur (CofA 25
Jul 46).   Canc Sep 49.   To military
markings.

C/n 10102/1216 F-BCCB.   To Air Transport (CofA 19 Oct
46).   Canc Nov 49.   To French AF.

C/n 10105/1219 F-BCCC.   To STARO (CofA 10 Sep 46).
Canc Nov 49.

C/n 10108/1222 F-BCCJ.   To Rapides Cote d'Azur (CofA 30
Sep 46).   Canc Feb 48.   To military
markings.

C/n 10109/1223 F-BCCK.   To STARO (CofA 14 Oct 46).
Written off Aug 47 (possibly 13 Aug 47 nr
Pau).

C/n 10110/1224 F-BCCE.   To CADAF (CofA 19 Sep 46).
Canc late 1947.   To French AF as F-UGAA
I/1 Aug 51.

C/n 10113/1227 F-BCCG.   To TAM (CofA 02 Oct 46).   Canc
Nov 49.   To military markings.

C/n 10116/1230 F-BCCI.   To SNCAN (CofA 05 Nov 46).
Canc late 1946.   To OO-CCR.   Written off
24 Aug 48.

C/n 10122/1236 F-BCCH.   To TAM (CofA 02 Oct 46).   Canc
late 1946.

C/n 10125/1239 F-BCCF.   To Louis Dolfuss, Tlemcen (CofA
26 Oct 46).   WFU May 63.

C/n 10133/1247 F-UHBF.   ER.33 1951.

C/n 10135/1249 F-BCCM.   To Aerotaxi (CofA 18 Apr 47).

To Navifrance Jun 48.   Canc late 1950.
To French Navy.

C/n 10137/1251 F-BCCV.   To SGACC (CofA 10 Dec 48).
Loaned to ONERA.   Canc Aug 53.

C/n 10138/1252 F-BCCO.   To SOTAM (CofA 21 Nov 46).
Canc Jan 48.

C/n 10139/1253 F-BCCN.   To TAM (CofA 22 Nov 46).   Canc
late 1951.

C/n 10142/1256 F-BCCT.   To Air Transport (CofA 28 Dec
46).   Canc Nov 49.

C/n 10143/1257 F-BCCR.   To Air Sud (CofA 11 Apr 47).
Canc late 1948.

C/n 10145/1259 F-BCCP.   To SOTAM (CofA 04 Dec 46).
Canc Jan 48.

C/n 10146/1260 F-BCCQ.   To SOTAM (CofA 04 Dec 46).
Canc Jan 48.

C/n 10153/1267 F-TEBF.   BE.701 1951.

C/n 10155/1269 F-BCCS.   To Air Sud (CofA 24 Jan 47).
Canc Feb 49.

C/n 10157/1271 F-SCCG.   ELA.44 1952.

C/n 10162/1276 F-SCCD.   ELA.44 1950/51.

C/n 10170/1284 C.447 F-SCEG.   ELA.46.

C/n 10172/1286 C.447 F-UGMT.   EC.5 1954.

C/n 10174/1288 C.447 F-SCEC.   ELA.46 1951.

C/n 10177/1291 C.447 F-SCRM.

C/n 10185/1299 F-UGMZ.   EC.5 1954.

C/n 10193/1307 F-ZJGG.   To CEV.

C/n 10195/1309 F-UGCK.   To II/7 Apr 59(?).

C/n 10196/1310 F-SCCF.   ELA.44 Aug 51.

C/n 10199/1313 F-UGAC.   EC.1.   To F-UGCQ II/7.

C/n 10204/1318 F-UGDD.   ELA.41.

C/n 10205/1319 F-BDXA.   To Air Sud (CofA 04 Jun 47).
Canc Dec 48.   To French Navy.

C/n 10214/1328 F-BDXH.   To French Govt, Cairo (CofA 17
Nov 48).   Canc Mar 51.   To French AF.

C/n 10215/1329 F-BCCX(1).   To SOTAM (CofA 24 Apr 47).
Canc Jan 48.

C/n 10216/1330 F-BCCY.   To TAM (CofA 06 May 47).   To
French AF Apr 48.   To French Navy.

C/n 10217/1331 F-BCCZ.   To SATI (CofA 30 May 47).
Ferried to Saigon 18 Jun 47-06 Jul 47.
Canc Jan 52.

C/n 10218/1332 F-SCCA.   ELA.44 Jun 47.   To F-UGMY EC.5
1951.

C/n 10225/1339 F-SCCC.   ELA.44 1950.

C/n 10226/1340 F-UGOB.   EC.5 1951.   To F-UGNY EC.5

C/n 10230/1344 F-BCCD(2).   To TAM (CofA 14 May 47).
Canc Mar 48.   To French Navy.

C/n "10231/1335" F-BDXI.   To Préfecture de Police (CofA
28 Mar 49).   Canc late 1951.   This c/n
is wrong and should be either 10221/1335, ex French AF,

to French Navy, or 10231/1345, ex French AF.

C/n 10235/1349 F-UGMW.   EC.5 Mar 52.

C/n 10239/1353 F-BAVO.   To Aero Club de la Lys, Lille
(CofA 10 Nov 54) (ex-military).   Canc
Sep 57.

C/n 10240/1354 F-BDXG.   To ONERA (CofA 31 Mar 48).
Canc Aug 53.

C/n 10241/1355 F-UGOA.   EC.5 1951.

C/n 10243/1357 F-SCCB.   ELA.44 1952.   To F-BEHX Aéro Club
Rhône et Sud Est, Lyon Bron (CofA 28 Nov
56).   To Soc Lyonnaise d'Import Export 1960.   WFU Dec
60 at Lyon.

C/n 10244/1358 F-BDXB.   To Air Sud (CofA 03 Jul 47).
Canc Dec 48.   To French Navy.

C/n 10245/1359 F-BDXC.   To Air Transport (CofA 03 Jul
47).   Canc Jun 48.

C/n 10246/1360 F-BDXD.   To SERAC (CofA 28 Jul 47).   Canc
late 1949.   To French Navy?

C/n 10248/1362 F-SCEH.   ELA.46 Dec 49.

C/n 10255/1369 F-UGGE.   3/6 Feb 50.

C/n 10257/1371 F-BDXE.   To Air Sud (CofA 29 Jul 47).
Canc Dec 48.   To French Navy?

C/n 10259/1373 F-UGDF.   EC.2.

C/n 10262/1376 F-BCAI.   To ENAC (CofA 02 Jan 52) (ex-
military).   Canc Sep 54.

C/n 10274/1388 F-UGNZ.   EC.5 1952.

C/n 10279/1393 F-BDXF.   To Rapides Cote d'Azur (CofA 28
Oct 47).   Canc Nov 49.   To French Navy.

C/n 10288/1402 F-SCAW.   ELA.41 1951.

C/n 10289/1403 F-BCAK.   To ENAC (CofA 25 Jan 52) (ex-
French AF).   Canc Sep 54.

C/n 10290/1404 F-SCCB.   ELA.44 1950/51.   To F-UGGU
EC.3 1952.

C/n 10291/1405 F-UGMV.   EC.5.

C/n 10293/1407 F-BCAL.   To ENAC (CofA 24 Jan 52) (ex-
French AF).   Canc Sep 54.

C/n 10296/1410 F-SCCA.   ELA.44 1950/51.

C/n 10297/1411 F-TFKN.   CERO.303 May 57.   Last in
service with French AF.   SOC 01 Jul 57.

C/n 10300/1414 F-UGMU.   EC.5 1952.

C/n 10303/1417 F-UGJO.   EC.4 Sep 50.

C/n 10304/1418 F-UGNZ.   EC.5 1951.

C/n 10305/1419 F-BCCX(2).   To ENAC (CofA 25 Jan 52 (ex-
military).   Canc Sep 54.

C/n 10308/1422 To French AF, del 13 May 48.

MS-700

The MS-700 was one of the few twin-engined
light aircraft designed in the late forties,
hardly a challenger for commercial operation,
but promoted as a colonial twin-engined aircraft (the so-called colonial aircraft appears
to have been a typical French aeronautical
fancy - as was also the case with postal aircraft).   Also intended as a communications and
medical evacuation type, several power-plants
were considered for the MS-700.   Three prototypes were built, but in the event the new
aircraft from the famous Morane-Saulnier stable
was not ordered into production.   Only the
first prototype was used for several years by
the manufacturers.   Also powered by two 160 hp
Potez 4D-33s, the second aircraft was much
publicised in 1950.   First, on 8 September, it
won the Rallye de Biarritz when flying 2,813 km
mainly around southern France in less than 17
hours with four on board.   Later in that year
(16 November to 20 December) an extensive
demonstration tour to French West Africa was
undertaken.   One hundred and one hours were
flown successfully, considering that both the
aircraft and the power-plants were prototypes,
but no contract was awarded (the SCAN-30 and
Castel-Mauboussin were also part of this tour,
which had been organised to promote light
transport aviation in remote areas).

Variants with more powerful engines were to
follow.   In 1951 the new MS-703 was displayed
at the Salon du Bourget, but development was
halted before the MS-704 (actually a conversion
of the first prototype) was flown.   The main
explanation for the failure of this Morane-
Saulnier venture seems to lie in the small
capacity of the aircraft, despite its promotion
of low fuel consumption and cheap operation.

Specifications

MS-700/MS-704

Span: 14.2m
Length: 9.92m
Height: 3.4m
Wing area: not quoted
Empty weight: 1,645 kg (unqualified)
Loaded weight: 2,220 kg
Power-plants: 2 x Potez 4D-33, 160 hp

Payload: 255 kg
Maximum speed: 290 km/hr at 1,500m
Cruising speed: 260 km/hr at 1,000m
Take-off distance to clear 66 ft/20m: 430m
Service ceiling: 6,200m
Range (with maximum load): 1,200 km (unqualified)
Passenger configurations: four-seat liaison transport,
                           ambulance, light freighter,
                           twin-engine trainer

MS-701

Engine change to 180 hp Mathis 8G-20s.

MS-702

No details available.

MS-703

Six-seater aircraft with longer fuselage and 240 hp
Salmson 8.AS.00 engines.

MS-704

As MS-703 but with 220 hp Potez engines.

Production

Three prototypes were built.

MS-700-01 F-WFDC.   F/f 08 Jan 49.   To F-BFDC Morane-
          Saulnier (CofA 16 Feb 56 after overhaul).   WFU
Jun 59 (due for conversion to MS-704).

MS-700-02 F-BFDE.   To DTI (CofA 08 Aug 50).   Demonst-
          ration tour to Africa.   Flown back to CEV
Brétigny May 51.   Canc late 1952 and converted to
MS-701-01.

MS-703-01 F-WFDD.   F/f 03 Jan 51.

SO-7010

The SO-7010 was an unorthodox venture into
the light transport field, and it could well
have been described as a twin-engined aircraft
that looked like a single-engined one!
Reportedly intended as a replacement for the
Caudron Goëland and other similar types (the
Caudron was then widely used), the SO-7010 was
a six-seater built around two Mathis G-8
engines combined into a single power-plant
driving the propellor through a double free-
wheel reducing gear, which allowed the propellor to be driven by one engine only if
necessary.   This, it was hoped, was a safety
device in case of an engine failure.   (Years
later, but unrelated to this design and in a
much different way, another French design - the

Breguet 940/941 - was also to use linked
propellors).   It was also intended to use the
aircraft on one engine only for economical
flight.

The unusual tandem location of the two
engines in the fuselage of the SO-7010 was
definitely surprising, and this was claimed to
improve performance owing to a generally clean
wing design.   It is also of interest to note
that the SO-7010 (originally referred to as the
Cassiopée) was mentioned as a twin-engined
postal aircraft, a programme which was then so
fashionable in France.   It was of course also
offered as a feeder-liner (bimoteur d'appoint
pour lignes secondaires).

The low-wing nosewheel undercarriage SO-7010, dubbed Pégase (Pegasus), was a generally clean-looking aeroplane and was first shown at the 1946 Paris exhibition. But, with extensive test-bench work being undertaken to solve technical difficulties, the aircraft registered F-WEAG did not make its maiden flight until 1948.   There was however apparently no future in such a design, and the Pégase remained in prototype form only - not to mention the fact that third-level operators and purchasers of executive aircraft (words which had still to be invented in the late forties!) could find much cheaper and more easily maintained aircraft on the market.

The SO-7010 had long been forgotten when, during 1971, the wreck of the prototype was discovered in a scrapyard near Mormant (Dept de Seine et Marne) by local aviation historians. Only the fuselage remained in poor condition, but by 1977/1978 a growing interest in this rare wreck had developed, culminating in steps taken by Les Ailes Anciennes to salvage the aircraft.   This volunteer organisation has been doing splendid work for the benefit of the national Musée de l'Air at Le Bourget, and an agreement has been reached with the owner of the scrapyard for the SO-7010 to be taken back to Le Bourget for storage and, when time allows, to be restored to become an exhibit at the Musée de l'Air.   This unique aircraft will then be another interesting item within the collection and also the witness of another unsuccessful attempt to compete with foreign aircraft in the light transport aircraft market.

Specification

Twin-engined light transport aircraft (taxi aérien).

Span: 14.7m
Length: 11.12m
Empty weight: 1,850 kg
Maximum loaded weight: 2,875 kg
Payload: 800 kg
Power-plants: 1 x Mathis 16 G-21 assembly (in fact 2 x 200 hp Mathis G-8R driving a single propellor)
Cruising speed: 275-300 km/hr
Service ceiling: 5,000m
Range: 800-1,250 km

Performance above possibly estimated, as very little is known about the flight tests and the results obtained. Crew of two (pilot plus radio operator) and accommodation for six passengers or equivalent freight load.

Prototype

SO-7010-01 F/f 27 Feb 48 F-WEAG.   Wreck derelict at railway station Mormant (Dept de Seine et Marne) and first spotted there 1971.   In the process of being restored for eventual exhibition at the Musée de l'Air, Le Bourget.

MAUBOUSSIN M.300

The Mauboussin M.300 is one of the least known French prototypes to the extent that it has not been possible, at the time of going to press, to find out if the only prototype built did in fact undergo flight trials and with what results.

It was a twin-engined aircraft of wooden construction designed by Pierre Mauboussin and built in the Fouga works at Aire-sur-Adour as a communications, air taxi or postal aircraft. In the military role it could have been used as a multi-engine trainer and for staff transport. Surprisingly this 1943 wartime design was nearly complete by May 1944 with engines already fitted but still without all the necessary equipment.

A low-wing cantilever monoplane with twin-fin tail and rudder assembly and conventional landing-gear, the M.300 was of a generally clean design, and if comparison in layout were to be found, the aircraft may be described as reminiscent to some extent at least of a scaled-down Messerschmitt Bf 110.   Construction was completed by April 1948, and the maiden flight was announced to be imminent by mid 1948.   It has not however been possible to trace the air-craft during flight trials, and no registration is known to have been allocated.

Specification

Originally announced as a three-seater communications aircraft, then as a passenger transport with crew of two and four passengers.

Span: 13.5m
Length: 10.65m
Height: 2.75m
Wing area: 24.56 sq m
Empty weight: 1,670 kg
Loaded weight: 2,500 kg
Payload: 830 kg
Power-plants: 2 x 240 hp Renault 6.Q.10/11
Maximum speed: 325 km/hr
Cruising speed: 282 km/hr
Ceiling: 6,200 km
Range: 1,300 km

All performance figures are estimated.

SIPA S-70

La Société Industrielle pour l'Aéronautique (SIPA) was a small firm established in 1938 with works in the Paris suburbs at Suresnes and Neuilly. Sub-contract work on components for various major aircraft manufacturers was undertaken, and during the occupation of France an Arado 396 production line under German control was launched. This resulted in post-war French derivatives of the German trainer, SIPA S-10, S-11 and S-12 (nearly 240 aircraft being manufactured for the French Air Force). After the Liberation of France SIPA designed various light aircraft (the S-50, a single prototype of 1946, followed by the SIPA 90-110 range in various variants for Aero Club use, and the only light jet aircraft for joyriding use built in the early fifties, the SIPA 200). More projects in various fields of aviation were also designed.

In 1947 the firm began work on another light transport prototype. This was at a time when an excessively optimistic mood prevailed, when the French aeronautical industry had often failed to realise that excellent British and American aircraft could be purchased brand-new or second-hand for light commercial operation. The SIPA S-70 was intended as an inter-city light transport, especially on the so-called French 'transversal' routes, ie east-west flights or from Paris to the small provincial cities. It is now known that what we call third-level airlines today were hardly feasible at that time. Work proceeded on the S-70 nevertheless. The general appearance of the aircraft was reminiscent of a clumsy-looking Beech 18, Caudron Goëland or NC-702 Martinet,

which is enough to show that it would have come into an already crowded market. Of wooden construction, the prototype, powered by two 240 hp engines, was registered F-WCZI. The aircraft is known to have been carrying out flight trials by July 1949, but the outcome of those trials has not been traced, and the aircraft is one of the least-known French prototypes. A wooden mock-up was on display at Paris Grand Palais in November 1946.

Specification

Span: 13.6m
Length: 10.2m
Height: 3.65m
Wing area: 24.5 sq m (estimated)
Empty weight: 1,055 kg (1,294 kg + oil)
Loaded weight (8 passengers + freight/luggage): 2,428 kg
(also quoted as 2,075 kg)

Payload: not quoted
Take-off distance: not quoted
Service ceiling: not quoted
Range: 1,000 km (unspecified)
Passenger configurations: 6-8 passengers + centrally-seated pilot
Power-plants: 2 x 150 hp Potez 4D or
2 x 190-200 hp Mathis
Cruising speed: 230 km/hr (Potez)
240 km/hr (Mathis)
Maximum speed: 270 km/hr (Mathis)

No photographic evidence of the S-70 was known of at the time of going to press, and the three-view drawing alongside provides one of the very few illustrations of the type in existence.

MILITARY PROTOTYPES

The approach to military transport in France as elsewhere is two-fold. While there is clearly a specific military need for tactical transports with an associated possible outlet for commercial use (especially in under-developed areas), basically commercial aircraft are also ordered and used by the military services.

The French Nord 2501 and derivatives are a typical example of military transports which, in a few instances, were delivered to civil operators. The Franco-German Transall is another example of a possible if limited use as a commercial freighter. The French-built Junkers 52 was used mostly in the military role, but one must not forget that the original pre-war German aircraft had been launched originally as a commercial transport.

On the other hand, no fewer than five post-war French-designed airliners have been delivered to the services, namely the SE-161, SO-30, Bréguet 763/765, Caravelle and Nord 262.

This is a reminder that it is often difficult to give a clear answer to the question as to whether an aircraft is a military or commercial transport. The aircraft in this chapter offer a useful variation on this theme, while the types quoted above are dealt with in other parts of this monograph.

The Bréguet 730/731 flying-boat which, after some hesitation, has been included in this publication, is a good example. This was a military flying-boat, and commercial variants were envisaged. In fact the Breguet 731 was widely reported as a commercial type but was only to see military service with the French Navy. The Bréguet 500 Colmar was a contender for what was obviously a commercial specifica-tion, and the winner of the contest, the SO-30, was an airliner. However, with the single prototype being used by the French Air Force, the Colmar is traditionally referred to as a military aircraft. The Castel-Mauboussin CM-100 and 101 were designed to meet both military and commercial applications, and yet they were only powered versions of a military glider.

The Breguet 890/892 prototypes were built both for commercial and military use, but had the aircraft been selected doubtless they would have been ordered mostly for military use. The Breguet 941 development was widely agreed to be a promising STOL transport, and besides wide-spread expected military use could well have been a successful commercial inter-city aircraft had the evolution of air transport taken a different turn.

On the other hand, Dassault ventures into the light twin-engined transport field were only tentative. The ill-fated SNCAC Cormoran deserves special mention as an aircraft for the military market for which France had no use at the time.

Lastly, a note on flying-boats. As is well-known, France continued to press for the use of flying-boats for some years after World War II (eg the SE-200 and SE-1200 projects), one even entering limited commercial service (the Laté 631). As mentioned in the chapter devoted to aircraft of German origin, the Dornier 24T was built for use by the French Navy. Another comparatively large flying-boat was also designed and built for the Marine Nationale. Originally an amphibian, the Nord 1400 and 1401 Noroit and the 1402 production variant were delivered in small numbers to the French Navy during the early fifties. With limited use only and restricted to flying-boat operation, the Noroit was short-lived in French Navy use, and with apparently no commercial variants in project it has been omitted from this monograph. A couple of prototypes were allocated civil registrations but for SNCAN test purposes only.

CM-100

Curiously the story of the CM-100 transport prototype began with a large transport glider, the CM-10, built by Etablissements Fouga. In the immediate post-war years the Technical Director of Fouga was Robert Castello who, as a former member of the Dewoitine and SNCASE design teams, had designed several sailplanes during the occupation of France including the C.25S, C.30S and C.300S, which were built in quantity both before and after the Liberation of France.

Fouga had also manufactured light touring aircraft built to the designs of Pierre Mauboussin, whose office was entirely taken over by Fouga by the end of the war.

This resulted in a close co-operation between the two men under the name of Castel-Mauboussin. After designing the CM-7 and CM-8 gliders (the latter a single-seat aerobatic sailplane), they produced the large CM-10 troop-carrying and freight glider. The prototype, towed by a French-built Ju-52, was flown by Mauboussin for the first time after some delays in June 1947. Construction was of mixed wood and light alloy, and a second CM-10 followed, with five more production gliders. Large-scale production was considered by SNCAN at the Sartrouville and Caudebec factories, and although in 1948 the

French forces were still impressed with the use of assault gliders the order was cancelled; indeed, one may wonder to what use France would have put a large military glider which could accommodate 35 troops or various freight combinations such as two vehicles of Jeep size.

The unwanted glider therefore gave way to a powered version, the CM-100. Power-plants selected were the Renault (later SNECMA) 12S, widely used on various post-war French light and medium transport aircraft.

Using CM-10 airframes, two prototypes were built. Although in late 1946 completion was hoped for by April 1947, flight trials did not in fact begin until 1949, and even then the first CM-100 was found to have a poor perform-ance, as exemplified during a demonstration tour in French Africa under tropical conditions. This resulted in the CM-101R, retaining the same engines but with performance boosted by two small auxiliary turbojets, the 110 kg st Turboméca Piméné (a retractable landing gear was also considered to reduce drag). Both types of this, the only French venture into the large transport glider field and then converted to a powered variant, were abandoned as they had no commercial prospective sales. A military project, known as the CM-103R, with the more powerful Turboméca Marboré jets was not developed.

The Castel-Mauboussin team turned to more realistic if less glamorous activities: light gliders in connection with auxiliary turbojet use, the introduction of the butterfly-type tail and experience which soon found its way into the world's first military jet trainer, the Fouga CM-170 Magister.

Specifications

Castel-Mauboussin CM-10

Large transport glider. The entire nose section of the fuselage including the freight compartment was hinged to open to starboard.

Span: 26.7m
Length: 17.9m
Wing area: 71.9 sq m
Empty weight: 2,833 kg
Loaded weight: 6,420 kg
Useful load: 3,500 kg (with crew of 1 and 35 armed
                       troops)
                4,050 kg (freighter, with maximum weight
                       being raised to 7,000 kg)
Cargo hold capacity: 32 cu m
Maximum speed: 280 km/hr

Castel-Mauboussin CM-100

Powered version of the original glider, retaining the opening nose. Dimensions as for CM-10; figures apply to freight operation, those between brackets to 15 passenger operation, both with crew of two.

Empty weight: 4,543 kg (4,865 kg)
Loaded weight: 7,300 kg (7,300 kg)
Useful load: 2,033 kg (1,200 kg + 525 kg baggage or
                      freight)
Power-plants: 2 x 580 hp Renault 12S

Cruising speed: 246 km/hr at 1,500m
Range: from 500 to 1,000 km (latter figures with 14
                      passengers and 182 kg baggage)

Castel-Mauboussin CM-101R

As for CM-100 but with two added auxiliary 110 kg thrust Turboméca Piméné turbojets, giving increased take-off and climb performance.

Empty weight: 4,920 kg (5,245 kg)
Loaded weight: 7,600 kg (7,600 kg)
Useful load: 2,010 kg (1,200 kg + 502 kg)

Cruising speed was improved, and take-off to 15m was reduced from 700m on CM-100 down to 450m.

Production

CM-10-01 F/f 07 Jun 47 at Mont de Marsan. To CEV for trials 21 Jan 48. Crashed 05 May 48 at Brétigny on test flight, pilot baling out to safety.

CM-10-02 F/f late 1948. To CEV Oct 50. Military marks. Production order for 100 cancelled after completion of five gliders (including two reported converted and fitted with engines to become CM-100s).

CM-100-01 F/f 19 Jan 49 as F-WFAV. Later modified to CM-101R-01, still as F-WFAV.

CM-101R-02 F/f 23 Aug 51 as F-WFAX.

CM-103R Military variant but not built.

Note The nose section of an unidentified CM-100 was still kept in a hangar at Chavenay, on the out-skirts of Paris, as late as August 1976.

BREGUET 890

Soon after the end of World War II there were considered to be prospects for a new market for twin-engined cargo aircraft, both for military and commercial short- and medium-range operation. Of course this did not take into account the fact that an incredibly large stock of surplus military aircraft fell into the same category; it was so much cheaper to buy the proven DC-3!

As early as 1948 however the French Armée de l'Air had drawn up a specification for a C-47 and AAC-1 (Ju-52) replacement, and the Bréguet 891 Mars was proposed, together with the Nord 2500. One military prototype and two commercial variants of the Bréguet design were ordered. The first prototype, flown in 1949, was in fact a four-engined derivative of the original design, known as the Br-892S (S denoting the Renault 12S power-plants, themselves evolved from the original German Argus AS411, left over from the German occupation of France). Later that year it was followed by the Bréguet 891R (SNECMA 14R engines) military transport prototype.

Both types were derived from the Br-890 basic design called Mercure (Mercury), which was not flown until April 1950, two years after it had been ordered alongside the 892S and 891R prototypes. Power-plants under consideration for the Mercure had included the French Arsenal-built Jumo 211 (hence the Br-890J designation), but more reliable British engines were preferred and, with the ensuing delay, the aircraft was then re-designated Br-890H (Bristol Hercules). The Br-892S was later converted to become the second Bréguet 890H. Both types (890 Mercure and 891 Mars) were structurally identical except for power-plants. Both were high-wing cantilever aircraft with the rear part of the fuselage hinged to fold to port, thus giving direct access to a capacious hold of 57 cubic metres. The Nord 2500 Noratlas was however preferred by the military services, a choice possibly influenced by the success of the contemporary twin-boom Fairchild designs.

As could have been expected, no commercial orders were obtained and development of the Bréguet 89 line was abandoned, together with the projected Hercules-powered Br-893H and also the Br-894 of 1950, the latter to be powered by Pratt & Whitney R-2800 CA-18 engines. A larger variant was announced in 1953 as the Br-895H, but this was also dropped.

Specifications

Bréguet 890H

Initially 890J with Jumo power-plants.

Span: 30.62m
Length: 21.6m (20.6m on 890J project)
Height: 7.2m (6.75m on 890J project)
Wing area: 101 sq m
Power-plants: 2 x Bristol Hercules 739 (2,030 hp each)
Empty weight: 12,040 kg
Loaded weight: 20,000 kg
Payload: 6,210 kg
Maximum speed: 380 km/hr at 1,500m
Cruising speed: 350 km/hr at 1,500m
Range: 1,000 km with 6,000 kg payload

Bréguet 891R

Dimensions as for Br-890H.

Power-plants: 2 x 1,600 hp SNECMA 14R 200
Empty weight: 11,500 kg
Loaded weight: 16,850 kg
Payload: 5,350 kg
Maximum speed: 420 km/hr
Cruising speed: 316 km/hr
Range: 1,500 kg with 3,000 kg payload

In the passenger transport tole the 890 and 891 could have been arranged to accommodate 32-36 passengers.

Bréguet 892S

Dimensions as for Br-890H and 891R.

Power-plants: 4 x 570 hp Renault 12S-02
Empty weight: 10,460 kg
Loaded weight: 16,000 kg
Payload: 4,000 kg
Maximum speed: 350 km/hr
Cruising speed: 285 km/hr
Range: 1,000 km with 4,000 kg payload

Production

Three prototypes only.

Bréguet 892S Commercial transport prototype. F/f 01 Mar
Mercure   49 as F-WFDO at Toulouse-Blagnac.   Regn
               retained when converted as Bréguet 890H No 02.

Bréguet 891R Military freighter prototype. F/f 15 Nov
Mars   49 as F-WFRB.

Bréguet 890J Commercial transport prototype. F/f as
Mercure   Bréguet 890H F-WFRF 05 Apr 50.

BREGUET 500

The Bréguet 500 was a twin-engined commercial monoplane and owed much to the four-engined Br 482 bomber designed in 1940 and a subsequent derivative planned as a twin-engined long distance staff and VIP transport projected as the Br 483T shortly after the collapse of France. The designation of the project was soon changed to Br 500 with the use of 1,600 hp Gnôme-Rhône power-plants in place of the 900 hp Gnôme-Rhône 14Ns under consideration.

Two prototypes were ordered during the occupation of France by the Vichy Government, construction being undertaken (with German agreement) at Montaudran, near Toulouse. Since the factory was engaged in overhaul work for the Luftwaffe it became a target for RAF bombers, and in April 1944 the first Bréguet 500 airframe which was nearing completion was destroyed during a raid over Montaudran, notwithstanding removal of the prototype to a hangar a few hundred yards away from the main buildings.

The other airframe was completed some time after the French Liberation. In accordance with plans drawn up in 1943, the name of the Alsace town Colmar had been made official for the Bréguet design, the initial letter C denoting all Bréguet types. The name was retained after the Liberation of France, not surprisingly as much fighting had taken place during the later stage of the war in Alsace, which had been severed from France and annexed by Germany.

With test pilot Jean Gonord at the controls the Colmar flew on 27 February 1945, taking as many as ten passengers on its very second flight. The Br 500 was of an exceptionally clean design and embodied several technical features such as double-slotted flaps. On the other hand, the cabin had to be divded into two sections because of the main spar going right through it, with accommodation for six passengers in the front and seventeen more in the rear section. Total capacity for the eight baggage holds was 11 cubic metres.

The challenger for production orders was the SNCASO SO-30, and after completion of official trials of both types Bréguet held strong hopes of winning the contract. In fact the SO-30 was preferred, possibly on the grounds that a product of a nationalised industry should be given priority. The Colmar was abandoned in 1946, and in August of the same year the sole prototype was delivered to the French Air Force as a VIP and staff transport with the GLAM, ironically serving as the personal aircraft of the Air Minister. It was withdrawn from use after an uneventful career.

The early project for a lengthened version, known as the Br 510 (with 32 passengers instead of 23) and more derivatives with British or American power-plants (reported with unconfirmed designations of Br 501 and 550) were also dropped.

Specification

Twin-engined airliner, all-metal construction with cantilever mid-wing monoplane. Crew of five and 23 passengers.

Span: 24.12m
Length: 19.95m
Height: 4.8m
Wing area: 67 sq m
Empty weight: 9.320 kg
Loaded weight: 14,380-14,560 kg (lowered to 13,610 kg in Armée de l'Air use)
Power-plants: 2 x Gnôme-Rhône 14 R-4/5 (1,210 hp at sea level, 1,600 hp on take-off)
Maximum speed: 460 km/hr
Cruising speed: 400 km/hr
Cruising range: 2,000 km
Payload: 2,620 kg

The Bréguet 510 project would have featured an increase in length to 22.95m, and its weight would have been circa 15,500 kg with 32 passengers, still with GR 14R engines. (Bristol Hercules and P&W Twin Wasp Juniors were also considered as unconfirmed Br 501 and Br 550 variants.)

Production

Two prototypes only. One uncompleted airframe destroyed by RAF bombing 1944. The other (whether 01 or 02 is not clear) became Br 500-01. F/f 27 Feb 45 at Toulouse-Montaudran. Military marks with Lorraine cross on fins. After official trials, to Armée de l'Air/GLAM Aug 46. WFU and scrapped.

BREGUET 730

From the start the Bréguet 730 was designed as a four-engine hydravion de croisiere flying-boat for military operations. It was a pre-war design, and the type was not to see commercial service; commercial development projects and experience obtained for post-war commercial transports plead however in favour of its inclusion in this monograph. The prototype for the French Navy had flown in 1938 but had sustained heavy damage shortly after trials had begun. Further tests were delayed, but four production flying-boats were ordered in

1939.   After the fall of France and following the Wiesbaden agreement of 1941, the French were allowed by the German and Italian authorities to resume manufacture of the type, and a rebuild of the original prototype was undertaken.   After the German invasion of southern France in late 1942, progress of flight trials was slow enough to prevent a German take-over of the flying-boat.   These trials were resumed on Berre Lake after the liberation of southern France, and the Br-730 entered service with the French Navy in the long-range communications role.   A typical flight from Toulon to French West Africa took some 16 to 17 hours with stop-overs at Port Lyautey and Port Etienne.   A second Br-730 followed.

A commercial development of the Br-730 appeared post-war as the Br-731.   The main feature was a redesigned streamlined fuselage; modified floats and more powerful engines were also fitted to this variant designed purely as a transport flying-boat, with strengthened structure.   Two hulls which had survived the 1944 allied bombings while under construction were used, and both flying-boats were delivered to the French Navy as there was no potential use on the civil side.

Many commercial projects evolved from the original Br-730 design including three pre-war derivatives for transatlantic operation with manufacturer's references Br-731, 731bis and 731ter depending on power-plants, a reworked wartime design (the Br-830) which was a double-deck commercial transport with considerably increased weight, the larger Br-732 (Pratt & Whitney R-1830s) and Br-733 (Wright R-1820s) with a payload across the Atlantic which would have been around five metric tons or 34 passengers, the commercial 26-passenger Br-740 which was soon abandoned, and the improved Br-741.

However, experience gained with the Br-730/731 and its projected derivatives was transferred to a new double-deck landplane design which was to become the well-known Bréguet Deux-Ponts series.

Specifications

Bréguet Br-730

Four-engined military reconnaissance flying-boat, known officially during World War II as the Cherbourg.

Span: 40.36m
Length: 24.37m
Height: 8.6m
Wing area: 172 sq m
Empty weight: 16,300 kg
Loaded weight: 26,900 kg
Payload: 10,600 kg

Power-plants: 4 x 1,050-1,150 hp Gnôme-Rhône 14N 44 and 45
Maximum speed: 330 km/hr
Cruising speed: 230 km/hr
Range: 2,500 km with 4,400 kg payload
Ceiling: 5,000m
Endurance: 30 hrs

Performance applies to 1944 variant; initial prototype performance was somewhat inferior during 1942 trials.

Bréguet Br-731

Commercial development of the Br-730.

Span: 40.36m
Length: 24.35m
Height: 8.18m
Wing area: 172 sq m
Empty weight: 17,000 kg
Loaded weight: 35,000 kg
Disposable load: 16,000 kg
Power-plants: 4 x 1,350 hp Gnôme-Rhône 14R 200 and 201
Maximum speed: 375 km/hr
Cruising speed: 250 km/hr
Time to 2,000m (loaded): 13 mins
Range with 9,000 kg payload: 2,500 km

Production

Br-730-01 F/f 04 Apr 38 at Le Havre.   Damaged Jul 38 and WFU.   Wings mated to new production hull and redesignated Br-730 No 1.   F/f early 1942 on Berre Lake. Taken over by the Germans Nov 42 but not flown again until salvage and further flight tests Dec 44.   To French Navy Feb 45, with Transport Flotilla 9FTr, named Vega. Destroyed in French North Africa (probably 09 Jan 49 at Arzew).

Br-730 No 2 F/f 14 Jun 46.   To French Navy 9FTr as Sirius.   Written off after accident at Port Lyautey (Morocco) 27 Jun 51.

Br-731 No 1 F/f 02 Sep 47 at Biscarosse.   To French Navy (Escadrille 33S), named Bellatrix.   WFU during early fifties.   Aircraft kept in good condition, still with engines, as late as 1956 at joint AF/Navy base Rochefort.

Br-731 No 2 F/f 22 Mar 49.   To French Navy (Escadrille 33S), named Altair.   WFU during early 1950s.

CORMORAN

The SNCAC (Aérocentre) NC-211 Cormoran originated in a four-engine military transport designed in 1945 as the NC-210 but not built. Change of engines from Gnôme-Rhône/SNECMA 18R to SNECMA 14R resulted in the change of designation to NC-211.   The Cormoran was a very large aircraft by contemporary French standards, featuring a tricycle undercarriage and front loading door with a 12 metric ton payload.   Optimistic plans called for large-scale production, and it was planned to order as many as 160 aircraft for the Armée de l'Air.

Unfortunately the prototype crashed on its first flight in July 1948, killing all five on board.   As was to be found out much later, the incorrect design of the tail surfaces in relation to the wing-flap effect had caused the disaster.

A second NC-211 was flown the following year but by that time both the Air Force and the Ministry of Defence were growing increasingly sceptical about the unrealistic design.   In many successive cuts, orders had been drastically reduced, and now only ten aircraft were to be built.   As flight tests with F-WFKH proved extremely disappointing and following much parliamentary dispute over the NC-211 programme, the entire order was finally cancelled.   This also brought to an end the ill-fated history of the Aérocentre concern. The huge and totally inadequate Cormoran was thus the swan-song of the SNCAC.   The design team had also considered two more versions: the NC-212 with Bristol Hercules 730 engines, and a Jumo-powered NC-213.   A Pratt & Whitney R2000 version was also under consideration.

Specification

High-wing four-engined all-metal transport aircraft. Cargo and military application (76 paratroops or 60 stretchers for casualty evacuation) or civil transport with accommodation for 100-150 passengers.

Span: 44m
Length: 30.5m
Height: 10.7m
Empty weight: 25,265 kg
All-up weight: 42,250 kg
Payload (freight): 12,000 kg
Power-plants: 4 x Gnôme-Rhône 14R (1,600 ch on take-off)
Maximum speed at 3,000m: 390 km/hr
Cruising speed at 3,000m: 288 km/hr

Service ceiling: 8,000m
Range with full-payload: 1,000 km
Range with 9,600 kg payload: 2,000 km

Production

One prototype, one pre-production aircraft and about half a dozen uncompleted airframes.

NC-211-01 Fuselage on display Nov 46 at first post-war Salon de l'Aéronautique in Paris, but aircraft not completed until 1948 at Toussus-le-Noble.   F/f 20 Jul 48 at Toussus, crashing 20 mins later while on approach to Villacoublay.   No regn.

NC-211

C/n 1   Completed at Billancourt, transferred by road to Villacoublay Dec 48 where assembled.   F/f 09 Apr 49 as F-WFKH.   Shown at Orly Air Show 29 Apr 49.   To Orléans-Bricy end May 49 for further Aérocentre flight testing.   Last flown 07 Jul 49 with approx 30 hrs TT, following canc of programme.   Dismantled at Villacoublay, fuselage being used for many years to house radio-transmitter with aerials on roof.   Last reported 1970/71, scrapped circa 1972/73.

C/n 2   Uncompleted airframe at Billancourt.   Scrapped.

C/n 3   Uncompleted airframe.   Was to become NC-213 with 2100 hp Junkers Jumo 213 engines.   Scrapped.

C/n 4-10 Components and some airframes under construction at Bourges.   Scrapped when Aérocentre plant taken over by SNCAN.

BREGUET 941

The Bréguet company had always been a pioneer in technical advances and showed early interest in high-lift research.   Gradual steps were taken, from the double-slotted flaps of the 1940 Br-482 series through the Br-500 Colmar and culminating in the Br-940 family.

By the mid fifties the STOL concept was being defined, and the French company began appropriate research on the use of propellor slipstream deflected into full-span flaps, the so-called "Integral" principle.   This, it was thought, would radically improve the aircraft lift coefficient.   In 1955 a twin-propellor research test-bed was ordered by the French Government, converted in the following year into a four-engined installation with a whole flapped wing.   The results were so impressive that the next obvious step was to start flight research with an experimental aircraft which could be built around the blown wing already under ground-tests.

The experimental vehicle was designated the Br-940 and flown in May 1958 with Turmo II power-plants.   Three years of test flying and some 300 flights provided a great deal of experience and data, and performance was judged unexpectedly remarkable.   US support for the evaluation programme had been granted by Piasecki in 1957.   Design innovations included triple-slotted inner flaps and double-slotted outer flaps, which gave a very high full-span blown wing, the propellor slipstream being deflected down into a lift component.   Also of interest was the simple engine connection

system to link all four turboprops, thus any loss of power was minimised in case of one engine failure, all four propellors still being able to provide continuous wing lift.

This successful STOL prototype gave birth to a number of derivative projects.   First, a tactical assault-transport ordered in 1961 was called the Br-941.   Built at Villacoublay, it was first flown at Toulouse in June 1961. Powered by four 1,250 hp Turmo IIDs, it was a larger aircraft than the original 940. Extensive trials were conducted, and the Br-941 was also used widely as a demonstrator, including two visits to the United States, where the Piasecki licensee was planning production of the type as the PA-94, as well as McDonnell a few years later as the MDC-188. Publicity shots of the Br-941 taking off from the heart of large cities were widespread at that time, as inter-city STOL commercial operation of the type was under serious consideration.   The only problem was, of course, to obtain production orders!

Unfortunately the French Air Force was reluctant to order the new type into service, having already spent funds on the much larger Transall programme.   Several military customers were showing interest in the Br-941 but cost was high, and it was difficult to boost sales of a type not even ordered at home.   The French Government at long last however released sufficient funds for a production batch of four Br-941Ss (Série, or production aircraft), a slightly enlarged version with a maximum payload of 9.8 metric tons.

The first Br-941S was flown in April 1967 and was delivered initially to the CEAM at Mont de Marsan for military trials before joining the 62° Escadre de Transport at Reims.   The third and fourth aircraft followed in 1968, while the second Br-941S was not delivered until 1970, having been diverted for an extensive demonstration programme in the USA; in 1968-69 it was demonstrated to the FAA and American commercial operators such as Eastern Airlines and American Airlines, and at that time there were strong hopes that McDonnell production of the type would soon start.

Unfortunately those promising prospects soon faded away, and there is every reason to believe that this was mostly sheer misfortune. The first blow occurred in the late sixties when the DHC Buffalo was preferred by the US Army to the McDonnell-sponsored Breguet transport in the Army Tactical Transport competition.

At home, the Armée de l'Air had very little use for their small batch of aircraft, and the whole Br-941S line was withdrawn from use in 1974.   One has been preserved by the Musée de l'Air at Le Bourget, a tribute to a sound but unlucky design which ended with failure owing to a lack of orders.

Numerous projects had been drawn up, including (in order of designation):

Br-941B (1967) - production freighter version (both military and civil)

Br-941C (1961) - feeder-liner with 52 passengers and an auw of 22 tons

Br-941M (1961) - military transport

Br-942 (1960) - basically a 941 with pressurised fuselage of circular section for commercial use (40-60 passengers)

Br-943 (1961) - a 941 derivative with more powerful 1,800 hp turbo-props (the Turmo VI project) meeting the NATO NMBR.4 specification of 1960-61. Boundary layer control would have been introduced, and a Br-943R variant was also announced with auxiliary pivoting Rolls-Royce RB.162 jet engines added

Br-944 (1964) - a high-density project for 150 passengers, with various power-plants considered (British-built General Electric T-64, Pratt & Whitney JFTD-12 or derated Rolls-Royce Tynes, with optional Rolls-Royce RB.1538 jet engines used as boosters

Br-945 (1960) - a promising light tactical transport with secondary role as co-operation aircraft powered by four Turmo IIIDs, competing with the Hurel-Dubois 150, Max

Holste MH-261 and Dassault Spirale III in an Armée de l'Air programme.   The Spirale III was chosen in February 1962 but eventually not ordered.

Other projects for larger aircraft (still in the 94. designation range) were also planned but soon abandoned.

American interest in the Bréguet aircraft also led to several McDonnell-Douglas projects, starting with the MDC 188 (basically the French Br-941 retaining Turmo IIID power-plants), a military variant known as the MDC 941 (losing competitor for a US Army programme) and a 1970 international co-operation project (AIT 941) involving the USA, France, Belgium and Italy - again a military variant.

Other MDC-188 sub-types were offered by the American manufacturers, as well as an entirely different and larger project (MDC 210).   None progressed further than the drawing-board stage. BEA also showed interest in the Bréguet design, and Short Bros & Harland held a licence agreement for the UK.

"Thinking of STOL?" the advertisements in the aeronautical publications of the early sixties enquired, "Then think of BREGUET, the manufacturers of the world's ONLY true STOL transport."   Perhaps the Br-941 venture was too far in advance of its time?

Specifications

Br-940

Experimental STOL research aircraft.

Span: 17.85m
Length: 12.09m
Height: 4.41m
Wing area: 47.48 sq m
Empty weight: 5,260 kg
Loaded weight: 6,500 kg
Power-plants: 4 x 400 hp Turboméca Turmo IID

Take-off run was 40m, or take-off to 15m at 7,000 kg weight 180m.

Br-941 Prototype

Span: 23.4m
Length: 22.73m
Height: 9.35m
Wing area: 82.6m
Empty weight: 11,530 kg
Loaded weight: 24,000 kg
Payload: 9,800 kg
Power-plants: 4 x 1,250 hp Turboméca Turmo IIID
Cruising speed: 400 km/hr
Maximum speed: 520 km/hr
Range with 4,000 kg payload: 1,400 km

Br-941S Production Aircraft

Span: 23.4m
Length: 24.35m
Height: 9.7m
Wing area: 84 sq m
Empty weight: 14,160 kg
Maximum loaded weight: 26,500 kg
Payload: 9,800 kg
Power-plants: 4 x 1,500 hp Turboméca Turmo IIID3
Cruising speed: 425 km/hr
Maximum speed: 520 km/hr
Ground run on take-off to 10m: 440m at 24,000 kg weight
Range with 4,000 kg payload: 2,800 km

Production

Br-940-01 Experimental STOL aircraft. F/f 21 May 58 as
        F-ZWVF "F". WFU after three years' testing
and some 300 flights.

Br-941-01 F-WJSD initially but to military marks F-ZWVZ
        "Z" prior to f/f 01 Jun 61. To CEV
for further tests. Demonstration tour in USA as
McDonnell 188 Jun 64 but seriously damaged in landing
accident due to pilot error. Repaired in St Louis by
McDonnell and re-engined with more powerful 500 hp
Turmo IIID3s. Flown again early 1965. WFU after
1,200 hrs TT on test flights and demonstration tours.

Br-941S (all French AF)

C/n 1 F/f 19 Apr 67 F-ZJRK "K". To CEAM as 118-IX
        (F-SDIX). To 62-NA (F-RANA) 62^e Escadre. WFU
Apr 74.

C/n 2 F-ZJRL "L". Extensive demonstration tour in USA
        Jun 68-Jun 69. To French AF 1970 as 62-NB
(F-RANB). WFU 1974.

C/n 3 F-ZJRM "M". To CEAM as 118-IH (F-SDIH), later
        62-NC (F-RANC). WFU Jul 74.

C/n 4 F/f 25 Apr 68 as 118-IY then 62-ND (F-RAND). WFU
        Jul 74 and preserved by Musée de l'Air, Le
Bourget, where currently on display.

PROJECTS

As often as possible projected derivatives
and progenitors of aircraft actually built have
been mentioned within each individual aircraft
history. However, mention of a few projects
ranging from the early post-war SE-1200 flying-boat to the early French airbus proposals are of
interest, and a selection of major representative
undeveloped aircraft is included herewith.

SE-1200

The SE-1200 was another unrealistic project
of the late forties. It was a giant flying-boat for transatlantic operation with 125
passengers, generally similar in outline to its
British contemporary, the Saunders-Roe SR.45
Princess, but slightly longer. Had it been
built, it would have come second in size only
to the giant Hughes Hercules, but still by a
wide margin.

The project was to be powered by eight 3,000
hp Arsenal 24H engines grouped in four tandem
pairs with contra-rotating propellors. During
the early stage of design, turboprops were also
considered, the best choice at that time being
the British Armstrong Siddeley Python. Con-sidering the fate of other French large flying-boats, namely the Latécoère 631 and the SNCASE
SE-200 which were both commercial failures, it
was perhaps fortunate that the SE-1200 was not
built.

A flying scale model of the SE-1200 was
however planned as the SE-1210. Approximately
one third the size of the SE-1200, it was of
wooden construction, and power-plants were four
240 hp Renault 6Qs.

The SE-1210-01 made its maiden flight on 9
June 1948 with the provisional test registration
F-WEPI, but trials were discontinued after the
cancellation of the SE-1200 programme. The
aircraft was scrapped in 1952.

Another flying-boat project was the 1946
SE-1300 for 167 passengers. The weight of the
flying-boat was to have been 180 metric tons,
but no other details have been traced.

Specifications

SE-1200

Giant 125-passenger transatlantic flying-boat.
Dimensions would have been as follows:

Span: 61m
Length: 47.9m
Height: 15.6m
Empty weight: 78,480 kg
Maximum weight: 140,000 kg
Payload: 22,000 kg
Power-plants: 8 x 3,000 hp Arsenal 24H 24-cylinder H-type
               liquid cooled (turboprops also considered)
Maximum cruising speed at 10,000m: 695 km/hr (estimated)
Economical cruising speed at 10,000m: 535 km/hr
                               (estimated)

Operational range was estimated at 6,000 km, maximum
range at 10,000 km.

SE-1210

Flying scale model of the SE-1200 transatlantic flying-boat built for preliminary aerodynamic investigations and
to obtain data for the larger flying-boat.

Span: 21.75m
Loaded weight: 5,461 kg
Power-plants: 4 x 240 hp Renault 6Q six-cylinder in-line
               inverted

Aerocentre NC.211 Cormoran F-WFKH (J Delmas)

Top   : **Breguet 940, No.01, coded F, at Le Bourget**
          in June 1959 (JMG Gradidge)

Bottom : Production form, the Breguet 941S in Armee
          de l'Air service with the 62eme escadron.
          In this 1972 photo, probably taken at Solen-
          zara in Corsica, the aircraft is coded 62-NB
          (Archives M Rostaing)

A fine shot of the Nord Noroit in Aeronavale service
with 33.S escadrille (via J-P Dubois)

Top   : French-built DO.24T in Aeronavale service on
           the River Seine (E.C.P.Armees)

Bottom : Another view of an Aeronavale DO.24, coded
           53.S.4 (via J-P Dubois)

MS-785/861/880

During the mid and late forties the famous Morane-Saulnier firm (established in 1911) had designed and flown a four-seat communications and training jet aircraft, the MS-760 Paris (prototype F-WGVO making its first flight on 29 June 1954).   Production was undertaken at Tarbes-Ossun close to the Pyrenees, and the MS-760 entered service not only with the Armée de l'Air, French Navy and CEV, but also with the Forca Aerea Brasileira and Fuerza Aerea Argentina (more being built or assembled under licence by the FMA at Cordoba).   The Paris was no doubt one of the very first executive jets, sold to civilian customers all over the world. Total home production reached around 120 aircraft.

A larger version also appeared in 1963-64 in prototype form only as the MS-760C Paris III, a light executive aircraft registered F-WLKL.

By then, the future of the company was doomed.   Most of the last Morane-Saulnier designs were executive jet projects and deserve mentioning here.   First was the MS-785 Comté de Nice project of 1957, with four podded Turboméca Gabizo jet engines.   The aircraft, which resembled a scaled-down Convair CV-880, would have accommodated eight passengers with a crew of two.   These initial studies soon led to another version, the MS-880.   With a similar capacity, the project retained the four 1,100 kg st Gabizo jets, two being slung under the outer wings and two moved back on to the fuselage in Caravelle fashion.   All-up weight was planned at 14,200 kg, with a cruising speed of 900 km/hr which would have taken the MS-880 at full load over a 3,500 km typical range. Development of this small high speed high-altitude transport was however abandoned, although the MS-880 designation later became world-renowned with an entirely different aircraft, the successful SEEMS (and later SOCATA) Rallye series, more than 3,000 having already been built.

In 1961 one of the last projects by the original Morane-Saulnier design office was the MS-861 business seven-seater (with two ST-12 engines).   The firm also entered the Armée de l'Air contest for a light military transport with short-field capability.   There were many competitors, and the Morane-Saulnier project was rejected along with Potez, Nord and SIPA designs, as were the Dassault Spirale III, Hurel-Dubois HD-150 and Max Holste MH-261. The Breguet 945 was selected, but the project was later abandoned with the termination of the Algerian War of Independence.

Meanwhile Morane-Saulnier was getting into financial difficulties, bankruptcy being declared in November 1962.   The revered name was taken over by the now also ailing Potez group as the Société d'Exploitation des Etablissements Morane-Saulnier (SEEMS) and later by Sud Aviation, the MS Rallye line nowadays being produced by the SOCATA division of the nationalised industry.

MYSTERE 30

The Mystere 30 project of 1964 was another Dassault venture into the air transport field. Although generally similar in layout to the Mystere XX (Roman numerals for this biz-jet often being turned into Mystere 20 - and the American name of Falcon now being preferred), the Mystere 30 was in fact an entirely different aircraft.   It was to be powered by two Rolls-Royce RB-172 jet engines, located Caravelle style.   Designed as a short- and medium-haul transport and with a span and length of circa 22 metres, the Mystere 30, loaded at 16,000 kg, would have carried about 40 passengers at a speed of 800 km per hour.

International co-operation was by this time almost essential for aircraft manufacturers, and this led to an agreement between Dassault (or GAMD as it was then called) and Siebel of Germany.   The agreement was signed in March 1964, providing the German company with about one third of the design and construction work. Manufacture of the prototype was planned for late 1965.   The target market was in the replacement of piston-engined airliners such as the Viscount or DC-4, but after some preliminary work the project was dropped.

BREGUET 944

With the experience gained with the blown wing principle (see Bréguet 940-942) employing slipstream deflected into flaps to obtain higher lift, Bréguet were tempted to design a heavy transport aircraft retaining this technical innovation.   Apart from projects which can be directly related to the Bréguet 941 and 942, a much larger high-density Bréguet 944 was designed in 1964.   With four turbo-props (3,500 hp General Electric T-64s) and optional auxiliary wing-tip pods each housing a 3,100 kgp Rolls-Royce RB-153, the 944 would have carried 120-150 passengers in a six-abreast layout.

The general outline of the Bréguet 944 project was very similar to that of the Breguet 941 and 942 but was considerably scaled up;   length was 35 metres, span approximately 35 metres.   With an all-up weight of some 60,000 kg, disposable load varied between 15,000 and 18,000 kg.   Estimated cruising speed mentioned was 500 km per hour or about 650 km per hour with the auxiliary jets.

But inter-city short-haul air transport, considered with such optimism by Bréguet, soon proved an elusive goal, as evidenced by the failure of the Bréguet 941/942 to attract customers for commercial operation.   Development of the 944 was abandoned.

The Bréguet 946 designation was given to another high-density transport project which was in fact a variant of the early Bréguet 124 airbus proposal (qv), fitted with four turbo-props and a blown wing, and bore little relation to the 944.

MISCELLANEOUS

Other projects worth mentioning are the following (arranged chronologically in order of development):

Bréguet 1011 Capricorne

A 1946 project in the so-called strato-spheric category for a non-stop Paris-New York flight with 80 passengers on board.

Speed: 550 km/hr
Range: 5,560 km
Ceiling: 10,000m
Span: circa 32m
Length: 22.43m
Loaded weight: 25,425 kg
Power-plants: not traced

Construction of both the fuselage and wings was undertaken, but by late September 1946 all work was stopped as no fundings could be obtained.

SO-3030

A SNCASO project of late 1946 for a long-range transatlantic airliner.

Bréguet 870

Also a late 1946 project, quoted with six Arsenal 24H engines and 90 metric tons, and with eight engines, 100 metric tons and 650 km per hour at 11,000 metres.

SNCAC NC-800 Cab

Abandoned project for a grand tourisme twin engined aircraft, also planned as a fast light transport (circa 1947).

SNCASO SO-5000

Construction also intended in 1947 and called a Rapide Aerien mondial.   For 250 passengers (!) with a speed of 700 km per hour. Followed by the SO-5100 in 1949.

SNCAC NC-290

March 1949.   With four Rolls-Royce Nene engines, carrying 60 passengers.   Span 44 metres, length 31.8 metres and weight of 45 tons.

Dassault MD-600

March 1949.   Cargo/freighter aircraft with four Hercules engines and a payload of 9,000 kg over 1,000 km or 5,000 kg over 2,000 km.

SNCASO SO-60C

April 1952.   A twin-jet in the current 737 category, this was to be fitted with two large/powerful jet engines and probably two auxiliary jets in an unknown arrangement.

APPENDICES

French Départements

France is administratively a highly centralised country with a system of division into Départements.   There is a growing tendency to use the département numbers (which are also to be found in the last two digits of car registration numbers and as the first two numbers in the French post code), and they are given in the list below for the benefit of the researcher on current French aviation or the would-be visitor to the country.

A map showing the various départements is given on the inside back cover.

01	Ain		50	Manche
02	Aisne		51	Marne
03	Allier		52	Haute-Marne
04	Alpes-de-Haute-Provence (formerly Basses-Alpes)		53	Mayenne
			54	Meurthe-et-Moselle
05	Hautes-Alpes		55	Meuse
06	Alpes-Maritimes		56	Morbihan
07	Ardèche		57	Moselle
08	Ardennes		58	Nièvre
09	Ariège		59	Nord
10	Aube		60	Oise
11	Aude		61	Orne
12	Aveyron		62	Pas-de-Calais
13	Bouches-du-Rhône		63	Puy-de-Dôme
14	Calvados		64	Pyrénées Atlantiques (formerly Basses-Pyrénées)
15	Cantal			
16	Charente		65	Hautes-Pyrénées
17	Charente-Maritime		66	Pyrénées-Orientales
18	Cher		67	Bas-Rhin
19	Corrèze		68	Haut-Rhin
20	Corse (see remarks)		69	Rhône
21	Côte-d'Or		70	Haute-Saône
22	Côtes-du-Nord		71	Saône-et-Loire
23	Creuse		72	Sarthe
24	Dordogne		73	Savoie
25	Doubs		74	Haute-Savoie
26	Drôme		75	Ville de Paris (formerly Seine - see remarks)
27	Eure			
28	Eure-et-Loire		76	Seine-Maritime (formerly Seine-Inférieure)
29	Finistère			
30	Gard		77	Seine-et-Marne
31	Haute-Garonne		78	Yvelines (formerly Seine-et-Oise - see remarks)
32	Gers			
33	Gironde		79	Deux-Sèvres
34	Hérault		80	Somme
35	Ille et Vilaine		81	Tarn
36	Indre		82	Tarn-et-Garonne
37	Indre-et-Loire		83	Var
38	Isère		84	Vaucluse
39	Jura		85	Vendée
40	Landes		86	Vienne
41	Loir-et-Cher		87	Haute-Vienne
42	Loire		88	Vosges
43	Haute-Loire		89	Yonne
44	Loire-Atlantique		90	Territoire de Belfort
45	Loiret		91	Essonne
46	Lot		92	Hauts-de-Seine
47	Lot-et-Garonne		93	Seine-Saint-Denis
48	Lozère		94	Val-de-Marne
49	Maine-et-Loire		95	Val-d'Oise

Overseas there are four more départements:

971	Guadeloupe		973	Guyane
972	Martinique		974	Réunion

In this monograph, covering nearly 35 years, it has been felt of some use for the historian to mention the few changes in names that have taken place over the years (see 04, 64, 76 for instance).

The most important change took place after July 1964 when the area around Paris was completely reorganised.   The former Département of Seine-et-Oise was split up into three new Départements (78, 91 and 95), and the Département de la Seine (basically Paris and near suburbs) also split up into the new Départements scheme (75, 92, 93 and 94), a few boroughs being reshuffled from one Département to another in the process.   This has been taken into account throughout the monograph, an aircraft accident for instance being located in the Yvelines or Seine-et-Oise depending on the date.

Recently the Mediterranean island of Corsica (Corse) was divded into two new Départements, Corse-du-Sud with Préfecture at Ajaccio, and Haute-Corse with Préfecture at Bastia (2A and 2B respectively, replacing the former 20).

The only other recent change (not reflected on the map) is the transfer of the Préfecture du Var (83, on the Mediterranean coast) from Draguignan to the larger city of Toulon (see T on the map).

Please note that the period of the late sixties has been taken as significant for the map, hence the above explanations, a few of them not appearing on the map.

FRENCH AERONAUTICAL GEOGRAPHY

First, it is necessary to remind all aviation enthusiasts that access to civil airfields is NOT always permissible and that photography of aircraft is not always authorised. When in doubt, it is expressly recommended to enquire on the spot about possible restrictions of access.

Even at major airports visitors' areas have been severely limited or have even in some cases disappeared owing to terrorist actions in recent years. Security patrols by the Gendarmerie and French Police in and around major airports, runways and taxiways are part of the daily scenery at most airports, and no-one should interfere with the local regulations or commit any act in contravention of the French laws.

With this clearly in mind, France is an interesting country for the aviation enthusiast with a civil bias, with so many small airports and airfields.

The French countryside is richly endowed with small airfields where, with a growing development of air transport at regional level, third-level transport aircraft and even short-haul and medium-haul airliners will be seen.

The airfields will be promptly spotted on the 1 cm:2 km Michelin maps which are the standard reference for the motorist travelling in France (map reference numbers 51 to 84, 90 for Corsica, plus special regional maps at the same scale such as 230 for Brittany or 1 cm:1 km map 195 for the French Riviera. No 97 "150 km autour de Paris" is very convenient for the area around Paris).

In recent years the Institut Géographique National (IGN) has also promoted sales of a wide range of excellent maps at various scales including a 1 cm:1 km series (map reference numbers 1 to 74, Corsica included), which usually include a general layout of airfield runway(s). There are also many special IGN maps of local interest, particularly useful to hikers and motorists alike.

Basic geographical information is given below on the main aspects of French aviation.

<u>Military Aviation</u>

Most French Air Force bases are located in north and north-east France, with training bases in southern France. Of interest to readers of this monograph at least are the transport units centred around Paris (Villacoublay, where HQ COTAM is located, and Le Bourget) and the major transport units at Evreux-Fauville (27), Orleans-Bricy (45) and until recently near Reims (51), before 62.ET was disbanded and shared between Toulouse (31) and Evreux. Obviously Armée de l'Air transport aircraft can be seen at various bases both at home and overseas, not to mention of course communications and transport aircraft such as the Nord 262 attached to regional headquarters.

French Navy bases are mainly in Brittany and close to the Mediterranean coast, and the French Navy no longer operates a transport unit of its own. French Navy Nord 262s can however been seen flying on transport and communications duties at all the naval air stations, with aircraft permanently based for training purposes at Ajaccio-Aspretto (Corse-du-Sud, 2A). The C-47 fleet at Nîmes-Garons (30) is soon to disappear, and the future of the single DC-6 at Le Bourget (93) remains to be seen.

The French Army (Aviation Légère de l'Armée de Terre), with only helicopters and a dwindling number of light aircraft in operation, is beyond the scope of this monograph.

<u>Civil Aviation</u>

The three major airports of Orly, Le Bourget and Charles-de-Gaulle (Roissy-en-France) close to Paris are well known and conveniently located near the main French motorways. Le Bourget, north of Paris, is now no longer the major airport it used to be but is worth a visit all the same, being also a repair and technical centre with unusual visitors and an interesting location for biz-jet enthusiasts.

Of course, a visit to the Musée de l'Air at Le Bourget is a must for any foreign visitor to the Paris area, including a visit to the Concorde prototype at a small extra charge.

The two other French airports which may be called international airports are both located on the Mediterranean shores: Marignane, not far from Marseilles, with heavy traffic to Mediterranean countries, particularly North Africa, and Nice-Côte-d'Azur on the French Riviera, with an important domestic and international traffic.

The foreigner must always be reminded that east-west land communications in France have been continually hampered by the lack of so-called "transversales" - trunk roads and railways. The Massif Central area is very hilly and even mountainous, and the only convenient and speedy north-south route either by road or train, is down the Rhône valley, a narrow corridor between the Massif Central and the Alps.

This has been a serious drawback but is now gradually being remedied. As far as air transport is concerned, the post-war era has seen the development of an efficient air postal night service and the growth of the national domestic airline, Air Inter, since the sixties.

Last but not least, third-level air transport has considerably increased in recent years with

operators such as Air Alpes, Touraine Air
Transport (TAT), Union Aéronautique Regionale
and several other small fleets.   This means
that transport aircraft can now be seen at many
regional airports all over the country.

Undoubtedly the main turning point is Lyon,
conveniently located both geographically and
economically, and a major airport opened
recently at Satolas.   Regional capital cities
such as Bordeaux (33), Strasbourg (67),
Toulouse (31), Lille (59), Nantes (44),
Clermont-Ferrand (63) and others have airports
of interest.

Even comparatively small towns may prove to
be aeronautically interesting, depending on
local conditions and the operators based there.
Such is the case for instance with Dinard-
Pleurtuit (35), Nîmes-Garons (30) and
Perpignan-Llabanere (66) with its unique EAS
Vanguard fleet, many of them derelict, or
Beauvais-Tillé (60) north of Paris, a well-
known air gateway to the UK.

Scheduled air transport at third level at
least has now reached many medium- or small-
sized towns in rural or remote areas of the
country.   Other places of interest include the
special status airport of Mulhouse (68) with
Swiss connections and the two airports in
Corsica, Bastia (2B) and Ajaccio (2A), with
seasonal peak traffic in the summer.

Aerospace Industry

Gone are the days when aircraft manufacturers
were located at various places all over the
country.   Originally there were several
regional nationalised concerns, but concentra-
tion gradually took place.

With the disappearance of SNCAC only four
industries remained after a law passed on 2
August 1949:   SNCAN for northern France, SNCASE
for the south-east, SNCASO for the south-west
plus the manufacturers of power-plants, SNECMA.

During late 1956 and early 1957 more mergers
took place, resulting in Nord Aviation and Sud
Aviation being the only airframe manufacturers
until the decision was taken to form a single
Société Nationale Industrielle Aérospatiale
(SNIAS), which has been the sole nationalised
aerospace industry since the beginning of the
seventies, except for SNECMA.   Of course there
are still many Aérospatiale industrial
facilities all over the country, often with a
long historical tradition.   The Aérospatiale
facility at Meaulte (80) goes back to the
famous pre-war Potez works.   At Cannes-
Mandelieu (06) the aviation historian may
remember the pre-war Romano aircraft or the wartime Groupe Technique de Cannes.   But aircraft
are no longer assembled there.   With the
development of sub-contracting and sub-
assemblies, only two places of current interest
remain as the birth-place for assembled airframes, both in southern France: the busy
helicopter division at Marignane (13), close to
Marseilles, and Toulouse (31), now often
referred to as the capital of the French aerospace industry.

Both places are of great interest to air-
minded visitors, and both have a rich historic
past.   There are several aerodromes around
Toulouse where the Caravelle and Concorde
prototypes were first flown, where the European
Airbus is currently built and where links
between past, present and future are too many
to be detailed here.

Marseille-Marignane and the nearby salt
water Berre Lake also have historic
connections, with the first seaplane flight
there in 1910.   Still bearing witness to
aeronautical archaeology are the seaplane and
flying-boat hangars at nearby Vitrolles and
Berre (SNCASE and French Navy respectively).
Post-war outsized hangars were built at
Marignane to accommodate the large transatlantic flying-boats but have long since been
scrapped, although the buildings remain,
housing the helicopter overhaul and servicing
department.

On the private side of industry manufacturers
are no longer extant (except for light aviation)
but there is the well-known exception of
Dassault (now linked to the famous pre-war name
of Bréguet).   Activities and facilities are
centred around Bordeaux-Mérignac (33), another
centre of aeronautical interest and busy
producing jet fighters and business jets.

Going south from Bordeaux one can approach
Cazaux and Biscarosse (33 and 40) - Biscarosse
is another location connected with transatlantic
flying-boats - but unfortunately access to both
is restricted.   Further south, close to the
Spanish border, Biarritz (64) with facilities at
Parme and Anglet was the home of the now defunct
Avions Bréguet (other Bréguet works and
facilities being located in the suburbs of
Toulouse, at Colomiers and Montaudran and at
Villacoublay, south-west of Paris).

Avions Marcel Dassault had also established a
brand-new assembly line for the Mercure airliner
at Istres (13), but owing to lack of sales the
large building has been diverted to the assembly
of other types.

Istres, by the way, on the western side of
the Berre Lake, is the 'sunshine branch' of the
CEV, the French test centre located at Brétigny-
sur-Orge (91) and also a military airfield,
making it a very busy place indeed.

Clearly this short guide to French aeronautical places of interest has to be sketchy.
Light aviation, including manufacturers, has
been omitted, and it was not possible to record
all SNIAS facilities, some of which are not of
particular interest to the aviation enthusiast.
Besides, changes do occur.   For instance,
since autumn 1978 the factory at Nantes (44)
has become the airfield to which Caravelles are
flown for wing strengthening.

Sketch-Map of France with Places of Aero-
nautical Interest

On the accompanying map many, but not all,
airfields of interest are given, and generally
only airports (including the small regional
airfields where third-level traffic is to be
encountered) and places mentioned in this monograph (eg factories, industrial facilities etc)
are included.   Place-names where crashes
occurred are not included.

When necessary a few additional remarks are
added after each name: distance from town
centre for airports with scheduled traffic,
military or Government agencies in the case of
aerodromes not open to civil traffic.   But it
must be kept in mind that in some instances (eg
Hyères or Metz) scheduled air transport operates
into military airfields.

The letters IF denote 'industrial facility',
be it a major plant with aircraft production
line or only repair shops.   In some cases a

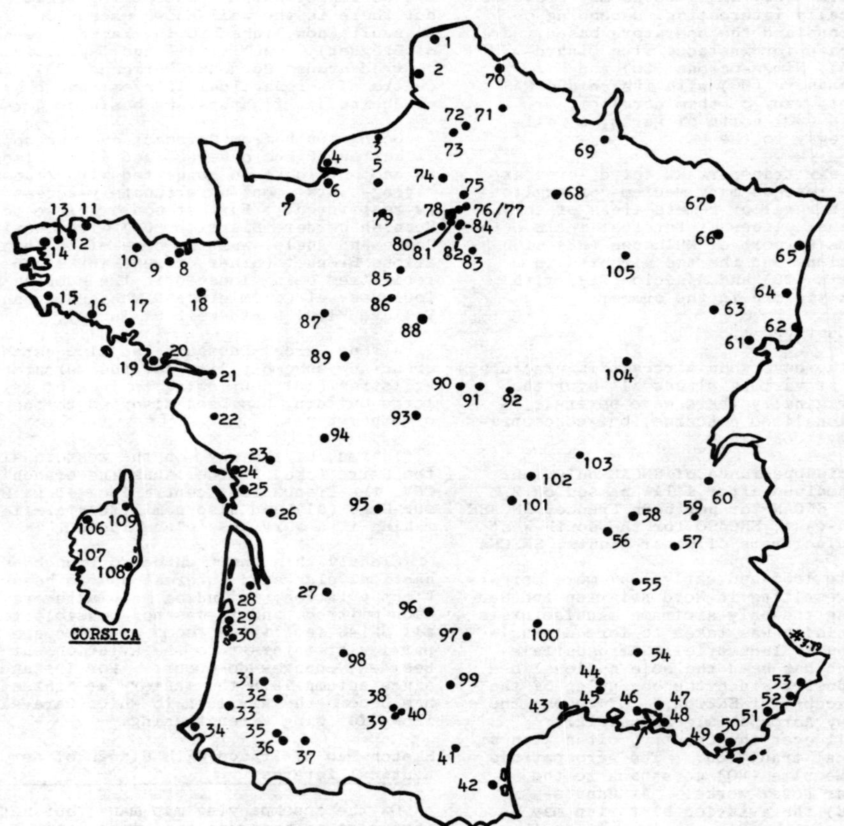

Locations of aeronautical interest

factory or workshop mentioned on the map is no longer extant but has been mentioned for historical interest.   The same remark applies to seaplane base facilities.

1 Calais-Marck
2 Le Touquet-Paris Plage
3 Dieppe-St Aubin
4 Le Havre-Octeville (Caudebec-en-Caux, 50 km E, was a seaplane IF)
5 Rouen-Boos (10 km)
6 Deauville-St Gatien (8 km)
7 Caen-Carpiquet
8 Dinard-Pleurtuit (7 km IF)
9 Dinan (small IF)
10 St-Brieuc-Trémuson
11 Lannion-Servel (home of CNET fleet, government agency for telecom research)
12 Morlaix-Ploujean (traffic to Channel Islands and the UK)
13 Landivisiau (French Navy)
14 Brest-Guipavas (11 km)
15 Quimper-Pluguffan (8 km)
16 Lann-Bihoué (French Navy)   Also Lorient airport (8 km)
17 Vannes-Meucon
18 Rennes-St Jacques
19 La Baule-Escoublac
20 St-Nazaire-Montoir (20 km, also IF)   St-Nazaire also former seaplane IF
21 Nantes-Chateau Bougon (9 km) Nantes-Bouguenais (IF)
22 La Roche-sur-Yon
23 Niort
24 La Rochelle/La Pallice (La Rochelle also former seaplane IF, see SCAN)
25 Rochefort (French AF and French Navy)   Base currently being moved south from Soubise/La Beaune to new location
26 Cognac-Châteaubernard (French AF)
27 Bergerac-Roumanière
28 Bordeaux-Mérignac (IF AMD-BA, French AF, AIA etc)   Former Dassault works located southern suburbs of Bordeaux at Talence.   No airfield there
29 Cazaux (French AF and Govt agencies firing range)
30 Biscarosse (former seaplane base and IF, Test Centre.   Currently located near Centre d'Essais des Landes - CEL - the French Space Centre)
31 Mont-de-Marsan (French AF, CEAM)
32 Aire-sur-Adour (IF)
33 Dax (French Army tri-service helicopter school)
34 Bayonne-Anglet (and nearby Turbomeca IF) Biarritz-Parme (IF)
35 Pau-Pont Long Uzein (NW of Pau) and Pau-Idron (E).   Pau is the main French paratroop training centre
36 Bordes (IF, Turboméca, no airfield there)
37 Tarbes-Ossun (IF light aircraft).   Also Lourdes Airport 10 km (Aéroport de Lourdes-Les Pyrénées).   Small airfield of Tarbes-Laloubère S of town
38 Toulouse-Blagnac (IF, Airbus production line; other IF at nearby Colomiers and St Martin du Touch) (10 km)
39 Toulouse-Francazal (French AF, transport school CIET)
40 Toulouse-Montaudran (IF)
41 Carcassonne (CN)
42 Perpignan-Llabanère (7 km, IF, home of large French Vanguard fleet)
43 Montpellier-Fréjorgues
44 Nîmes-Courbessac (French AF)
45 Nîmes-Garons (11 km - French Navy including training C-47 fleet)
46 Istres-Le Tubé (CEV, French AF, IF, AMD-BA etc)
47 Salon-de-Provence (French AF Academy)

48 Marseille-Marignane (30 km, IF, SNIAS helicopters, Sécurité Civile Canadair CL-215 fleet)   Former seaplane base on Berre Lake and another former French Navy seaplane station on the other side of lake
49 Cuers-Pierrefeu (French Navy repair and IF)
50 Hyères-le Palyvestre (French Navy, also Toulon Airport, 18 km)
51 Fréjus-St Raphaël (French Navy test centre and civil airfield)
52 Cannes-Mandelieu (IF, also heavy private flying to wealthy Riviera)
53 Nice-Côte d'Azur
54 Orange-Caritat (French AF)
55 Valence (also French Army helicopter centre)
56 St Etienne-Bouthéon (17 km)
57 Grenoble-(St Etienne de) St Geoirs (39 km)
58 Lyon-Bron (10 km) and Satolas (25 km)
59 Le Bourget-du-Lac (Aix-les-Bains and Chambéry Airport, also French AF helicopter school)
60 Annecy-Meythet
61 Belfort-Fontaine (15 km)
62 Bâle (Basel)-Mulhouse (5 km)
63 Luxeuil-les-Bains/St Sauveur (French AF)
64 Colmar-Meyenheim (French AF)
65 Strasbourg-Entzheim (13 km)
66 Nancy-Essey(-les-Nancy) (4 km, also French Army) Nancy-Ochey (French AF)
67 Metz-Frescaty (6 km - French AF)
68 Reims-Champagne (also French AF, IF etc)
69 Mézières-Charleville
70 Lille-Lesquin (also French Army)
71 Cambrai-Epinoy (French AF)
72 (Albert-)Méaulte (IF)
73 Amiens-Glisy
74 Beauvais-Tillé
75 Creil (French AF, IGN)
76 Le Bourget (IF, Musée de l'Air)   The small French Navy station and former French AF depot on the other side are referred to as Dugny or Dugny-Le Bourget
77 Roissy-en-France/Charles-de-Gaulle (IF)
78 Paris suburbs: throughout the monograph factories and workshops located in the immediate suburbs of Paris have been mentioned.   Some of them no longer exist. They were (or still are) at Argenteuil, Billancourt, Chatillon-sous-Bagneux, Colombes, La Courneuve, Issy-les-Moulineaux (still nowadays a busy heliport), Les Mureaux, Neuilly-sur-Seine, Sartrouville, Suresnes.   (Les Mureaux and Sartrouville also had seaplane facilities on the Seine.)
79 Evreux-Fauville
80 Toussus-le-Noble (busy light aviation centre)
81 (Vélizy-)Villacoublay (CEV)
82 Brétigny-sur-Orge (CEV)
83 (Melun) Réau-Villaroche (IF, SNECMA etc)
84 (Paris-)Orly.   Currently split between Orly-Ouest and Orly-Sud
85 Chartres-Champhol
86 Châteaudun (French AF storage base)
87 Le Mans
88 Orléans-Bricy (French AF)
89 Tours-St Symphorien (also French AF)
90 Bourges (IF)
91 Avord (French AF)
92 Nevers-Fourchambault (IF)
93 Châteauroux (IF)
94 Poitiers-Biard
95 Limoges-Bellegarde (8 km)
96 Figeac (IF, propellors)
97 Rodez-Marcillac (5 km)
98 Agen-La Garenne
99 Albi-le Sequestre
100 Mende-Brenoux (4 km)
101 Clermont-Ferrand/Aulnat (also French AF, IF, AIA)
102 Vichy-Charmeil
103 St-Yan (CN St-Yan)

104 Dijon-Longvic (French AF)   Also Dijon-
    Darois (IF, Robin light aircraft)
105 St Dizier (French AF)

<u>Island of Corsica</u>

106 Calvi-Ste-Catherine (7 km)

107 Ajaccio-Campo dell'Oro (8 km - also French
    Navy with Nord 262 squadron)
108 Solenzara (military base with NATO training
    commitments)
109 Bastia-Poretta (25 km)

To locate a place on the map, follow the numbers anti-clockwise, starting from the extreme north of France, Calais being number 1.

Agen - 98	Limoges - 95
Aire sur l'Adour - 32	Lorient - 16
Aix-les-Bains - 59	Lourdes - 37
Ajaccio - 107	Luxeuil - 63
Albert - 72	Lyon - 58
Albi - 99	Le Mans - 87
Annecy - 60	Marignane - 48
Argenteuil - 78	Marseille - 48
Aulnat - 101	Méaulte - 72
Avord - 91	Melun-V - 83
Bâle - 62	Mende - 100
Bastia - 109	Mérignac - 28
La Baule - 19	Metz - 67
Bayonne - 34	Mézières - 69
Beauvais - 74	Mont-de-Marsan - 31
Belfort - 61	Montpellier - 43
Bergerac - 27	Morlaix - 12
Biarritz - 34	Mulhouse - 62
Biscarosse - 30	Les Mureaux - 78
Blagnac - 38	Nancy - 66
Bordeaux - 28	Nantes - 21
Bordes - 36	Neuilly-sur-S - 78
Bourges - 90	Nevers - 92
Le Bourget - 76	Nice - 53
Le Bourget-du-Lac - 59	Nîmes - 44
Brest - 14	Niort - 23
Brétigny - 82	Orange - 54
Bron - 58	Orléans - 88
Caen - 7	Orly-Sud & Ouest - 74
Calais - 1	Ossun - 37
Calvi - 106	La Pallice - 24
Cambrai - 71	Paris & suburbs - 78
Cannes - 52	Pau - 35
Carcassonne - 41	Perpignan - 42
Caudebec-en-C - 4	Poitiers - 94
Cazaux - 29	Quimper - 15
Chambéry - 59	Réau-Villaroche - 83
Charles-de-Gaulle - 77	Reims - 68
Charleville - 69	Rennes - 18
Chartres - 85	Reims-Champagne or Prunay - 68
Châteaudun - 86	Rochefort - 25
Châteauroux - 93	La Roche-sur-Yon - 22
Chatillon-sous-B - 78	La Rochelle - 24
Cognac - 26	Rodez - 97
Colmar - 64	Roissy-en-France - 77
Colombes - 78	Rouen - 5
La Courneuve - 78	St-Brieuc - 10
Creil - 75	St-Dizier - 105
Cuers - 49	St-Etienne - 56
Dax - 33	St-Geoirs - 57
Deauville - 7	St-Nazaire - 20
Dieppe - 3	St-Raphaël - 51
Dijon - 104	St-Yan - 103
Dinan - 9	Salon-de-Provence - 47
Dinard - 8	Sartrouville - 78
Dugny - 76	Satolas - 58
Evreux - 79	Solenzara - 108
Figeac - 96	Strasbourg - 65
Fréjus - 51	Suresnes - 78
Grenoble - 57	Talence - 28
Le Havre - 4	Tarbes - 37
Hyères - 50	Toulon - 50
Issy-les-Moulineaux - 78	Toulouse - 38, 39, 40
Istres - 46	Le Touquet - 2
Landivisiau - 13	Tours - 89
Lann-Bihoué - 16	Toussus-le-Noble - 80
Lannion - 11	Valence - 55
Lille - 70	Vannes - 17
	Vichy - 102
	Villacoublay - 81
	Villaroche - 83

REGISTRATION–C/N CROSS-REFERENCES

Types are listed in their order of appearance in the monograph, with only those included of which a reasonable number of examples have been completed.

Column 1

A300

Registration	C/N
AP-BAX	96
AP-BAY	98
AP-BAZ	99
D-AHLA(2)	64
D-AHLB	83
D-AIAA	21
D-AIAB	22
D-AIAC	26
D-AIAD	48
D-AIAE	52
D-AIBA	53
D-AIBB	57
D-AIBC	75
D-AIBD	76
D-AIBF	77
D-AMAP	9
D-AMAX	12
D-AMAY	20
D-AMAZ	25
EP-IBR	61
EP-IBS	80
(F-BDHC)	8
F-BUAD	3
F-BUAE	4
F-BUAF	8
F-BUAG	15
F-BUAH	27
F-BUAI	62
F-BUAJ	97
F-BVGA	5
F-BVGB	6
F-BVGC	7
F-BVGD	10
F-BVGE	11
F-BVGF	13
F-BVGG	19
F-BVGH	23
F-BVGI	45
F-BVGJ	47
F-BVGK	70
F-BVGL	74
F-BVGM	78
F-BVGN	100
F-GBEA	50
F-GBEB	
F-GBEC	
F-GBNA	65
F-GBNB	66
F-GBNC	67
F-GBND	68
F-GBNE	86
F-GBNF	87
F-GBNG	91
F-GBNH	92
F-OCAZ	1
(F-ODCX)	3
F-ODCY	9
F-ODHC	8
F-ODHY	49
F-ODHZ	51
F-ODJU	29
F-WLGA	9, 32, 35
F-WLGB	14, 16, 27, 34, 36, 39, 46
F-WLGC	12, 15, 27, 31
F-WNDA	21, 25
F-WNDB	8, 26, 30, 44, 48, 50
F-WNDC	22, 28, 33, 52?
F-WNDD	24, 29
F-WUAA	4
F-WUAB	1

Column 2

Registration	C/N
F-WUAC	2
F-WUAD	3
F-WUAT	36, 43
F-WUAU	37, 42
F-WUAV	38, 49
F-WUAX	40, 44, 47
F-WUAY	31
F-WUAZ	41
F-WVGA	5
F-WVGB	6
F-WVGC	7
F-WVGH	23
F-WZEA	51, 68
F-WZEB	52, 69
F-WZEC	55, 71, 85
F-WZED	54, 72, 88
F-WZEE	53, 73
F-WZEF	56, 89
F-WZEG	57, 75, 90
F-WZEH	58
F-WZEI	59, 76, 93
F-WZEJ	60, 77, 94
F-WZEK	61
F-WZEL	63
F-WZEM	64, 95
F-WZEN	65, 79
F-WZEO	66, 80
F-WZEP	67, 81, 96
F-WZEQ	31
F-WZER	46, 98
F-WZES	49, 83
F-WZET	50, 84, 99
HK-2057(X)	29
HL-7218	14
HL-7219	16
HL-7220	18
HL-7221	24
HL-7223	28
HL-7224	30
HL-7238	31
HL-7246	81
HS-TGH	33
HS-TGK	35
HS-TGL	54
HS-TGM	55
HS-TGN	71
HS-TGO	72
HS-TGP	84
HS-TGR	85
HS-VGD	8
(HS-VGF)	9
JA	82
JA	89
JA	90
LN-RCA	79
N201EA	41
N202EA	42
N203EA	43
N204EA	44
N205EA	65
N206EA	66
N207EA	67
N208EA	68
N209EA	86
N210EA	87
N212EA	91
N213EA	92
N291EA	49
N292EA	51
OO-TEF	2
OO-TEG	17

Column 3

Registration	C/N
PH-TVL	8
RP-C3001	63
RP-C3002	69
SE-DFK	94
SU-AZY	25
SU-BBS	17
SX-BEB	46
SX-BEC	56
SX-BED	58
VT-EDV	34
VT-EDW	36
VT-EDX	38
VT-EDY	59
VT-EDZ	60
VT-EFV	88
ZS-SDA	32
ZS-SDB	37
ZS-SDC	39
ZS-SDD	40
9M-MHA	73
9M-MHB	93
9M-MHC	95

Concorde

Registration	C/N
F-BTSC	3
F-BTSD/N94SD	13
F-BVFA/N94FA	5
F-BVFB/N94FB	7
F-BVFC/N94FC	9
F-BVFD/N94FD	11
F-BVFF	15
F-WJAM	13
F-WJAN	15
F-WTSA	02
F-WTSB	1
F-WTSC	3
F-WTSS	001
G-BBDG	02
G-BFKW	014
G-BFKX	016
G-BOAA	06
G-BOAB	08
G-BOAC	04
G-BOAD	010
G-BOAE	012
G-BSST	002
G-N81AC	04
G-N94AA	06
G-N94AB	08
G-N94AD	010
G-N94AE	012
N81AC	04
N94AA	06
N94AB	08
N94AD	010
N94AE	012

SO-30 Bretagne

Registration	C/N
EP-AAM	14
F-DAAC	14
F-DAAX	28
F-DABA	31
F-DABB	32
F-DABC	33
F-DABD	34

Reg	No
F-DABE	35
F-WALY/BALY	30N.01
F-WAYA/BAYA	30R.01
F-WAYB	30R.02
F-WAYC	30P.1
F-WAYD	2
F-WAYI/BAYI	7
F-WAYJ/BAYJ	8
F-WAYK/BAYK	9
F-WAYL	10
F-WAYM/BAYM	11
F-WAYN	30C.01
F-WAYO/BAYO	12
F-WAYP	13
F-WAYQ/BAYQ	14
F-WAYR	15
F-WAYS	16
F-WAYT/BAYT	17
F-BAYU	18
F-BAYV	19
F-BAYX	20
F-WAYY	21
F-BAYZ	22
F-BEHA	23
(F-BEHB)	24
(F-BEHC)	25
F-BEHD	26
(F-BEHE)	27
(F-BEHF)	28
F-WEHG	29
F-WEHH	30
(F-BEHI)	31
(F-BEHJ)	32
(F-BEHK)	33
(F-BEHL)	34
(F-BEHM)	35
F-WEHN	36
F-WEHO	37
F-BEHP	38
F-BEHO	39
F-WEHR/BEHR	40
F-WEHS/BEHS	41
F-WEHT	42
F-BEHU	43
F-WEHV	44
F-WEHX	45
F-OAIT	13
F-OAIX	15
F-OAIY	12
(F-OAKS)?	40
F-OALG	16
F-OALH	17
F-OALI	21
F-OALJ	27
F-OAMA	14
F-OAQG	7
F-OAQH	9

SE-161 Languedoc

Reg	No
EC-AGU	10
EC-AGV	60
EC-AHT	38
EC-AKV	26
EC-AMH	49
EC-ANP	3
EC-ANQ	4
EC-ANR	28
EC-ANS	47
F-ARTV	01
F-BATA	1
F-BATB	2
F-BATC	3
F-BATD	4
F-BATE	5
F-BATF	6

Reg	No
F-BATG	7
F-BATH	8
F-BATI	9
F-BATJ	10
F-BATK	11
F-BATL	12?
F-BATM	13
F-BATN	14
F-BATO	(15)/32
F-BATP	16
F-BATQ	17
F-BATR	(18)/60
F-BATS	19
F-BATT	20
F-BATU	23
F-BATV	34
F-BATX	24
F-BATY	25
F-BATZ	26
F-BAYY	26
F-BCUA	27
F-BCUB	28
F-BCUC	29
F-WCUD	66
F-BCUE	35
F-BCUF	36
F-BCUG	37
F-BCUH	38
F-BCUI	39
F-BCUJ	40
F-BCUK	41
F-BCUL	42
F-BCUM	43
F-BCUN	61
F-BCUO	45
F-BCUP	46
F-BCUQ	47
F-BCUR	48
F-BCUS	49
F-BCUT	31
F-BCUU?	33
OD-ABJ	20
OD-ABU	14
OD-ABY	26
SP-LDA	21
SP-LDB	22
SP-LDC	18
SP-LDD	15
SP-LDE	44
SU-AHG	35
SU-AHH	41
SU-AHX	46
SU-AHZ	19
SU-AIA	24

SE.210 Caravelle

Reg	No
B-1850	121
B-1852	122
B-1854	38
B-1856	170
B-2501	108
B-2503	110
B-2505	197
CC-CCO	140
CC-CCP	164
CC-CCQ	160
CN-CCT	254
CN-CCV	32
CN-CCX	57
CN-CCY	154
CN-CCZ	195
CS-TCA	117
CS-TCB	125
CS-TCC	137

Reg	No
D-ABAF(1)	21
D-ABAF(2)	263
D-ABAM	214
D-ABAP	235
D-ABAV	243
D-ABAW	239
D-ANYL	247
EC-ARI	107
EC-ARJ	108
EC-ARK	109
EC-ARL	110
EC-ATV	163
EC-ATX	165
EC-AVY	173
EC-AVZ	159
EC-AXU	138
EC-AYD	197
EC-AYE	198
EC-BBR	171
EC-BDC	176
EC-BDD	202
EC-BIA	226
EC-BIB	223
EC-BIC	225
EC-BID	228
EC-BIE	230
EC-BIF	232
EC-BRJ	250
EC-BRX	261
EC-BRY	264
EC-CAE	176
EC-CIZ	247
EC-CMS	238
EC-CPI	236
EC-CUM	212
EC-CYI	263
EC-DCN	199
EC-DFP	257
EI-ATR	110
EI-AVY	108
F-BHHH	01
F-BHHI	02
F-BHRA	1
F-BHRB	2
F-BHRC	5
F-BHRD	8
F-BHRE	9
F-BHRF	12
F-BHRG	13
F-BHRH	16
F-BHRI	17
F-BHRJ	23
F-BHRK	26
F-BHRL	31
F-BHRM	37
F-BHRN	39
F-BHRO	41
F-BHRP	45
F-BHRQ	46
F-BHRR	50
F-BHRS	54
F-BHRT	55
F-BHRU	58
F-BHRV	59
F-BHRX	60
F-BHRY	61
F-BHRZ	52
F-BJAK	219
F-BJAO	42
F-BJAP	62
F-BJAQ	19
F-BJAU	70
F-BJGY	258
F-BJSO	143
F-BJTA	53

Registration	Page(s)
F-BJTB	68
F-BJTC	83
F-BJTD(1)	84
F-BJTD(2)	162
F-BJTE	111
F-BJTF	113
F-BJTG	115
F-BJTH	124
F-BJTI	105
F-BJTJ	119
F-BJTK	141
F-BJTL	142
F-BJTM	144
F-BJTN	145
F-BJTO	148
F-BJTP	152
F-BJTQ	177
F-BJTR	22
F-BJTS	27
F-BKGZ	83
F-BLCZ	51
F-BLHY	158
F-BLKF	42
F-BLKI	136
F-BLKJ	169
F-BLKS	176
F-BNFE	200
F-BNGE	10
F-BNKA	206
F-BNKB	208
F-BNKC	217
F-BNKD	220
F-BNKE	224
F-BNKF	227
F-BNKG	229
F-BNKH	248
F-BNKI	214
F-BNKJ	252
F-BNKK	256
F-BNKL	260
F-BNRA	201
F-BNRB	222
F-BOEE	212
F-BOHA	242
F-BOHB	244
F-BOHC	245
F-BRGU	237
F-BRGX	234
F-BRIM	193
F-BRUJ	209
F-BSEL	167
F-BSGZ	83
F-BSRD	38
F-BSRR	21
F-BSRY	258
F-BTDL	136
F-BTOA	274
F-BTOB	277
F-BTOC	278
F-BTOD	279
F-BTOE	280
F-BTON	97
F-BUFC	161
F-BUFF	101
F-BUFH	123
F-BUFM	209
F-BUOE	170
F-BUZC	94
F-BVPU	196
F-BVPY	271
F-BVPZ	218
F-BVSF	241
F-BVTB	270
F-BX00	76
F-BYAI	139
F-BYAT	205
F-BYAU	192
F-BYCA	66
F-BYCB	175
F-BYCD	67
F-BYCY	233
F-GAPA	99
F-GATZ	175
F-GBMI	19
F-GBMJ	149
F-GBMK	180
F-OBNG	18
F-OBNH	20
F-OBNI	28
F-OBNJ	51
F-OBNK	73
F-OBNL	75
F-OCKH	263
F-OCPJ	258
F-WBNG	18
F-WHHH	01
F-WHHI	02
F-WHRA	1
F-WHRB	2
F-WHRJ	23
F-WHRK	26
F-WJAK	19, 20, 21, 64, 67, 70, 71, 120, 135, 219, 249, 263
F-WJAL	32, 38, 65, 78, 107, 153, 161, 175, 196, 203, 215, 275
F-WJAM	24, 33, 38, 42, 66, 108, 235, 243, 247
F-WJAN	69, 117, 126, 156, 174, 263, 269
F-WJAO	35, 63, 76, 118, 131
F-WJAP	10, 62
F-WJAQ	19, 140, 167, 204, 223
F-WJSO	143
F-WLGA	21, 78, 134
F-WLGB	231, 238
F-WLGC	239
F-WLHY	158
F-WLKF	42
F-WLKI	136
F-WLKJ	128, 169
F-WLKR	171
F-WLKS	176
F-WTOA	274
HB-ICK	200
HB-ICN	253
HB-ICO	255
HB-ICP	234
HB-ICQ	222
HB-ICR	119
HB-ICS	121
HB-ICT	122
HB-ICU	123
HB-ICV	147
HB-ICW	33
HB-ICX	38
HB-ICY	43
HB-ICZ	48
HC-BAD	35
HC-BAE	40
HC-BAI	82
HC-BAJ	117
HC-BAT	125
HC-BDS	146
HC-BFN	137
HK-1709X	133
HK-1778	140
HK-1779	164
HK-1780	160
HK-1810	165
HK-1811	138
HK-1812	109
HK-2212	131
HK-2287X	168
HS-TGF	56
HS-TGG	49
HS-TGH	29
HS-TGI	25
HS-TGK	34
HS-TGL	30
I-DABA	71
I-DABE	72
I-DABF	179
I-DABG	205
I-DABI	74
I-DABL	132
I-DABM	143
I-DABP	192
I-DABR	81
I-DABS	106
I-DABT	85
I-DABU	77
I-DABV	146
I-DABW	150
I-DABZ	82
I-DAXA	35
I-DAXE	36
I-DAXI	40
I-DAXO	44
I-DAXT	80
I-DAXU	79
I-STAE	93
JY-ACS	199
JY-ACT	200
JY-ADG	236
LN-KLH	3
LN-KLI	7
LN-KLN	209
LN-KLP	24
LN-KLR	30
LV-HGX	19
LV-HGY	127
LV-HGZ	149
LV-III	180
LV-PBJ	180
LV-PRR	19
LV-PVT	127
LV-PVU	149
LX-LGE	234
LX-LGF	166
LX-LGG	156
N45SB	19
N46SB	149
N49SB	180
N98KT	102
N210G	138
N420GE	42
N555SL	102
N777VV	87
N901MW	62
N902MW	88
N903MW	89
N904MW	93
N905MW	95
N907MW	129
N1001U	86
N1002U	87
N1003U	88
N1004U	89
N1005U	90
N1006U	91
N1007U	92
N1008U	93

Registration	No.
N1009U	94
N1010U	95
N1011U	96
N1012U	97
N1013U	98
N1014U	99
N1015U	100
N1016U	101
N1017U	102
N1018U	103
N1019U	104
N1020U	114
(N2001U)	62
N2296N	102
OD-ADY	83
OD-ADZ	51
OD-AEE	153
OD-AEF	157
OD-AEM	23
OD-AEO	174
OE-LCA	161
OE-LCE	156
OE-LCI	166
OE-LCO	167
OE-LCU	136
OH-LEA	21
OH-LEB	22
OH-LEC	27
OH-LED	116
OH-LER	162
OH-LSA	181
OH-LSB	182
OH-LSC	185
OH-LSD	187
OH-LSE	189
OH-LSF	188
OH-LSG	169
OH-LSH	211
OH-LSI	259
OH-LSK	212
OO-CVA	97
OO-SBQ	123
OO-SRA	64
OO-SRB	65
OO-SRC	66
OO-SRD	69
OO-SRE	67
OO-SRF	76
OO-SRG	70
OO-SRH	78
OO-SRI	175
OO-SRK	196
OY-KRA	6
OY-KRB	14
OY-KRC	29
OY-KRD	47
OY-KRE	49
OY-KRF	170
OY-KRG	191
OY-SAA	270
OY-SAB	271
OY-SAC	269
OY-SAD	272
OY-SAE	273
OY-SAF	275
OY-SAG	276
OY-SAH	88
OY-SAJ	104
OY-SAK	99
OY-SAL	89
OY-SAM	95
OY-SAN	98
OY-SAO	101
OY-SAP	90
OY-SAR	103
OY-SAY	255
OY-SAZ	263
OY-SBV	91
OY-SBW	93
OY-SBY	94
OY-SBZ	114
OY-STA	183
OY-STB	186
OY-STC	212
OY-STD	238
OY-STE	249
OY-STF	257
OY-STG	259
OY-STH	262
OY-STI	265
OY-STK	266
OY-STL	267
OY-STM	268
PH-TRH	96
PH-TRM	21
PH-TRN	191
PH-TRO	33
PH-TRP	43
PH-TRR	48
PH-TRS	100
PH-TRU	102
PH-TRX	92
PH-TRY	87
PH-TVT	93
PH-TVV	44
PH-TVW	36
PH-TVZ	91
PI-C969	103
PI-C970	90
PP-CJA	129
PP-CJB	133
PP-CJC(1)	162
PP-CJC(2)	62
PP-CJD	168
PP-PDU	118
PP-PDV	120
PP-PDX	126
PP-PDZ	131
PP-VJC	10
PP-VJD	15
PP-VJI	20
PT-DUW	86
RP-C970	90
RP-C123	257
SE-DAA	4
SE-DAB	11
SE-DAC	25
SE-DAD	34
SE-DAE	56
SE-DAF	112
SE-DAG	172
SE-DAH	193
SE-DAI	210
TL-AAI	10
TL-ABB	249
TR-LWD	114
TS-IKM	84
TS-ITU	246
TS-MAC	207
TS-TAR	178
TT-AAD	100
TT-AAM	100
TU-TCN	199
TU-TCO	215
TU-TCY	219
TU-TXQ	201
TU-TXR	78
VT-DPN	155
VT-DPO	128
VT-DPP	130
VT-DSB	134
VT-DUH	203
VT-DUI	204
VT-DVI	213
VT-DVJ	216
VT-DWN	231
VT-ECG	70
VT-ECH	78
VT-ECI	237
XU-JTA	145
XU-JTB	53
XV-NJA	10
XV-PNH	83
YK-AFA	184
YK-AFB	190
YK-AFC	183
YK-AFD	186
YU-AHA	139
YU-AHB	135
YU-AHD	151
YU-AHE	194
YU-AHF	218
YU-AHG	233
YU-AHK	237
YU-AJE	209
YU-AJG	191
YV-C-AVI	20
5A-DAA	158
5A-DAB	162
5A-DAE	221
5T-CJW	91
5T-MAL	91
5T-RIM	91
6V-AAR	5
(6V-ACP)	5
7T-VAE	51
7T-VAG	18
7T-VAI	28
7T-VAK	73
7T-VAL	75
9Q-CLC	240
9Q-CLD	251
9U-BTA	144
9XR-CH	209

Military

Argentine

Registration	No.
T-91	19
T-92	149
T-93	180

France

Registration	No.
F-RAFA	158
F-RAFG	141
F-RAFH	201
F-RBPR	240
F-RBPS	251
F-RBPT	264
F-ZACE	116
F-ZACF	193

Sweden

Registration	No.
85172	172
85210	210

Yugoslavia

Registration	No.
7601	241
74101	241

N-262

Registration	
CF-BCR	16
CF-BCS	23
CF-BCT	23
CF-BCU	9
D-CADY	37
(D-CAMY)	34
D-CIFG	54
D-CIMA	34
D-CIMB	25
F-WJDA	250.001
F-WJDV	260.01
F-WJSN	260.1
F-WKRB/BKRB	260.2
F-BKRH	260.3
F-BKRS	260.4
F-WKVR	262.01
F-BLEA	260.5
F-WLGP/BLGP	260.6
F-BLHE	262.3
F-BLHN	260.7
F-BLHO	260.8
F-WLHP	260.9
F-WLHQ/BLHQ	262.2
F-WLHR/BLHR	3
F-WLHS/BLHS	4
F-BLHT	5
F-BLHU	6
F-BLHV	7
F-WLHX	9
F-BLHX	20
F-WLKA/BLKA	1
F-WLKE/BLKE	1
F-WNDA	30, 54, 56?, 96
	100
F-WNDB	31, 35, 103
F-WNDC	37, 50, 82
F-WNDD	29, 33
F-BNDE	8
F-BNGB	2
F-BNKX	29
F-WNLI/BNLI	21
F-BNMO	27
F-WNMP	28
F-WNTT/BNTT	26
F-WOFA	38
F-WOFB	
F-WOFC	41
F-WOFD?	42
F-BOFQ	21
F-WOFX	49
F-WOFY	
F-WOFZ	48
F-BOHH	49
F-BPNS	30
F-BPNT	35
F-BPNU	38
F-BPNV	39
F-BPNX	40
F-BPNY	42
F-WPXA/BPXA	36
F-BSTN	48
F-BSUF	74
F-BTDQ	29
F-BVPP	41
F-BVRV	100
F-BYCT	104
F-OCNQ	50
F-OCOG	48
F-ODBT	104
G-AYFR	29
I-SARL	25
I-SARP	34
JA8646	8
JA8652	15
JA8663	22
LN-LMB	260.5
LN-LME	260.8
LN-LMG	260.6
N87TC	8
N88TC	15
N89TC	22
N274A	33
N417SA	15
N418SA	41
N419SA	22
N420SA	8
N481A	25
N486A	84
N487A	54
N488A	34
N7885A	21
N7886A	47
N26201	9
N26202	10/102
N26203	11
N26207	12
N26208	13
N26209	14
N26210	16
N26211	17
N26212	18
N26213	19
N26215	23
N26217	24
N26222	69
N26224	31/100
N26225	32
N26226	56
N26227	29
N26228	42
N29802	31
N29808	29
N29811	42
N29812	17
N29813	56
N29814	99
N29816	101
N29817	50
N29824	48
OY-BCO	33
OY-BDD	21
OY-BDL	25
OY-BDM	34
OY-BDR	84
OY-BKR	47
OY-BLV	37
OY-IVA	57
OY-TOV	54
RP/PI-C966	22
RP/PI-C967	8
RP/PI-C968	15
SE-CCR	31
SE-CCS	32
SE-CCT	56
SE-FUA	69
TN-ACS	48
TS-LIP	50
ZS-IZX	6
4R-ACL	29
5R-MCC	2
5R-MCU	41
5Y-DCA	96
7T-VSQ	14
7T-VSR	10
7T-VSS	12
7T-VST	18
7T-VSU	19

AAC-1 Toucan

Registration	
CS-ADA	148
F-BAJA	1
F-BAJB	2
F-BAJC	3
F-BAJD	4
F-BAJE	5
F-BAJG	6
F-BAJH	7
F-BAJI	8
F-BAJJ	13
F-BAJK	14
F-BAJL	15
F-BAJM	16
F-BAJN	17
F-BAJO	18
F-BAJP	19
F-BAJS	20
F-BAJT	44
F-BAJU	36
F-BAJV	37
F-BAJX	38
F-BAKK	101
F-BAKL	102
F-BAKM	55
F-BAKN	56
F-BAKO	57
F-BAKP	58
F-BAKQ	59
F-BAKR	60
F-BAKS	71
F-BAKT	72
F-BAKU	73
F-BAKV	74
F-BAKX	75
F-BAKY	76
F-BAKZ	77
F-BALA	78
F-BALB	79
F-BALC	80
F-BALD	103
F-BALE	90
F-BALF	91
F-BALG	92
F-BALH	93
F-BALI	94
F-BALJ	95
F-BALK	96
F-BALL	97
F-BALM	98
F-BALN	99
F-BALO	100
F-BAMO	104
F-BAMP	105
F-BAMQ	106
F-BAMR	107
F-BAMS	108
F-BAMT	109
F-BAMU	110
F-BAMV	111
F-BAMX	112
F-BAMY	113
F-BAMZ	114
F-BANA	115
F-BANB	136
F-BANC	137
F-BAND	138
F-BANE	139

Reg.	No.	Reg.	No.	Reg.	No.
F-BANF	140	F-BDYA	382	F-BDLP	280
F-BANG	141	F-BDYB	385	F-BDLQ	233
F-BANH	142	F-BDYC	400	F-BDLR	281
F-BANI	143	F-BDYD	390	F-BDLS	300
F-BANJ	144	F-BDYE	404	F-BDLT	325
F-BANK	145	F-BDYF	398	F-BDLU	350
F-BANL	146	F-BDYG	399	F-BDVS	324
F-BANM	147	F-BDYH	401		
F-BANN	148	F-BDYI	180	SE-KAE	264
F-BANO	149	F-BYDJ	388	SE-KAG	265
F-BANP	150	F-BDYK	148	SE-KAL	159
F-BANQ	151	F-BDYM	389	SE-KAM	172
F-BANR	152	F-BEPD	288	SE-KAN	241
F-BANS	153				
		LR-AAC	73	SP-LFA	185
F-BBOF	7	LR-AAI	113	SP-LFB	227
F-BBYA	221	LR-AAJ	109	SP-LFC	228
F-BBYB	222			SP-LFD	229
F-BBYC	223	L-RAMT	109	SP-LFE	231
F-BBYD	224	L-RAMY	113	SP-LFF	232
F-BBYE	225				
F-BBYF	226	YU-ACE	316	**SO-95 Corse**	
F-BBYG	227				
F-BBYH	228	**NC-701/702 Martinet**		F-WBIA/BBIA	42
F-BBYI	229			F-BBIB	26
F-BBYJ	195	F-BAIP	1	F-WBIC	43
F-BBYK	231	F-BAIQ	20	F-BBID	45
F-BBYL	232	F-BAJY	4	F-BBIE	44
F-BBYM	233			F-BBIF	28
F-BBYN	234	F-BAOF	7	F-BBIG	29
F-BBYO	235	F-BAOG	8	F-BBIH	27
F-BBYP	236	F-BAOH	9	F-BBIQ	38
F-BBYQ	237	F-BAOI	10	F-BBIR	39
F-BBYR	238	F-BAOJ	11		
F-BBYS	239	F-BAOK	12	VT-DBW	28
F-BBYT	240	F-BAOL	13	VT-DDI	45
F-BBYU	241	F-BAOM	14		
		F-BAON	16	**SN-601 Corvette**	
F-BBZA	246	F-BAOO	121		
F-BBZB	247	F-BAOP	68	F-WIFA	15
F-BBZC	248	F-BAOQ	69	F-WIFU/BIFU?	11
F-BBZD	249	F-BAOR	70	F-WNGQ	17, 30
F-BBZE	250	F-BAOS?	71	F-WNGR	18, 32
F-BBZF	194			F-WNGS	20, 34
F-BBZG	196	F-BBFA	124/87	F-WNGT	22
F-BBZH	197	F-BBFB	125	F-WNGU	25
F-BBZI	202	F-BBFC	126	F-WNGV	26
F-BBZJ	203	F-BBFD	127	F-WNGX	28
F-BBZK	204	F-BBFE	128	F-WNGY	29
F-BBZL	230	F-BBFF	131	F-WNGZ	31
		F-BBFG	132/190		
F-BCAA	281	F-BBFH	133	F-WPTT	8
F-BCAB	282	F-BBFI	134	F-WRNZ/BRNZ	2
F-BCAC	283	F-BBFJ	135	F-WRQK/BRQK	9
F-BCAD	284	F-BBFK	136/267	F-WRSN	01
F-BCAE	286	F-BBFL	137/268	F-BTTK	31
F-BCAF	287	F-BBFM	138/269	F-BTTL	28
		F-BBFN	139	F-BTTM	17
F-BCHA	312	F-BBFO	140	F-BTTN	20
F-BCHB	310	F-BBFP	141	F-BTTO	18
F-BCHC	313	F-BBFQ	142	F-BTTP	30
F-BCHD	311	F-BBFR	143	F-BTTQ	32
F-BCHE	308	F-BBFS	144	F-BTTR	9
F-BCHF	309	F-BBFT	145	F-BTTS	11, 36
F-BCHG	324	F-BBFU?	146	F-BTTT	33
F-BCHH	323	F-BBFX	123	F-BTTU	22, 37
F-BCHI	314	F-BBFY	129		
F-BCHJ	315	F-BBFZ	130	F-WUAS	1
F-BCHK	316			F-WUQN/BUQN	3
F-BCHL	317	F-BDLA	272	F-WUQP/BUQP	4
F-BCHM	318	F-BDLB	185	F-WUQR	6
F-BCHN	319	F-BDLE	227		
F-BCHO	320	F-BDLF	274	F-BVPA	5
F-BCHP	321	F-BDLG	275	F-BVPB	6
F-BCHQ	322	F-BDLH	278	F-BVPC	12
F-BCHR	384	F-BDLI	279	F-BVPD	13
F-BCHT	380	F-BDLJ	228	F-BVPE	21
F-BCHU	386	F-BDLK	229	F-BVPF	23
F-BCHV	391	F-BDLL	231	F-BVPG	25
F-BCHX	381	F-BDLM	232	F-BVPH	27
F-BCHY	387	F-BDLN	276	F-BVPI	24
F-BCHZ	383	F-BDLO	277	F-BVPJ	29
				F-BVPK	7

F-BVPL	19	N2811D	27	F-BAPX	1120, 1191	
F-BVPO	10	N2812D	28	F-BAPY	1121, 1195	
F-BVPS	14	N2813D	29	F-BAPZ	1122, 1148	
F-BVPT	16	N2814D	30			
F-BYCR	34	N3923	10	F-BAQA	1149	
		N3924	13	F-BAQB	1150	
F-OBZP	29			F-BAQC	1151	
F-OBZR	7	N4120A	35	F-BAQD	1152	
F-OBZV	25	N4121A	36	F-BAQE	1153	
F-OCJL	19	N4122A	41	F-BAQF	1154	
F-ODFE	22	N4451	33	F-BAQG	1155	
F-ODFQ	26	N4452	34	F-BAQH	1181	
		N4453	31	F-BAQI	1183	
N611AC	7	N4732V	4	F-BAQJ	1190	
N612AC	9	N7775C	21	F-BAQK	1182	
N613AC	11	N7911C	12	F-BAQL	1209	
N614AC	17	N7912C	16	F-BAQM	1176	
N615AC	18	N7913C	25	F-BAQN	1172	
N616AC	20	N7913L	15	F-BAQO	1175	
N617AC	22	N7916C	32	F-BAQP	1138	
N618AC	26	N7917C	20	F-BAQQ	1177	
		N7918C	11	F-BAQR	1178	
OO-MRC	30	N7921C	31	F-BAQS	1213	
				F-BAQT	1171	
OY-ARA	32	N48011	14	F-BAQU	1144	
OY-ARB	34	N68596	20	F-BAQV	1159	
				F-BAQX	1173	
PH-JSB	26	VP-KNV	19	F-BAQY	1212	
PH-JSC	35			F-BAQZ	1145	
		SFERMA-60 Marquis				
SE-DEN	15			F-BAVM	1188	
		D-ILCA	26	F-BAVO	1353	
TN-ADB	22	D-ILFA	4	F-BCAI	1376	
TN-ADI	9	D-ILFE	5	F-BCAK	1403	
		D-ILFI	7	F-BCAL	1407	
TR-LWY	11	D-ILFO	6			
				F-BCCA	1147	
5A-DCK	38	F-WJHC	01	F-BCCB	1216	
		F-WJSI/BJSI	2	F-BCCC	1219	
6V-AEA	8	F-WJSJ/BJSJ	3	F-BCCD	1191, 1344	
		F-BKOO	8	F-BCCE	1224	
SCAN-30		F-BKOP	9	F-BCCF	1239	
		F-BLKU?	7	F-BCCG	1227	
AP-AMW	2	F-BLLP	10	F-BCCH	1236	
		F-BLLQ	11	F-BCCI	1230	
CF-LFQ	14	F-BLLR	12	F-BCCJ	1222	
CF-MLC	15	F-BLLS	13	F-BCCK	1223	
CF-ODR	28			F-BCCM	1249	
CF-SPA	2?	F-WLLT/BLLT	14	F-BCCN	1253	
		F-WLLU/BLLU	15	F-BCCO	1252	
EI-ALE	19	F-WLLV/BLLV	16	F-BCCP	1259	
		F-BLLX	17	F-BCCQ	1260	
F-WFDM	01	F-BLLY	18	F-BCCR	1257	
F-WFHA	1	(F-BLLZ)	19	F-BCCS	1269	
F-WHFB	2			F-BCCT	1256	
F-BFHC	3	HB-GBB	5	F-BCCV	1251	
F-BFHD	4	HB-GBF	6	F-BCCX	1329, 1419	
F-BFHE	5			F-BCCY	1330	
F-BFHF	6	**C-449 Goëland**		F-BCCZ	1331	
F-BFHG	7					
F-BFHH	8	F-BAPA	1098	F-BDXA	1319	
F-BGTD	19	F-BAPB	1099	F-BDXB	1358	
		F-BAPC	1100	F-BDXC	1359	
F-OALL	38	F-BAPD	1101	F-BDXD	1360	
F-OALM	39	F-BAPE	1102	F-BDXE	1371	
F-OALN	40	F-BAPF	1103	F-BDXF	1393	
F-OARB	36	F-BAPG	1104	F-BDXG	1354	
		F-BAPH	1105, 1141	F-BDXH	1328	
G-ARIX	19	F-BAPI	1106	F-BDXI	"1335"	
		F-BAPJ	1107, 1142			
N10BR	35	F-BAPK	1108, 1143	F-BEHX	1357	
N50G	27	F-BAPL	1109	F-BFAA	1121	
N57LM	4	F-BAPM	1110	F-BFAB	1122	
N58LM	14	F-BAPN	1111	F-BFAC	1125	
N58Q	27	F-BAPO	1112			
N60LM	15	F-BAPP	1113, 1130	LV-FGI	1195	
N62G	9	F-BAPQ	1114			
N62L	3	F-BAPR	1115	OO-CCJ	1155	
N63LM	2	F-BAPS	1116	OO-CCK	1181	
		F-BAPT	1117	OO-CCR	1230	
N115WB	4	F-BAPU	1118	OO-CEA	1121	
N2810D	26	F-BAPV	1119, 1146	OO-CEB	1122	
				OO-CEC	1125	

Alphabetical Index